Ecological Principles
of Agriculture

Join us on the web at

www.Agriscience.Delmar.com

Visit www.thomson.com for information on 35 Thomson Publishers
and more than 25,000 products! or email: findit@kiosk.thomson.com

Ecological Principles of Agriculture

Dr. Laura E. Powers
Dr. Robert McSorley

Africa • Australia • Canada • Denmark • Japan • Mexico • New Zealand • Philippines
Puerto Rico • Singapore • Spain • United Kingdom • United States

NOTICE TO THE READER

Publisher does not warrant or guarantee any of the products described herein or perform any independent analysis in connection with any of the product information contained herein. Publisher does not assume, and expressly disclaims, any obligation to obtain and include information other than that provided to it by the manufacturer.

The reader is expressly warned to consider and adopt all safety precautions that might be indicated by the activities herein and to avoid all potential hazards. By following the instructions contained herein, the reader willingly assumes all risks in connection with such instructions.

The Publisher makes no representation or warranties of any kind, including but not limited to, the warranties of fitness for particular purpose or merchantability, nor are any such representations implied with respect to the material set forth herein, and the publisher takes no responsibility with respect to such material. The publisher shall not be liable for any special, consequential, or exemplary damages resulting, in whole or part, from the readers' use of, or reliance upon, this material.

Delmar Staff:
Business Unit Director: Susan Simpfenderfer
Executive Editor: Marlene McHugh Pratt
Acquisitions Editor: Zina Lawrence
Development Editor: Andrea Edwards Myers
Executive Marketing Manager: Donna Lewis
Executive Production Manager: Wendy Troeger
Production Editor: Carolyn Miller

COPYRIGHT © 2000
Delmar is a division of Thomson Learning. The Thomson Learning logo is a registered trademark used herein under license.

Printed in the United States of America
1 2 3 4 5 6 7 8 9 10 XXX 05 04 03 02 01 00 99

For more information, contact Delmar, 3 Columbia Circle, PO Box 15015, Albany, NY 12212-0515; or find us on the World Wide Web at http://www.delmar.com

All rights reserved Thomson Learning 2000. The text of this publication, or any part thereof, may not be reproduced or transmitted in any form or by any means, electronic or mechanical, including photocopying, recording, storage in an information retrieval system, or otherwise, without prior permission of the publisher.

You can request permission to use material from this text through the following phone and fax numbers. Phone: 1-800-730-2214; Fax 1-800-730-2215; or visit our Web site at http://www.thomsonrights.com

Library of Congress Cataloging-in-Publication Data
Powers, Laura E.
 Ecological principles of agriculture/Laura E. Powers, Robert McSorley.
   p. cm.
 Includes bibliographical references.
 ISBN 0-7668-0653-7
 1. Agricultural ecology. I. Title. II. McSorley, R. (Robert)
S589.7 .P69 1999
630'.277 21-dc21   99-043250

Table of Contents

Preface — ix

Chapter 1 Introduction to Agroecology — 1
  Studying Agroecosystems — 2
  The History of Agriculture — 4
  Experimentation in Agriculture — 6
  Information Sources in Agricultural Ecology — 12

Chapter 2 Carbon and Energy in the Agroecosystem — 15
  Introduction — 15
  Basic Chemical Processes: Photosynthesis and Respiration — 16
  Carbon — 17
  Energy in Agroecosystems — 20

Chapter 3 Community Ecology — 34
  Above-Ground Food Webs — 34
  Soil Organisms and Soil Biology — 36
  Benefits and Problems of Soil Organisms — 45
  Below-Ground Food Webs — 46
  Interconnection of Food Webs — 50

Chapter 4 Nutrient Cycling and Decomposition — 55
  Essential Nutrients for Plant Growth — 55
  Nitrogen in Agroecosystems — 57
  Macronutrients and Micronutrients in Agroecosystems — 64
  Decomposition — 68

Chapter 5 Physical Factors and the Agroecosystem — **76**
Climate — 76
Limiting Factors — 83
Water and the Water Cycle — 83
Soils and the Soil Environment — 85

Chapter 6 Growth and Dynamics of Populations — **101**
Dynamics of Plant Growth — 102
Plant Population Dynamics — 107
Exponential Growth — 111
Logistic Growth — 117
Modeling and Forecasting Population Growth — 121

Chapter 7 Niches and Competition — **126**
Ecological Niches — 126
Plant Competition and Allelopathy — 127
Migration, Dispersal, and Colonization — 131
Plant Succession — 134

Chapter 8 Adaptation of Plant Cultivars — **148**
Biodiversity and Ecosystem Health — 148
Genetic Variation and Agriculture — 149
Genetic Resistance to Pests and Diseases — 159
Herbivory and Plant Defense Mechanisms — 162

Chapter 9 Predation and Parasitism — **166**
Interactions — 166
Predation — 167
Parasitism — 173
Biological Control — 176

Chapter 10 Agricultural Pests — **180**
Weeds — 180
Plant Diseases — 187
Plant-Parasitic Nematodes — 196
Insects — 201
Vertebrates — 208
Pests of Domestic Animals — 209

Chapter 11 Principles of Pest Management — **214**
Pest Management Strategies — 215
Tactics and Issues in Pest Management — 216
Deciding When to Manage Pests — 222

Ecological Basis for Pest Management	230
Multiple Pests	232
Examples and Case Studies	234

Chapter 12 Agriculture in the Landscape — 243
Diversity in the Agroecosystem and the Landscape	243
Agriculture-Environment Interactions	250
Agrochemicals and the Environment	252

Chapter 13 Cropping Systems — 266
Monoculture versus Polyculture	266
Principles of Multiple Cropping Systems	268
Examples of Multiple Cropping Systems	268
Shifting Cultivation	280
Multiple Cropping Systems as an Integrated Farming Method	283
Problems with Polycultures	283

Chapter 14 Conservation of Nutrients — 287
Conservation Tillage	287
Mulches	295
Organic Amendments	296

Chapter 15 Water Conservation and the Politics of Irrigation — 308
Sources of Fresh Water	309
Agricultural Problems Associated with Water	310
Irrigation and Water Conservation	313
Salinity and Sodicity: The Chemistry of Land in Trouble	320
Waterlogging and the Importance of Drainage Systems	323
The Politics of Water	324

Chapter 16 Energy Conservation and Animal Agriculture — 328
Shortening the Food Chain	328
Livestock Production in Industrialized and Developing Nations	330
Desertification	333
Integrating Animals Into Agroecosystems	334
Livestock Production Systems	335
Grazing Management Systems	336

Chapter 17 Political and Socioeconomic Issues of Agroecology — 344
Agricultural Support Policies	346
Agriculture and the Environment	354
Social Issues and International Justice	355
Human Aggression: Competition, Conflict, and War	357

Chapter 18 Issues in Tropical Agriculture   360
Introduction   360
The Difficulties of Managing Tropical Agroecosystems   362
Tropical Crops   364
Tropical Cropping Systems That Work   381

Chapter 19 Human Population Growth   384
Population Growth—Past, Present, and Future   384
Pressures From Population Growth   386
Feeding a Growing Population   390
Sustainable Agriculture   392

Appendix I Organizations Related to Agricultural Ecology   398

Appendix II Scientific Names of Some Common Crop Plants   401

Appendix III Conversions Between English and Metric Systems of Measurement   405

Appendix IV Glossary of Common Terms   407

Index   417

Preface

Agriculture and ecology are taught separately in many institutions, and often in separate departments. The two subjects, however, are never independent of one another. Whether recognized as such or not, every site or field where agricultural products are grown is an ecosystem subject to the same ecological processes and principles that affect the function of any ecosystem, natural or otherwise. Taken as a whole, agriculture represents the largest group of managed ecosystems in the world—a significant portion of the earth's land surface. Because of this, the impact of agriculture reaches beyond these agricultural ecosystems into other ecosystems as well. This influence extends not only as populations expand and seek additional land, but also through the interactions of different types of ecosystems in the regional landscape. Fortunately, the recognition of this close connection between agriculture and ecology is growing, as indicated by the increasing development of agricultural ecology or agroecology courses on college campuses and the increasing numbers of publications and articles on this subject.

It was our intention to develop a college textbook for the subject of agricultural ecology, one that would be versatile enough to serve as an introduction to ecology and ecological principles for agricultural students with no prior coursework in ecology, as well as for students familiar with the ecology of natural or urban systems, but with little previous exposure to agriculture. Our primary audience consists of agricultural students at the first-year or sophomore levels in college. We assume that students have little or no background in ecology, and, thus, we have included many basic concepts in this area. We believe that this approach to agriculture and the understanding of the ecological principles behind agricultural practices should be beneficial to most students of agriculture, and that this is an approach that is applicable to most courses that emphasize general agriculture.

Although intended as an introductory text, we believe that the text will be beneficial for more advanced students as well. Upper-level students or students with a strong background in ecology can move quickly through some of the basic principles and concentrate on the agricultural applications. However, even at these levels, backgrounds of students will vary, and some students may be unfamiliar with (or possibly may have forgotten!) certain basic concepts. It is our intent to provide a level background of principles and concepts for students with various levels of experience, so that all students can understand the application of these principles to agricultural systems. One of us even uses this text, supplemented by additional reading, as the basis for an agricultural ecology course for graduate students.

There are many ways to organize a text on agricultural ecology. In the early chapters, we introduce important ecological principles and many of the organisms that inhabit agricultural ecosystems. The latter half of the book focuses on key issues in agriculture, viewed with the background of the principles introduced earlier. Applications—boxed readings—throughout the book provide examples or additional insight into some of the topics.

Every agricultural ecosystem is unique, so no text could serve as a guide or manual for all practices and problems encountered in agriculture, and no specific method mentioned in this book will be practical for all circumstances. In any agricultural system, interactions among organisms and between organisms and their environment can determine the success of a particular practice. Experimentation and trial and error with various techniques will ultimately produce the best methodology for a particular situation. However, by recognizing the ecological principles that function in agricultural systems, and by understanding their implications in agricultural management, we hope that students will be able to apply these principles in agricultural design, management, and problem solving, even in very specific situations. Many of the examples in this book are drawn from agriculture in the United States, since agriculture in this country presents such a wide diversity of types, from home gardens to some of the largest and most intensive commercial operations in the world. Nevertheless, the principles examined here should be applicable in many agricultural ecosystems throughout the world.

While written primarily as a text in agricultural ecology, we hope that this book will have other important applications as well. We hope that it will be a useful reference for scientists who are specialists in certain areas of agriculture but wish to obtain an overview of other areas, since all study areas and specialties merge within the agricultural ecosystem. We have tried to keep these ideas in mind as well in writing this text. Finally, we should mention that many of the topics introduced in the text are controversial, and the various pros and cons of many practices must be weighed carefully before reaching a management decision. It is not our intention to advocate or endorse any particular practice or approach, but only to provide you, the reader and manager, with the principles and awareness to make an informed choice.

Acknowledgments

Preparation of a text such as this was a large task, and many people helped us significantly along the way. We would especially like to thank Nancy Sanders, Pam Howell, and Lila Collins for their help in preparing the text. John Frederick and Denise Bonilla helped with various preparations and proofreading.

A number of scientists reviewed portions of the book and offered helpful suggestions. For this, we are grateful to Kenton Brubaker, Ken Buhr, Larry Duncan, Bob Dunn, Jackie Greenwood, Tom Hewlett, Renato Inserra, Bill Johnson, Carroll Johnson, Monica Ozores-Hampton, Hugh Smith, and Phil Stansly.

Publication of this book would have been impossible without the excellent and efficient help of Andrea Myers, Judy Roberts, Carolyn Miller, Kathy Hans, and the other members of the Delmar staff.

The authors and Delmar Publishers would also like to thank those individuals who reviewed the manuscript and offered suggestions, feedback, and assistance.

John Bartell
Alfred State
SUNY College of Technology
Alfred, New York

Richard J. Brzozowski
University of Maine Cooperative Extension
Portland, Maine

Dr. Charles Francis
University of Nebraska
Lincoln, Nebraska

Len Harzman
Western Illinois University
Macomb, Illinois

Richard Olson
University of Nebraska
Lincoln, Nebraska

Dr. Gary A. Peterson
Colorado State University
Fort Collins, Colorado

Stephen E. Williams
University of Wyoming
Laramie, Wyoming

Finally, we thank many scientists for their wonderful work and examples that are cited throughout this text, for this is the information upon which a science like agroecology is built.

Laura E. Powers
Robert McSorley

Chapter 1

Introduction to Agroecology

Key Concepts

- Ecosystem structure
- The history of agroecology
- Obtaining and interpreting data

Agroecology has its roots in both traditional agriculture and ecology. While the term has come into vogue only within the past two decades, the concepts of agroecology date back to when agriculture first began, more than 10,000 years ago. Since early humans first started to settle in one area and actively plant and tend crops, they have paid close attention to the organisms within their systems, how they interact, and how they relate to their environment. In short, they have become familiar with the ecology of the system.

There are as many definitions of ecology as there are ecologists. Odum (1971) defined ecology as "the study of structure and function of nature." Krebs (1985) defined it as "the scientific study of the interactions that determine the distribution and abundance of organisms." Ehrlich and Roughgarden (1987) defined ecology as "the study of the relationship between organisms and their physical and biological environments." However we choose to define the term, we must consider that ecology encompasses all systems on the globe, that it deals with both physical and biological components, and that it includes the study of a range of organisms from molecular to complex at all levels of organization from individual to ecosystem.

Agroecology is the ecology of agricultural production systems and the natural resources required to sustain them. We can define it as the study of the interactions among organisms in an agricultural setting, and the interactions between these organisms and their environment. Agroecology includes all of the ecological components commonly studied within natural systems. While it may be convenient to examine these components within the well-defined boundaries of an agricultural field, the setting is not limited to the single field. Depending on the question involved, the context may range from only a small (even microscopic) portion of a field, to a much larger scale such as a whole farm, region, or world.

Agricultural systems, as well as all other kinds of ecological systems, are made up of both biotic and abiotic components. **Biotic factors** are the biological components of the system, or the living organisms. **Abiotic factors** are the nonliving or physical components, and include such factors as energy, water, climate, and nutrients. Both biotic and abiotic factors have significant effects on plant growth and productivity within agricultural systems.

Studying Agroecosystems

We can study agricultural systems from many different levels of organization (Figure 1-1). We can start at the **individual** level, perhaps looking at how an individual corn plant is adapted to a region. We may look at the physiology of the plant to determine that it grows best in full sun. We may examine the molecular structure of the plant to determine that it will have resistance to a particular pathogen. Groups of individuals of the same species make up a **population.** In an agricultural system, we may speak of a population of corn plants. If we study the system at this level, we might look at questions concerning competition among corn plants. All populations in a given area are referred to as a **community,** which includes many different types of organisms. Communities in a corn field not only include the corn plants, but also all of the weeds in the field, the insects that feed on the plants, the fungi below ground, the soil nematodes and earthworms, and all other living organisms. Agricultural research at the community level may examine how certain insects affect corn growth, or how soil organisms influence decomposition and subsequent plant growth. Finally, we can study an agricultural field at the **ecosystem** level, which includes the community and all abiotic factors such as energy, water, and climate. Most agroecological research is conducted, to some degree, at the ecosystem level, and the agricultural ecosystems are referred to as **agroecosystems.**

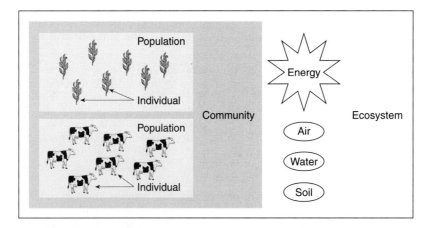

Figure 1-1 Relationship among individuals, populations, communities, and ecosystems. Ecosystems are made up of many different communities, which include many populations of many individuals.

Agroecosystems can be studied from many different viewpoints and within a variety of disciplines (Figure 1-2). The abiotic factors that make up an agroecosystem can be studied within such fields as chemistry, meteorology, soil science, and hydrology. The biotic components can be studied within the disciplines of entomology, weed science, plant physiology, plant pathology, agronomy, and horticulture. And, of course, the system as a whole must be examined from the viewpoints of economics, sociology, and political science, which are particularly important in managed systems since the choices that growers make are almost always driven by feasibility and economic profitability!

Management is one of the main characteristics of an agroecosystem that distinguishes it from a natural system. Natural systems, while almost always affected directly or indirectly by human activity, are not often managed by humans. Agricultural systems are directed by human activity and human choices. Humans decide what plants to grow in a field, whether or not to spray pesticides, how much water to introduce through irrigation, whether or not to add nutrients to the field, and a variety of other decisions. While many of these choices are influenced by nature (a farmer will seldom choose to irrigate after receiving a healthy amount of rain), most are directly influenced by economic or political factors.

Natural systems also differ from agricultural systems in regard to their boundaries. Boundaries of natural systems tend to blend together without well-defined edges. Forests blend into shrublands, which blend into grasslands, and the borders of each system tend to blur with the borders of adjacent systems. However, agricultural boundaries are often economically defined, and therefore have very distinct borders. The end of the managed area is usually considered to be the end of the agroecosystem, even though the natural systems around it may be the sources of beneficial insects, weed seeds, and vertebrate pests. For this reason, hedge rows and planned borders are often included within the realm of an agroecosystem, and are discussed in more detail in Chapter 12.

Perhaps one of the most important differences between natural and agricultural systems concerns the regulation of the system. Natural systems tend to regulate themselves. Nutrients that are removed from the soil by plants are returned to the system when the plants die and decompose. In agricultural systems, the plants are grown for the main purpose of

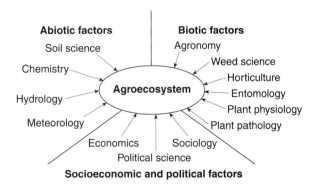

Figure 1-2 Agroecosystems can be studied from a range of viewpoints. From a scientific viewpoint, any number of biotic and abiotic factors can be studied within many different disciplines. Agroecosystems can also be studied from a socioeconomic or political viewpoint.

harvesting some particular product. This means that nutrients are necessarily removed from the system and are not cycled back through it. Because of this, special management is necessary to replace these nutrients and to keep the system productive.

The History of Agriculture

Early humans lived in hunter-gatherer societies that moved across a landscape in search of food and water. They knew their land well, since knowledge of the land meant survival, and were well-adapted to their environments. Their demand on natural resources was quite low since the population was very small. Consequently, as long as the environment remained healthy, there was plenty of food to meet the needs of all people. However, the early hunter-gatherers were not necessarily good conservationists. Their survival depended on exploiting the environment, and using any resources necessary to support them. They would move into an area, deplete it of the food resources that were usable, and then move on. This nomadic existence prevented the complete destruction of the ecology in a local area, but was not intended to conserve natural resources for the next stopover in the area.

Eventually, as civilizations grew and more people occupied any given area of land at a time, the nomadic existence no longer met the needs of all people. Division of labor began, and a select group of people within each tribe cultivated certain plants and animals to provide food for the entire group. The increasing cultivation of plants and animals allowed populations to become more sedentary. Agriculture improved nutrition and stability of the food supply. As agricultural development intensified, so did human population growth (Minc and Vandermeer 1990). The impact of humans on the natural environment increased. Forests and grasslands were cleared and cultivated, and pastures were heavily grazed. There was little concern for the environment, perhaps because the population was so small and the area of land available to people was so vast. Unfortunately, the overgrazing and overcultivation of land inevitably led to the degradation and desertification of the landscape.

As these agricultural societies continued to develop, environmental destruction became common, and overgrazing led to desertification. Overcultivation led to soil erosion and land scarring. Improper use and mismanagement of irrigation led to salinization in arid regions, and to depletion of regional groundwater. Sumerian agriculture faced the problem of land loss to salinization 4400 years ago (Gardner 1996). Lest we think that these were all problems of the distant past, we must remember that these same problems occurred more recently when Europeans colonized the Americas, and continue to occur today. Unfortunately, with increasing population, pollution, and the effects of climate change, the environmental destruction related to agricultural systems is even more serious.

Population growth is probably the most important defining force of agricultural development. With low populations and abundant resources, sustainability of ecosystems is not affected much by agriculture. But as populations increase, productivity is forced to increase beyond sustainable limits. To increase productivity, fertilizer and pesticide inputs, as well as irrigation, have all served to increase yields (Table 1-1). However, all of these inputs are dependent on fossil fuels, so they are not sustainable over the long term. As population continues to grow, new ways to increase productivity, while maintaining sustainability, must be found, or the system will be vulnerable to collapse.

Diversification of production systems, which can help to address some of these problems, will be examined later in this book. Diversification can lead to stability and sustainability, not only on small family-owned farms, but also on larger farms that are production-oriented and directed by economics. Diversification can also increase the population of natural ene-

Table 1-1 Strategies for increasing productivity and their ecological consequences.

Farming Strategy	Ecological Consequences
Intensive farming	• Loss of wildlife habitat • Soil erosion (despite improvements in tillage equipment and tillage management)
Intensive animal production	• Pollution • Desertification
Decoupling of animal and grain production	Manure is not easily recycled as fertilizer, so there is often too much (pollution) at feedlots, and too little at sites of grain production.
Pesticides	• Harm to humans, wildlife • Weeds and pests are harder to control than ever. • Short-term solutions have led to long-term problems since many pests have developed resistance to common chemicals.
Fertilizers	Water pollution: Both surface and groundwater sources of water contain nitrates, phosphates, and pesticide residues.
Irrigation	Salinization
Deforestation	Soil erosion
Goats (poor man's cow)	Denuded land, erosion

mies of pests, allow nutrients to be used more efficiently, allow recycling of nutrients, and keep the soil covered to prevent erosion. Multiple-crop production leads to stability in some agricultural systems, as we will see in later chapters. Integrating livestock and crop production on the same farm closes the nutrient loop, decreasing the need for fertilizers and eliminating problems with pollution. It is rather ironic that we are quite approving of money managers who diversify investments to minimize risk (Brummer 1998), yet we continue to support and even subsidize those cropping systems that are extremely undiversified, providing no security against damage from natural disasters or economic changes.

Unfortunately, agricultural research that is currently conducted in many parts of the world, including the United States, does not look at the benefits of diversification. Nor does it examine how to make diversified systems work with the technology currently available. A great amount of research money has been used to make a select group of crops (e.g., corn, soybeans, wheat) highly productive as monocultures, but very few research dollars have focused on the design of sustainable systems or the improvement of productivity in a diversified system. Even in developing nations, we are beginning to see a movement away from highly diversified feed-the-family systems toward the mass production of only one or a few crops in a large field.

This, perhaps, is why the field of agroecology is so important. There are very few natural systems in the world with only one species of plant present and no pests on that species. Conventional agricultural systems do not mimic natural systems in any way, and therefore must be managed carefully with a variety of expensive inputs that are often necessary to maintain ecosystem health and sustainability.

Many natural systems, on the other hand, are relatively stable over time without input or management. Nutrients are recycled. Natural enemies prey on pests that threaten plant species, so one species of insect rarely overwhelms the system. Nutrients are harvested efficiently by a variety of plants with slightly different requirements. Soil organisms enable decomposition to occur, providing a wealth of nutrients and organic matter to the system. If an agroecosystem can be designed to mimic a natural ecosystem, we can minimize expensive inputs of fossil fuels and other limited resources. The challenge with such a system is to maintain a level of agricultural production sufficient for an expanding human population while still conserving resources.

Agroecology, as a field of study, looks at how the agroecosystem functions naturally and examines how adjustments can be made to the agroecosystem without destroying other facets of the system. It acknowledges that changing one aspect of the system may have far-reaching consequences for other parts. It challenges scientists to look at an agricultural field as a system that must be respected, rather than conquered by chemicals and machines. Most importantly, it encourages us to remember that agroecosystems are most sustainable if they function as natural systems. By carefully observing both natural and agricultural systems, we can develop new ideas for increasing agricultural productivity, while maintaining the stability and sustainability that will allow that system to remain in production over the long term.

Experimentation in Agriculture

Since agricultural data can be obtained from a very wide variety of sources, it is not surprising that the quality and utility of the data may also vary. Advertising and testimonials may provide awareness of a particular product, but may be of little help in evaluating an agricultural product or procedure that we are thinking about using. Decision making based on scientific fact requires accurate information and data, detailed observations, or even experiments performed by ourselves or others.

Approaches to Obtaining Data

Much agricultural information and knowledge can be obtained by careful observation. Historically, many agricultural practices probably developed and improved as people observed which conditions appeared favorable for the growth of certain crop plants. These observations and recommendations were passed along by word-of-mouth and by demonstration, and refined by subsequent generations in an ongoing process that continues today.

Experimentation has often been used in evaluating and establishing agricultural practices. Experiments are inspired by some previous observation or question that needs additional study or testing. This question must be expressed in a format that can be tested or evaluated. One approach is to develop a **hypothesis,** or a guess about what we think will happen under a particular set of conditions. We can then design an experiment to test our hypothesis under these conditions. Data collected from the experiment provide measurable results by which we evaluate our hypothesis. Based on these results, we reach a conclusion about whether our hypothesis is supported or contradicted, and provide some evidence for our original question. These steps that occur in many experiments—observation, hypothesis, experiment, results, conclusion—are recognized formally as the **scientific method.**

Although we may think of the scientific method as an approach that is commonly used in laboratory tests or in formal field trials, it is probably used very often in a less formal way. We may observe that a certain weed species is common in apple orchards, hypothesize that it is more common in orchards than in alfalfa fields, examine several orchards and alfalfa fields, make mental notes about where the weed species occurs, and conclude whether it was more common in orchards or alfalfa fields. Although this example is not a particularly formal experiment and no data were written down, the approach follows the steps of the scientific method and allows us to reach a conclusion. Many people use this method of analysis without realizing it.

Obtaining Data From Field Experiments

Suppose we wanted to design a simple field experiment to evaluate whether the use of a new fertilizer resulted in the growth of taller corn than if we used our old fertilizer, Brand X. A very simple experiment would be to apply the new fertilizer to one row of corn and Brand X to the next row of corn (Figure 1-3A). *Everything else about these two rows of corn must be kept exactly the same; only the types of fertilizer can be different.* An individual row of corn is called a **plot,** the basic or smallest unit of an experiment. The size and composition of plots depend on the nature of the experiment; plots can consist of single rows, multiple rows, several trees, measured areas of land, etc. The two different fertilizer types are **treatments,** because they represent the only way (fertilizer treatment) in which the two plots have been treated differently. If we measured the heights of the corn plants in the two plots and found that the average height in the plot with new fertilizer was 2.5 m, and the average height in the plot treated with Brand X was 1.5 m, we can conclude that the use of the new fertilizer resulted in the growth of taller corn plants, *provided that everything else in these plots was treated exactly the same.*

Sometimes things are not exactly the same. What if there is a row of trees next to the plot that received Brand X fertilizer (Figure 1-3B)? Our original conclusion, that the new fertilizer improved plant height, may be incorrect. Maybe the plants treated with Brand X are shorter because they are shaded by trees. Maybe if the plot receiving the new fertilizer was located next to the trees, the results, and our conclusion, would be opposite. To avoid this type of problem, we could repeat the two treatments several times, at different locations in the field (Figure 1-4). Each location within the field is called a **replication,** or **block.** A

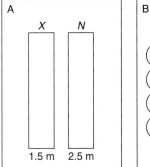
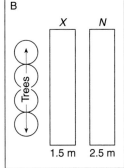

Figure 1-3 Diagrams of a simple field experiment in which a new fertilizer (N) is applied to one row of corn and old Brand X fertilizer (X) is applied to an adjacent row of corn. Each rectangle represents an individual plot, in this case, a row of corn plants. Numbers are average heights of corn plants in the plots.

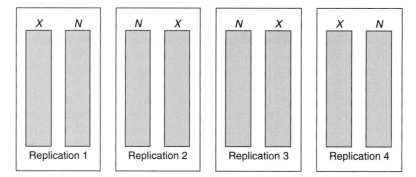

Figure 1-4 Diagram of a field experiment in which two treatments (N = new fertilizer; X = Brand X fertilizer) are applied to plots (represented by shaded rectangles) in four replications.

replication is an experimental unit containing several plots, each receiving a different treatment. In our example, each replication contains two plots, one with each of the two treatments. Both treatments are always included in each replication. The order of treatments within each replication is chosen at random, so, in Figure 1-3, sometimes treatment X is on the right and sometimes it is on the left. Replication is often used in agricultural field experiments because differences within fields may arise from many factors that are not easily observed, such as slight differences in soil type, topography, or concentrations of chemicals.

Analysis of Data From Field Experiments

The use of replications provides independent measurements of the treatment effects at several different locations. In the example (Figure 1-4), we could have a measurement of plant height from each of the two treatments in all four replications (Table 1-2). The average of the plant heights from the four replications gives a much more realistic impression of plant height in the field than does the measurement of plant height from just one replication. In our example (Table 1-2), the **mean,** or average, plant height is 2.35 m in plots treated with fertilizer N and 1.75 m in plots treated with fertilizer X (calculation in Application 1-1). It would seem as if the use of the new fertilizer (N) has resulted in taller plants, but there may be other factors to consider.

Table 1-2 Heights of corn plants in field plots treated with new fertilizer (N) or Brand X fertilizer (X) in a field experiment with four replications (see Figure 1-4).

	N	X
Replication #1	2.4 m	1.6 m
Replication #2	2.3 m	1.9 m
Replication #3	2.2 m	1.8 m
Replication #4	2.5 m	1.7 m

Application 1-1 *Computation of the Mean and Standard Deviation*

Suppose that four measurements are available, as was the case with the plant heights shown in Table 1-1. The four measurements are: 2.4, 2.3, 2.2, and 2.5.

The mean is the numerical average of the measurements (total divided by number of measurements):

$$(2.4 + 2.3 + 2.2 + 2.5)/4 = 9.4/4 = 2.35$$

The standard deviation is a measurement of precision, given by the formula:

$$\text{Standard deviation} = \sqrt{\frac{\Sigma y^2}{(n-1)}}$$

where y^2 is the sum of the squared differences between each measurement and the mean, and n is the number of measurements. To find the standard deviation, we first find the difference between each measurement and the mean, and then square each difference. In our example, the mean is 2.35, so:

	Difference	Difference Squared
2.4 − 2.35 =	0.05	0.0025
2.3 − 2.35 =	−0.05	0.0025
2.2 − 2.35 =	−0.15	0.0225
2.5 − 2.35 =	0.15	0.0225

Then we add up all of the squared differences:

$$0.0025 + 0.0025 + 0.0225 + 0.0225 = 0.050$$

We divide by the number of samples ($n = 4$ in this case) minus 1:

$$0.050/(4 - 1) = 0.050/3 = 0.0167$$

Finally, take the square root to get the standard deviation

$$\sqrt{0.0167} = 0.129 \approx 0.13$$

For the second column of data in Table 1-2 (1.6, 1.9, 1.8, 1.7), the mean is calculated as 1.75 and the standard deviation is 0.13.

It is often important to know how much variability occurs among a set of measurements. Natural variation and gradients in soils or other factors within a field may cause differences, or variability, among the measurements of plant height collected from different replications. The **standard deviation** is one statistic that describes how close measurements are to one another. The calculation of a standard deviation is illustrated in Application 1-1. The size of the standard deviation increases as variability among data increases. However, the standard deviation also has a tendency to increase as the size of the mean increases. Other statistics, such as the **coefficient of variation** (CV), may be used to adjust for this increase in standard deviation as the mean increases. CV is given by:

$$CV = \frac{\text{standard deviation}}{\text{mean}} \times 100\%$$

Both the standard deviation and the coefficient of variation are measures of **precision;** that is, they measure how close the various data or measurements are to one another. Precision is sometimes confused with accuracy, but **accuracy** refers to whether or not the measurements are correct. Problems with accuracy could result if plant heights were measured incorrectly or if the plants chosen for measurement were not really representative of most plants in the plot. However, even if all measurements are accurate, variability will still occur from uncontrolled factors (weather, soil types, plant genetics, etc.). Since not all plants will be *exactly* the same height, some measure of precision may be helpful for evaluating this uncontrolled variability.

In many cases, the experimental data and results do not show such clear differences as those observed between the two treatments in Table 1-2. Some common problems are illustrated by the two scenarios shown in Table 1-3. In the first case, the height measurements obtained from the two treatments are so close that we wonder whether the means of 1.85

Table 1-3 Heights of corn plants (in meters) in field plots treated with new fertilizer (*N*) or Brand *X* fertilizer (*X*) in a field experiment with four replications, for two hypothetical scenarios.

	Scenario #1: Similar Means	
	N	*X*
Replication #1	1.8	1.6
Replication #2	1.8	1.9
Replication #3	1.9	1.8
Replication #4	1.9	1.8
Mean	1.85	1.78
Standard deviation	0.06	0.13
	Scenario #2: Variable Data	
	N	*X*
Replication #1	2.0	1.8
Replication #2	2.8	2.5
Replication #3	1.3	2.6
Replication #4	3.1	1.7
Mean	2.30	2.15
Standard deviation	0.81	0.46

and 1.78 are really very different at all. In the second scenario, the mean from plots with treatment N is numerically greater, but the data are highly variable and erratic. Notice that the standard deviations are greater in this case (Scenario #2) than in the other examples shown, reflecting the greater variability in these data. Concerned about this variability, we may ask if the difference between the means of 2.30 and 2.15 is real or a result of chance.

Various kinds of **statistical tests** are used to evaluate whether observed differences among treatments are likely to have occurred as a result of the treatments or if they could have occurred simply by chance. We would not have much confidence in a result that could occur by chance 50% of the time. On the other hand, if we achieved a result that was likely to occur by chance only 5% of the time, we could be much more confident in our results. We could be extremely confident if differences among treatments were so great that they could occur by chance only 1% or even 0.1% of the time. However, we must remember that in agriculture we are dealing with living plants and animals rather than physical or chemical processes, so we usually expect more variability in agriculture than in engineering or other physical sciences. Most agricultural research is based on differences in results that could occur by chance only 5% of the time or less. In some cases, a 1% chance may be used, but in others, the standard may be related to a 10% chance. The point is that most things in agriculture do not happen the same way 100 out of 100 times; but 95/100 times is quite convincing, yet possible. Of the many statistical tests for evaluating differences among treatments, the analysis of variance procedure is probably the most widely used and applicable. Information on this and other statistical tests, as well as on the concepts of replication and field plot design, is presented by various statistics texts (e.g., Dyke and Grundy 1988; Peterson 1994; Steel, Torrie, and Dickey 1997).

Conclusions From Experimental Data

Provided that the observed results are statistically valid and not attributed to chance, we may reach conclusions about the experiment represented by the data shown in Table 1-2. A valid conclusion may be that, *under the conditions of this test,* the use of the new fertilizer resulted in the growth of taller corn plants than if Brand X fertilizer was used. Our conclusions must be rather conservative. Results may have been different if the test was conducted with a different corn hybrid or in a different year. If the test was conducted in Iowa, we cannot really make generalizations about what might happen in Georgia, the rest of the United States, the United Kingdom, or other locations—we don't know for sure if the test wasn't conducted in those locations. Of course, we may have a fairly good idea about what would happen in these locations, but technically, even a very reasonable idea would remain speculation until future testing confirmed it. From our example (Table 1-2), we can conclude nothing about corn yield, only plant height. To make conclusions about yield, plant weights, protein content, etc., these factors must be measured.

Ideally, a corn grower attempting to reach a decision to use the new fertilizer rather than Brand X could conduct a small experiment right on his or her own farm. However, time and opportunity do not always allow on-farm testing, so growers often need to depend on experiments conducted by other people at other sites. Growers and advisors must judge whether results of research conducted elsewhere are applicable to their particular conditions. Knowledge and understanding of ecological principles and the effects of their application on the agroecosystem are critical in judging whether experimental results can be transferred to different locations or whether more site-specific data will be required. If the situation is too uncertain, small-scale experiments or demonstrations may confirm whether a new method or product is useful for a specific site.

Information Sources in Agricultural Ecology

The vast information available on agricultural ecology ranges from general textbooks to highly specific advisories (circulars, bulletins, fact sheets, etc.) emphasizing production of a crop in a particular region or specific problems affecting that crop. The bibliography at the end of this chapter includes a number of books that provide additional information on agricultural ecology and related subjects. Many different scientific journals publish technical research articles on a variety of agricultural subjects. A small sample of the more general agricultural journals is listed in Table 1-4. Other journals may emphasize particular crops (e.g., *Peanut Science*), certain groups of pests or production problems (e.g., *Plant Disease, Environmental Entomology, Weed Science*), or topics of special interest to a state or region (e.g., *Soil and Crop Science Society of Florida Proceedings*). A wide variety of trade magazines and other periodicals can provide more specific information on certain topics, as well as interpretation of research results. Additional information on specific topics can be obtained through various scientific organizations that maintain active Internet sites. A number of these organizations and their Internet addresses (current at the time of publication) are summarized in Appendix I at the end of this book. Other Internet sources also provide valuable sources of information. However, all Internet information should be evaluated carefully because quality control on the Internet is highly variable.

The most specific information available is probably that obtained from local advisory services, either public or private. The nature and infrastructure of public advisory services vary from country to country, but some service of this type is available to growers in most regions. In the United States, the Cooperative Extension Service maintains agents and offices within individual counties who provide recommendations specific to agricultural production practices and problems for each county. County agents within a state are part of a network of agricultural scientists in that state, including specialists based at the land-grant university of a particular state. The objective of this agricultural network is to deliver specific information that is directly transferable, reducing the amount of experimentation and testing required of an individual grower. Unfortunately, the information available is some-

Table 1-4 Examples of scientific journals that include articles on topics in agricultural ecology.

Agricultural Water Management
Agriculture, Ecosystems and Environment
Agronomy Journal
American Journal of Alternative Agriculture
Applied Soil Ecology
Australian Journal of Experimental Agriculture
Biology and Fertility of Soils
Crop Science
Field Crops Research
HortScience
Plant and Soil
Soil Science
Soil Science Society of America Journal
Soil and Tillage Research

times limited to conventional production systems, and may not be particularly applicable to more diversified agroecosystems.

In addition to the list of scientific organizations, a few other reference items are included in the appendices at the end of this book. Appendix II shows the scientific names (Latin names) of common crop plants. Appendix III shows conversions of some common units between the metric and English systems, since many agricultural measurements in the United States are still reported in English units. A glossary of terms used throughout the text is included as Appendix IV.

Summary

Agroecology is a specific branch of ecology that examines agricultural systems that experience varying degrees of management inputs, rather than natural systems that may or may not be directly influenced by humankind. The growth of agriculture and the emphasis on increasing productivity have led to a series of changes in the way we feed people that has, unfortunately, caused a great deal of environmental destruction and degradation over time. In order to increase productivity in a sustainable way, scientists need to perform a variety of field tests and experiments, using the data they collect to draw conclusions about different management options. Knowledge of ecological principles is essential for determining the best practices for a particular region, and for discerning whether or not publicly disseminated information is useful and valid for any given situation.

Topics for Review and Discussion

1. Explain the difference between a population and a community.
2. Name several reasons for the changes in agriculture that have taken place over the past 10,000 years. Have systems become more or less sustainable? Why?
3. List some factors that may be different in various parts of an agricultural field that is planted to one crop.
4. Draw a plot plan for a field experiment in which the yields of three different bean varieties are to be evaluated, using four replications. (Hint: The different bean varieties are the treatments.)
5. Many kinds of claims and conclusions are presented by advertising. Choose a current advertisement that makes some claim about an agricultural product, and discuss how an experiment could be designed to test this claim.

Literature Cited

Brummer, E. C. 1998. Diversity, stability, and sustainable American agriculture. *Agronomy Journal* 90:1–2.

Dyke, G. V., and G. M. F. Grundy. 1988. *Comparative Experiments with Field Crops*. 2d ed. London: Charles Griffin and Company.

Ehrlich, P. R., and J. Roughgarden. 1987. *The Science of Ecology*. New York: Macmillan.

Gardener, G. 1996. *Shrinking Fields: Cropland Loss in a World of Eight Billion*. Worldwatch Paper 131. Washington: Worldwatch Institute.

Krebs, C. J. 1985. *Ecology: The Experimental Analysis of Distribution and Abundance*. 3d ed. New York: Harper & Row.

Minc, L. D., and J. Vandermeer. 1990. The origin and spread of agriculture. In *Agroecology*, eds. C. R. Carroll, J. H. Vandermeer, and P. M. Rosset, 65–111. New York: McGraw-Hill.

Odum, E. P. 1971. *Fundamentals of Ecology*. Philadelphia: Saunders.

Peterson, R. G. 1994. *Agricultural Field Experiments: Design and Analysis*. New York: Marcel Dekker.

Steel, R. G. D., J. H. Torrie, and D. A. Dickey. 1997. *Principles and Procedures of Statistics: A Biometrical Approach*. 3d ed. New York: McGraw-Hill.

Bibliography

Altieri, M. A. 1987. *Agroecology: The Scientific Basis of Alternative Agriculture*. Boulder, Colo: Westview Press.

———. 1994. *Biodiversity and Pest Management in Agroecosystems*. Binghamton, N.Y.: Food Products Press.

Campbell, K. L., W. D. Graham, and A. B. Bottcher, eds. 1994. *Environmentally Sound Agriculture*. St. Joseph, Mich.: American Society of Agricultural Engineers.

Carroll, C. R., J. H. Vandermeer, and P. M. Rosset, eds. 1990. *Agroecology*. New York: McGraw-Hill.

Collins, W. W., and C. O. Qualset, eds. 1998. *Biodiversity in Agroecosystems*. Boca Raton, Fla.: CRC Press.

Edwards, C. A., R. Lal, P. Madden, R. H. Miller, and G. House, eds. 1990. *Sustainable Agricultural Systems*. Delray Beach, Fla.: St. Lucie Press.

Gliessman, S. R. 1998. *Agroecology: Ecological Processes in Sustainable Agriculture*. Chelsea, Mich.: Ann Arbor Press.

Southwick, C. H. 1996. *Global Ecology in Human Perspective*. New York: Oxford University Press.

Spedding, C. R. W. 1996. *Agriculture and the Citizen*. London: Chapman and Hall.

Tivy, J. 1992. *Agricultural Ecology*. Harlow, U.K.: Longman Scientific and Technical.

Chapter 2

Carbon and Energy in the Agroecosystem

Key Concepts

- Photosynthesis as a source of carbon and energy
- The carbon cycle in agriculture
- Energy transfer through producers and consumers
- Energy efficiency of agricultural production

Introduction

Agricultural production requires many different physical **inputs**, such as water, light, soil, energy, and various chemical elements, which often must be supplied to the agroecosystem. Identification of the **sources** (where materials are obtained) of these agricultural inputs is critical in anticipating potential problems with shortages of supplies. The amount of a particular input used to avoid deficiencies or excesses will depend on the requirements of the crop and the environmental conditions. **Outputs,** which include anything removed from or leaving the agroecosystem, include not only the obvious harvested material, but also more subtle items such as dead leaves, crop residues, manure, or water and fertilizer that have not been used by plants. Knowledge of these outputs and their **sinks** (where materials go) can be helpful in recognizing potential pollutants or other future problems. Some basic questions about input and output of physical and chemical factors are summarized in Figure 2-1. In this diagram, the source and sink are shown apart from the agroecosystem to emphasize that they may be distant and physically separated. This is not always the case, since some sources and sinks may be within the agroecosystem. For example, the soil environment acts as both a source and a sink for some materials.

Inputs and outputs to agricultural systems include both matter and energy. Materials such as chemical elements and compounds may cycle between organisms and the physical environment, and to some extent, it is possible to follow their movements as they are

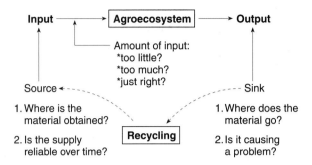

Figure 2-1 Basic questions for analysis of input and output of physical and chemical materials for an agroecosystem.

recycled. **Biogeochemistry** is the study of the movement of elements between organisms and different parts of the physical environment. Biogeochemistry is covered in more detail in Chapter 4, but the subject is introduced in this chapter with the cycling of carbon. Energy does not cycle, but many of its movements in agroecosystems are similar to those of carbon.

Basic Chemical Processes: Photosynthesis and Respiration

Photosynthesis

The major pathway of carbon and energy into the agroecosystem is through the photosynthesis reaction in green plants:

$$6CO_2 + 6H_2O + \text{energy} \rightarrow C_6H_{12}O_6 + 6CO_2$$

Carbon dioxide from the atmosphere is the source of carbon, which is incorporated into a six-carbon sugar, such as glucose. This sugar is an intermediate product that is a building block for more complex organic molecules. Energy is supplied by solar radiation when light energy is trapped by the chlorophyll of green plants. Some of this energy is stored as potential energy in the sugar molecule, which can be broken down later to provide energy for future cellular activities. Water provides a source of hydrogen atoms, and CO_2 provides a source of oxygen atoms for the organic molecules within the cells of plants and other living organisms.

Other types of photosynthesis reactions are possible in some situations. Photosynthetic bacteria use different inputs and produce different products, but are not of great importance in agricultural systems. The previous equation is really the most fundamental first step for agriculture. *The capture and storage of solar energy by plants makes agricultural production possible.* In a sense, plant agriculture can be thought of as simply photosynthesis on a grand scale! Agriculturists manage the efficient capture of solar energy and atmospheric carbon.

Respiration

The energy necessary for life functions and activities is released by the breakdown (oxidation) of energy-rich organic molecules. Many different organic molecules can be broken down. The oxidation of glucose is the reverse of photosynthesis:

$$C_6H_{12}O_6 + 6O_2 \rightarrow 6CO_2 + 6H_2O + energy$$

The glucose comes from the breakdown of larger organic molecules. Carbon dioxide and water are also released during the process, but the energy released is vital and is used in the cellular metabolism of plants and animals. A similar reaction results when organic materials are burned in fires, but the released energy is dissipated as heat.

The equation for respiration is oversimplified. The breakdown of large organic molecules is a complex process that involves many steps, although glucose or other sugars are common endpoints. Other materials can be oxidized as well, but any respiration that uses oxygen gas is called **aerobic respiration. Anaerobic respiration** can occur when oxygen is absent or limited, and is common in landfills, swamps, underwater sediments, and waterlogged soil. Methane (CH_4), carbon monoxide (CO), hydrogen sulfide (H_2S), and other gases can be produced by anaerobic respiration. Regardless of the pathway or type of respiration involved, respiration is the process by which all living organisms use the potential energy that was originally captured during photosynthesis and stored in organic compounds.

Carbon

The Carbon Cycle in Agriculture

Carbon is included in all organic compounds. Not only is it used to build energy-storing compounds, but it is an essential component of all cells, tissues, and organs. Carbon is incorporated into most agricultural systems through photosynthesis. Because green plants produce the organic carbon food source for the ecosystem, they are called **producers,** or **autotrophs.** Most other organisms, including animals, are referred to as **consumers,** or **heterotrophs.** They cannot produce organic carbon compounds from CO_2 and must obtain carbon from other sources. Growth of consumers, such as cattle or people, is limited by the amount of carbon in their food source (Figure 2-2). Much of the carbon fixed during

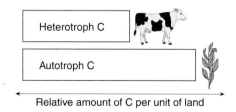

Figure 2-2 Amount of carbon available to consumers is limited by the amount produced by autotrophs.

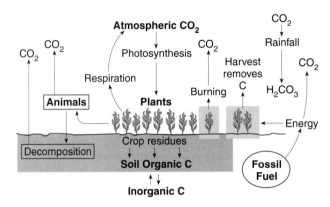

Figure 2-3 Major features of the carbon cycle in agriculture.

photosynthesis is unavailable to consumers. Some is lost as CO_2 during plant respiration; some is lost to decomposition of fallen leaves; some is tied up in plant parts like roots, which are not suitable for human or animal consumption; some is simply missed by cattle during grazing; and some of the carbon compounds consumed are not completely digested by the cattle, people, or other consumers.

The carbon cycle for an agroecosystem is summarized in Figure 2-3. The main input of carbon to the system is atmospheric CO_2 fixed by photosynthesis. A portion of this is returned by respiration of plants. Unharvested residues of plants or plant materials falling onto the soil are broken down and their carbon compounds are passed through a series of soil organisms in the process of decomposition, which is examined in more detail in the next two chapters. Respiration of these organisms releases CO_2 during the decomposition process.

The available pool of soil organic carbon is important to the long-term health of an agroecosystem. In systems in which it is maintained, organic carbon can provide a long-term source of food and energy to the soil organisms that are important in maintaining soil fertility (see Chapters 3 and 4). On the other hand, if organic carbon is rapidly depleted and decomposed to CO_2, it will be unavailable for these purposes. Organic material in soil provides a means for storing carbon and energy for future use in the system.

Three events may have particularly important impacts on the cycling of carbon in an agroecosystem. Harvest results in removal of a portion of the plant carbon to another location. This is, after all, the purpose of an agricultural system, and the main feature that sets agricultural systems apart from natural ecosystems, in which most plant carbon is recycled on the same site. Burning of crop residues, a common practice in some agricultural systems, simply releases the organic carbon as CO_2, preventing it from entering the decomposition process. Grazing by animals can remove a substantial portion of the organic carbon in plants, some of which ultimately becomes incorporated in animal tissue. The remainder of the organic carbon consumed by grazers is released as CO_2 during respiration or passed into the decomposition process as manure.

In agroecosystems, most of the cycling of carbon takes place among the atmosphere, producers, and consumers, including soil organisms (Figure 2-3). However, the largest

Application 2-1 *Fossil Fuels and CO_2 Balance*

Ultimately, carbon from organic compounds can be recycled, even if the cycle is interrupted for millions of years. It is believed that the high concentration of O_2 in the earth's atmosphere resulted from a net increase in O_2 and decrease in CO_2 from photosynthetic activity over a geologic time scale. Normally, the organic carbon from these early plants would have been recycled by decomposition, but many of the plant residues likely fell into prehistoric swamps during the Carboniferous and Permian periods. Oxygen and decomposition were limited in that environment, and so the organic compounds persisted, eventually being converted to oil, coal, or natural gas. When these materials are removed and burned as fuel, the carbon stored in prehistoric times is finally released as CO_2, completing the cycle. This carbon was originally stored in plants by photosynthesis over a very long period of time, and its release over the rather short (geological) time period of the past 100 years has resulted in measurable changes in the CO_2 concentration of the atmosphere.

---

reserves of carbon as inorganic carbon are found in the earth's crust, in the form of carbonates and other compounds. Some of these may be soluble to some extent in the soil water. Carbon dioxide dissolves in water to form carbonic acid:

$$CO_2 + H_2O \leftrightarrow H_2CO_3 \leftrightarrow H^+ + HCO_3^- \leftrightarrow 2H^+ + CO_3^{2-}$$

Depending on the pH and availability of positive ions such as Ca^{2+}, Mg^{2+}, or K^+, the H_2CO_3 may dissociate into bicarbonate (HCO_3^-) or carbonate (CO_3^{2-}) ions. The reaction is reversible, although limited, because some of the carbonate compounds are relatively insoluble. Although some of the soluble bicarbonates and carbonates may be taken up by plants, this is not of particular significance in meeting the carbon needs of plants (atmospheric CO_2 is the main source). Carbonates and bicarbonates may also be introduced into the agroecosystem in irrigation water pumped from wells containing these materials.

Finally, the burning of fossil fuels is indirectly involved in the carbon cycle in agriculture. The purpose of burning fossil fuels is to supply energy, not carbon, to the agroecosystem. Nevertheless, the process results in a release of CO_2 to the atmosphere (see Application 2-1).

Agriculture and CO_2 Production

The extent to which agriculture removes or produces CO_2 varies greatly, depending on the agricultural system involved. In general, plants remove CO_2 and release O_2, but many of the practices associated with plant production release CO_2. Fields that are idle for part of the year still release CO_2 from the decomposition of organic carbon compounds by soil organisms. Plowing disrupts soil aggregates, releasing some of the CO_2 that has accumulated in soil. Plowing also exposes soil organisms and organic carbon to O_2, stimulating decomposition processes that generate CO_2. Grazing or draft animals release CO_2, and the use of fossil fuels for agricultural operations is an important contributor of atmospheric CO_2. Additional CO_2 is produced by the manufacture of agrochemicals and by other agroindustrial processes. Deforestation and clearing of land for agriculture or other purposes increase CO_2 levels in two ways: First, the destruction of plants releases CO_2 slowly through decomposition or

Application 2-2 CO_2 Concentration and Global Warming

Air consists of about 78% N_2, 21% O_2, and 1% other gases, including CO_2. As of 1995, CO_2 concentration was about 0.036%, or 360 parts per million (ppm), and has been increasing at a rate of about 1.0–1.5 ppm per year (Figure 2-4). Worldwide, most of this increase has resulted from the burning of fossil fuels, although deforestation is a more important contributor in many countries.

Many scientists believe that increases in atmospheric concentrations of CO_2 and other gases may result in a **greenhouse effect,** or a gradual warming of the earth due to the increased absorption of heat by these gases. Although CO_2 is the most important "greenhouse gas," chlorofluorocarbons (CFCs), methane, and nitrous oxide are also significant contributors. Their concentrations have also been increasing, primarily from nonagricultural sources. It remains to be seen how much this greenhouse effect may increase the temperature of the earth, but many sources predict a rise of 1–4.5°C (2–8°F) by the second half of the twenty-first century. Opinions vary, and interpretation of temperature change is controversial since urbanization can increase temperature on a local scale. However, evidence of trends in temperature change over time is becoming more frequent. Global temperatures have increased about 0.5°C from the mid-1960s to 1993 (Wilson and Hansen 1994). The mean annual temperature of the forty-eight contiguous U.S. states has remained relatively constant during this century (Karl et al. 1994), although temperatures have fluctuated within individual regions. Coastal California appears to be the region showing the greatest temperature increase during this period.

Global warming and increased CO_2 concentrations could have important effects on agriculture. Crops may be more productive, since CO_2 for photosynthesis would be less limited. With an increase in global temperature, geographic ranges of some plant and animal species may shift, and agricultural production may slowly move toward the poles. Climatic changes or coastal flooding resulting from melted polar ice may eliminate agriculture from some locations. Erratic weather patterns may cause a variety of problems in crop production. Pest and disease problems would likely increase, because they are generally favored by higher temperatures. A variety of models have been developed to study and project the impact of global change on agriculture (Leemans 1997).

---

quickly by burning. Second, the destruction of a forest may result in a net loss or removal of green plants, which won't be available to remove CO_2 and release O_2. Burning of fossil fuels and deforestation are the two main contributors to increased CO_2 concentrations in the atmosphere (see Application 2-2 and Figure 2-4).

Energy in Agroecosystems

Solar Radiation and Photosynthesis

Both the quality and quantity of solar radiation affect the amount of energy absorbed by plants during photosynthesis. Most of the solar energy coming to earth is reflected, converted to heat, or used for evaporation and precipitation. Only a small amount (<1%) is actually used for photosynthesis. Solar radiation consists of a range of electromagnetic

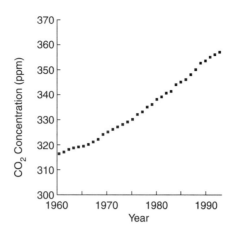

Figure 2-4 Atmospheric CO_2 concentration at Mauna Loa in January each year from 1960 to 1990. Based on data from Keeling and Whorf (1994).

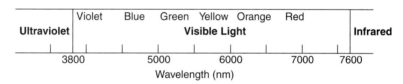

Figure 2-5 The visible portion of the electromagnetic spectrum.

waves, but only certain portions of the visible spectrum (Figure 2-5) are of suitable quality for photosynthesis. Plants reflect green light, and so its use in photosynthesis is limited. Most plant chlorophylls absorb light primarily in the violet/blue (400–500 nm) and red (650–700 nm) regions of the visible spectrum, and so these wavelengths are the most useful for photosynthesis. Other pigments such as carotenoids may absorb light from the yellow and orange regions of the spectrum, thus, some plants may use a wider range of wavelengths.

The quantity, or **intensity,** of light received by a plant depends on geographic location, time of day, season, cloud cover, shading, and many other factors. Crop species show remarkable variation in their adaptations to different levels of light intensity (see Application 2-3).

Thermodynamics and Energy Content

Physical scientists and engineers use the laws of thermodynamics to describe the transformation of energy from one form to another. The first law of thermodynamics deals with the conservation of energy, stating that while one kind of energy may be transformed into

Application 2-3 *C3 versus C4 Plants*

Plants differ in their **photosynthetic pathways,** the series of biochemical steps involved in photosynthesis. One large group of plants uses a three-carbon compound during this process (*C3 plants*), while many others use a four-carbon compound (*C4 plants*). C3 and C4 plants differ in their responses to light and temperature. C4 plants are better adapted to very intense sunlight and high temperatures. Both groups may perform well under moderate conditions, but C3 plants usually outperform C4 plants at low light intensities and cooler temperatures. Examples of C3 and C4 plants are found in a number of plant families, but many (though not all) tropical grasses (family Gramineae) are C4 plants. A third photosynthetic pathway, *CAM* (for crassulacean acid metabolism), is used by some succulent plants such as cacti. Pineapple is an example of an agricultural plant using this pathway.

Examples of C3 crop plants include alfalfa, barley,* bean, cassava, cotton, oat,* peanut, potato, rice,* rye,* soybean, sunflower, and wheat*

Examples of C4 crop plants include corn,* millet,* sorghum,* sugarcane*

*Family Gramineae

---

another form, energy itself cannot be created or destroyed. The second law of thermodynamics basically says that no energy transfer from one form to another is 100% efficient (often, the additional energy is lost as heat). These laws also apply to biological and agricultural systems, and can be illustrated by an important example. In photosynthesis, solar energy is converted into potential energy stored in organic compounds. This transfer is not 100% efficient, since some of the solar energy absorbed by the plant is "lost" as heat due to respiration by the plant. Overall, the total amount of energy is conserved because the incoming solar energy equals stored potential energy plus heat energy. This heat energy may not be the energy product of main interest or use to us, but it is an energy form nonetheless, and can be thought of as an energy "cost" that must be paid by a living organism to convert one form of energy into another.

All living and dead organisms and their products (wood, oil, etc.) contain potential energy stored in their organic compounds. If water is removed from an organism, the **biomass,** or weight of living tissue that remains, will contain some amount of energy that can be determined experimentally.

$$\text{Fresh weight} \rightarrow \text{dry weight (biomass)} \rightarrow \text{energy equivalent}$$

This amount of energy is available if an organism or other organic material is used as a food or fuel source. The energy equivalents of some common substances are presented in Table 2-1, and some of the references cited at the end of this chapter contain many more examples. Ranges are reported for several agricultural crops, representing lower values for hay and higher values for grain. Most values in this table are the gross energy contents of these substances if burned completely. Often, sources discussing animal feeds report digestible energy content rather than gross energy content. **Digestible energy** is the energy content of the material that is actually digested when used as a feed by an animal. It is always less than gross energy content since digestion is not 100% efficient (some energy

Table 2-1 Energy content of selected organic substances.

Material	Energy Content (kcal/g dry weight)	Reference
Soybean oil	9.2	Scott 1986
Fat (general)	9.0	Fluck 1992
Corn oil	9.0	Scott 1986
Animal fat, beef	7.0	Scott 1986
Vertebrates	5.6	Odum 1983
Insects	5.4	Odum 1983
Alfalfa hay	4.5	Fluck 1992
Corn	4.4–5.6[a]	Fluck 1992
Soybean	4.4–4.8[a]	Fluck 1992
Sorghum	4.2–4.4[a]	Fluck 1992
Starch	4.3	Scott 1986
Protein (general)	4.0	Fluck 1992
Carbohydrate (general)	4.0	Fluck 1992
Glucose	3.8	Fluck 1992
Corn grain	3.6[b]	Bennett, Tucker, and Maunder 1990
Wheat grain	3.5[b]	Bennett, Tucker, and Maunder 1990
Sorghum grain	3.5[b]	Bennett, Tucker, and Maunder 1990
Barley grain	3.1[b]	Bennett, Tucker, and Maunder 1990

[a]Higher number is for grain; lower number is for hay or other plant parts.
[b]Digestible energy content rather than gross energy content.

is lost in feces and other nondigestible material). The last four entries in Table 2-1 are digestible energies. Energy is a common denominator for all individual organisms (living or dead) in an ecosystem, and so measurements of its movement or flow through an ecosystem can provide a basis for comparing the relative importance or contribution of very different kinds of organisms.

Energy Use and Transfer: Respiration and Production

Herbivores obtain their energy from the potential energy that was originally stored in the plant tissue that makes up their food source. The energy **consumed** (C) by the herbivore may be used in several different ways (Figure 2-6). Some of the consumed energy may be **assimilated** (A) through digestion and is therefore available for use by the consumer. A portion of the assimilated energy is used for **respiration** (R), to maintain the life functions and activities of the consumer. The remainder of the assimilated energy is available for growth and is stored in the organic molecules that form the animal's tissues. The energy that has been stored in biomass over time is called **production** (P). Some material that is consumed may not actually be digested, and so the energy in this material simply passes through the consumer as feces. This energy is not really utilized or assimilated (NA) by the consumer. Note that

$$A = P + R \text{ and } C = P + R + NA$$

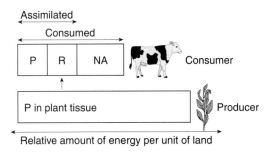

Figure 2-6 Partitioning of consumed energy into production (P), respiration (R), and nonassimilated (NA) energy.

Production, respiration, and other energy measurements are usually reported based on *area of land* and over some period of *time* (see Application 2-4). **Standing crop** is the energy equivalent of biomass available at one point in time. It is different from production, which would include all biomass produced on the site during the course of a year or some other specified time period. Standing crop is analogous to a single harvest of an agricultural crop, whereas production includes multiple harvests within a season and multiple crops within a year. Although the amount of crop biomass present on tropical and cool temperate sites may be similar at one point in time, production at a tropical site is usually greater due to the longer growing season.

Energy or Trophic Pyramid

The transfer of energy through several levels of consumers may be illustrated by the energy, or **trophic, pyramid** (Figure 2-7). Energy for a simple agroecosystem in which cattle are produced for human consumption originates from the solar energy that has been absorbed by plants. **Gross primary production** (GPP) is the total amount of energy captured by the plants during photosynthesis. As mentioned earlier, some of this energy is used for plant respiration and the remainder is stored in plant tissue as production. This production by plants is called **net primary production** (NPP), or net primary productivity. The plants themselves are called producers, since their NPP ultimately produces the energy for the consumers in the trophic pyramid. **Primary consumers** are the herbivores, which derive their energy directly from the NPP. The energy assimilated by cattle or other herbivores is partitioned into respiration and production. The production energy in cattle is available to humans, who are at the next level, the **secondary consumers,** in this simple system.

As energy is transferred upward through the layers, or **trophic levels,** of the pyramid, the amount of production energy available to the next level becomes smaller and smaller, defining the pyramid shape. This hierarchy of producers, primary consumers, secondary consumers, tertiary consumers, and so on, is also called the **food chain.** Notice the similarity in the flow of energy and carbon through a trophic pyramid or a food chain. This is because the production energy at each level is stored within organic (carbon-based) compounds.

Application 2-4 *Units for Energy Measurement*

Wavelength of light: 1 nanometer (nm) = 10^{-9} m
Energy measurement: Basic unit = 1 calorie (cal) = amount of heat needed to raise the temperature of 1 g of water 1°C

1 kilocalorie (kcal) = 1000 cal
1 cal = 4.184 joules (J)
1 kilojoule (kJ) = 1000 J
1 megajoule (MJ) = 10^6 J
1 gigajoule (GJ) = 10^9 J
1 British thermal unit (Btu) = 252 cal (English system)

Measurement of production and respiration:

Units are energy measured over area and time, such as kcal/m^2/yr or kcal/ha/yr.

Standing crop is measured in kcal/m^2 (one time).

Production is measured in kcal/m^2/yr (over time).

Notes on energy measurement: Although this text uses calories for energy measurement, many people, particularly those in agricultural engineering, use joules, so interconversion of the two measurements is important. Dieticians and various popular sources often use "calories" to measure the energy content of foods. This term is not a true calorie, but is actually a kilocalorie, sometimes written as Calorie (capital "C"), although often the capital "C" is dropped, causing confusion.

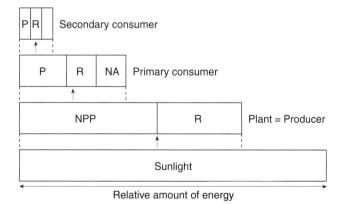

Figure 2-7 A typical trophic pyramid for an agricultural system. NPP = net primary production; P = production; R = respiration; NA = nonassimilated energy.

Ecological Efficiency

The relative amounts of energy available to each trophic level are shown to decrease in Figure 2-7, but these amounts are not drawn to scale. The capture of solar radiation by plants is particularly inefficient. On average, only about 1% or less of solar radiation is absorbed by plants during photosynthesis. Although photosynthetic efficiencies of up to 10% have been reported for some crops under optimum growing conditions, these are maximum daily rates that are not sustained over an entire season. Some C4 grasses such as sugarcane or corn can maintain efficiencies of about 2% over time. For most plants, about half of the absorbed energy is actually converted to production (50% for respiration), and so the actual conversion of solar energy to NPP is <1%. Energy is transferred from plants to higher levels of the trophic pyramid at about 10% efficiency; that is, 100 g of plant NPP produces 10 g of herbivore production, and so on.

The use and allocation of consumed energy varies among animal species, depending on their habits. Grazing herbivores such as cattle may consume a high proportion (30–50%) of NPP, but their assimilation efficiency is low. Much of the energy consumed is returned to the site in feces. While herbivores may assimilate only 20–50% of consumed energy, the rates for carnivores typically are much higher (80–90%). The amount of assimilated energy allocated for respiration increases as the activity of the organism increases. Active predators and animals that maintain a constant body temperature must divert particularly large amounts of energy to respiration. Birds and small predatory mammals may use 98% of their assimilated energy for respiration.

Energy Flow in Agroecosystems

The flow of energy through an agroecosystem follows the carbon cycle (see Figure 2-3) rather closely, since much of the energy is stored in carbon compounds. Unlike carbon, energy does not cycle because the pool of stored potential energy is continually decreased by energy that is "spent" on respiration and released as heat. These points of respiration energy release correspond to the points of respiratory CO_2 evolution shown in Figure 2-3. The cycling of materials like carbon is critical because the total amount of carbon in the earth and its atmosphere is essentially constant. Loss of one form of energy (potential energy in organic compounds) is not as critical since this is derived from solar radiation, which is a renewable resource.

Energy supplied to agroecosystems as fossil fuels, labor, or other means does not enter the trophic pyramid directly, but has significant indirect effects. Energy supplied as fuel for cultivation may result in removal of weeds that may be shading or competing with crop plants, improving the photosynthetic efficiency of the crop. Energy supplied to control insects feeding on a crop may result in less NPP lost to pests and more NPP available for harvest. Many different kinds of energy inputs may result in increased levels of production or in the conservation of existing production energy that has been incorporated into the plant or animal products harvested from agroecosystems.

Energy Equivalent of Agricultural Inputs

The energy equivalents of indirect (nonphotosynthetic) energy inputs into agricultural systems have been determined for many different crops and locations (see the References at the

Figure 2-8 Energy inputs to agriculture include not only tractor fuel and human labor, but also the energy used to produce pesticides, fertilizers, seed, and any other materials used in agriculture.

end of the chapter for many examples). Determining the energy input from fuel consumption is relatively straightforward, but theoretically it should be possible to determine or estimate an energy equivalent or cost for every material or process used in agriculture. This would include human or animal labor, any field operation involving machinery, irrigation, transportation, drying of some crops, etc. (Figure 2-8). The energy needed to produce various materials used in agriculture must also be considered, since the energy costs in producing synthetic fertilizers and pesticides can be considerable. When the energy equivalents of crop yield and other agricultural outputs are determined and compared to input energy, the energy efficiency of various agricultural production systems can be assessed.

For mechanized production of most crops in the United States, the largest energy inputs typically are for fertilizers, farm machinery, transportation and shipping, irrigation, crop drying, and pesticides. Pimentel and Burgess (1980) give detailed summaries of the energy inputs and outputs for nonirrigated corn production in twenty-five different U.S. states. In twenty-two of these (88%), the largest energy input is for nitrogen fertilizer. Total energy output is more than twice the total energy input in twenty-three states (92%).

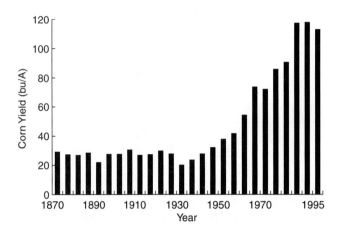

Figure 2-9 Average corn yield in the United States for 125 years from 1870 to 1995. Data shown in bu/A; 1 bu/A = 62.7 kg/ha, assuming 1 bu = 56 lbs.
Based on data presented by United States Department of Agriculture (1936, 1950, 1966, 1975, 1988, 1997).

Energy Input and Yield

The amount of energy input to agriculture has varied greatly through the course of history and into the present day. Three stages of energy intensification may be recognized. *Preindustrial* systems use low inputs of energy and labor, while providing only limited output. An extreme example is a hunting-gathering society, which may have a low harvest rate and minimal environmental impact, but may require a relatively large area to provide food for just one person. *Semi-industrial* agricultural systems use higher inputs of labor and energy, and often use supplemental energy in the form of draft animals to relieve human labor. Irrigation and other technologies may be involved and agriculture may support some nonagricultural population. *Full-industrial,* or **intensive,** agriculture relies on high inputs of fossil fuels and use of machinery to minimize human labor and maximize yield (energy output).

United States agriculture has shifted from a semi-industrial to full-industrial scale over the last 100 years. Crop yields have risen greatly over this time period, as illustrated by the trends in U.S. corn yields over the 125 years from 1870 to 1995 (Figure 2-9). Increases in corn yields in the 1940s were due mainly to the use of hybrid corn, while those of the 1950s and beyond were due mostly to an increase in mechanization and the use of agrochemicals, as well as hybrid improvement. As a result, use of fossil fuels and materials produced from fossil fuels (fertilizers, herbicides, insecticides) has escalated over this time period as well. For example, note the great increase in nitrogen fertilizer use in the United States during the fifty-year period from 1930 to 1980 (Figure 2-10).

In general, the overall energy efficiency of an agricultural system decreases with the increased use of fossil fuels since so much energy is input into agricultural production. Data from the energy summary of corn production mentioned previously (Pimentel and Burgess 1980) indicate that corn production in the United States had an energy efficiency (energy output:energy input) ratio of 2.47:1, with an average corn yield of 5160 kg/ha. Corn production in Mexico using oxen (semi-industrial scenario) resulted in an energy efficiency of

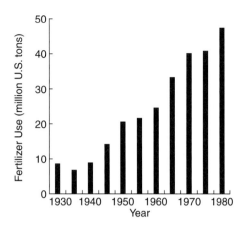

Figure 2-10 Commercial fertilizer use in the United States during fifty years from 1930 to 1980. Data shown in U.S. tons (1.0 U.S. ton = 0.91 mt).
Based on data presented by United States Department of Agriculture (1936, 1950, 1966, 1975, 1988).

4.25:1 and an average yield of 941 kg/ha, while production in Mexico by hand labor had an efficiency of 128.2:1, with a yield of 1944 kg/ha. However, energy efficiency of preindustrial agricultural systems can be very difficult to evaluate. For example, shifting ("slash and burn") agriculture is not efficient (energy efficiency ratio <1:1) if the energy lost in burning forests or other biomass is considered.

Less intensive agricultural production systems generally have less environmental impact and are more efficient in energy use than are intensive industrial systems. But mechanized intensive systems produce much higher overall yields. Much more land is required by less intensive systems to produce an equivalent amount of yield. This is critical, since a high level of yield (food production) is required to sustain the high population of the world today (see Chapter 18). Nevertheless, as mechanized agriculture strives to produce higher yields, there are increasing concerns about environmental problems and eventual depletion of the fossil fuel resources required by these systems. Interest in the development of alternative energy sources and production systems continues to increase. One possibility of particular interest to agriculture is the production of fuel from crop biomass (see Application 2-5).

Maximizing Crop Yield

There are several very general approaches for increasing the use of light energy and crop yield per unit of land:

Increasing plant size and weight. Larger plants result in more yield biomass per hectare, and the larger leaves increase the overall leaf area available for photosynthesis. Many agricultural practices are devoted to increasing the weight or mass of individual plants. Most of the yield increase resulting from increased energy input can be attributed to methods that increase plant weight.

Application 2-5 *Fuel from Biomass Crops*

The use of biomass for fuel is not a new concept—wood has been used as a major fuel source since ancient times. Many other organic materials can be used as fuel sources as well, including crop residues, manures, aquatic plants, exotic plants rich in hydrocarbons, or vegetable oils from crops such as sunflower, soybean, olive, coconut, canola, peanut, or palm. Anaerobic digestion of organic products can produce biogas consisting of methane and other combustible products. Crop residues can be sources of alcohols, such as ethanol and methanol.

Ethanol is useful as an engine fuel and its use in blends with gasoline is increasing. It can be produced by fermentation of sugars derived from crop residues. Residues from sugar-rich crops, such as sugarcane and sugarbeet, are particularly useful for this purpose. Crops high in starches such as cassava or potato and starchy cereals like corn or sorghum can also provide high yields of ethanol. Several of these crops (sugarcane, corn, sorghum) are C4 grasses with high conversion rates of solar energy. Although crops can be grown for ethanol production on a large scale, this requires tying up cropland that could be used to grow food crops.

- **Increasing harvest index.** The **harvest index** is the proportion of the crop yield (% by weight) that is actually used. When corn is harvested for grain, much of the plant is not used for human consumption. The harvest index is higher for lettuce, since most of the plant is eaten, or for potato, since the tubers are a particularly heavy part of the plant. Although harvest index can be affected by management, many methods aimed at increasing plant size affect both usable and nonusable plant parts, and so their relative proportions may remain unchanged. The most significant changes in harvest index are usually obtained over time through plant breeding and improvement.
- **Increasing plant population.** In some situations, yields may be increased easily by simply growing additional plants in what would normally be vacant space between rows, resulting in an overall increase in the plant population per hectare (see Application 2-6). Plant population is also increased over time through multiple cropping, cover cropping, or other practices that make use of available land for as much of the growing season as possible (see Chapter 13).

Application 2-6 *Optimum Plant Spacing*

Traditionally, wide spacing (76–91 cm = 30–36 in.) between plant rows has been maintained for ease in movement of equipment or animals during cultivation and other field operations. Frequently, higher rates of photosynthesis and plant production per unit of land can be achieved by using closer spacing between rows. A typical response of crop yield to plant spacing is shown in Figure 2-11. In this example, yield increased by 300 kg/ha when the distance between rows was reduced from 76 cm to 38 cm. Yield increases as plant population

(continued)

per hectare increases, up to the optimum spacing for the crop. Optimum spacings vary depending on plant species and cultivar, but are known for many crop plants. If spacing between rows is too low, plants become too crowded and yield or quality may be reduced due to plant competition (see Chapters 6 and 7) or possibly to increased incidence of plant disease (see Chapter 10). Plant populations may also be increased by decreasing distances between plants in rows or by increasing broadcast seeding rates for crops like small grains, which are often not planted in rows. The production system may impose limits on plant population. For instance, corn or sorghum cultivars grown for silage can be planted at much higher plant densities than those grown for grain yield, since optimum silage yields are achieved at higher densities.

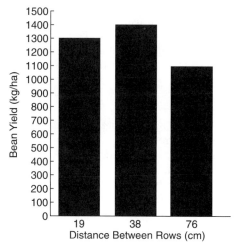

Figure 2-11 Relationship between marketable bean yield and spacing between crop rows. Based on data from Russo (1995).

Summary

Solar energy and carbon are supplied to agroecosystems through photosynthesis. Movement of carbon can be traced through the carbon cycle at the level of the agroecosystem, as well as on a world basis. The movement of energy parallels the carbon cycle to some extent because potential energy stored in carbon compounds is used by most organisms during respiration. However, since energy transfer is not efficient, the amount of energy available decreases greatly as it is transferred from producers to higher level consumers. The energy efficiency of an agricultural production system can be determined from the energy equivalent of yield produced and the energy equivalents of all input materials and operations. Modern mechanized agriculture has used high energy inputs to produce exceptionally high yields.

Topics for Review and Discussion

1. How can the efficiency of solar energy use in agriculture be increased?
2. How might the carbon cycle for an agroecosystem differ from that for a natural ecosystem?
3. Inputs such as CO_2 or solar energy are used directly by growing plants. But the energy used in applying herbicides or insecticides is indirect, since the plants do not actually use the energy from the application. Explain how, in terms of energy use and conservation, an indirect energy application, such as herbicide or insecticide use, may benefit plant growth and yield.
4. The organisms in an ecological community could be described in terms of energy (such as production or standing crop), or they could be described in terms of numbers or biomass. Discuss the advantages and disadvantages of describing the organisms in an agroecosystem in terms of numbers, biomass, or energy.
5. A semi-industrial agricultural system using relatively low energy input may require much more land to produce an amount of food equal to that produced by an intensive agricultural system with high energy input. Discuss the trade-offs involved in adopting production systems having various levels of energy intensity.

Literature Cited

Bennett, W. F., B. B. Tucker, and A. B. Maunder. 1990. *Modern Grain Sorghum Production.* Ames, Ia.: Iowa State University Press.

Fluck, R. C. 1992. Energy of agricultural products. In *Energy in Farm Production,* ed. R. C. Fluck, 39–43. Amsterdam: Elsevier.

Karl, T. R., D. R. Easterling, R. W. Knight, and P. Y. Hughes. 1994. U.S. national and regional temperature anomalies. In *Trends '93: A Compendium of Data on Global Change,* eds. T. A. Boden, D. P. Kaiser, R. J. Sepanski, and F. W. Stoss, 686–736. Carbon Dioxide Information Analysis Center, Oak Ridge National Laboratory, Oak Ridge, Tenn.

Keeling, C. D., and T. P. Whorf. 1994. Atmospheric CO_2 records from sites in the SIO air sampling network. In *Trends '93: A Compendium of Data on Global Change,* eds. T. A. Boden, D. P. Kaiser, R. J. Sepanski, and F. W. Stoss, 16–26. Carbon Dioxide Information Analysis Center, Oak Ridge National Laboratory, Oak Ridge, Tenn.

Leemans, R. 1997. Effects of global change on agricultural land use: Scaling up from physiological processes to ecosystem dynamics. In *Ecology in Agriculture,* ed. L. E. Jackson, 415–452. San Diego, Calif.: Academic Press.

Odum, E. P. 1983. *Basic Ecology.* Philadelphia: Saunders College Publishing.

Pimentel, D., and M. Burgess. 1980. Energy inputs in corn production. In *Handbook of Energy Utilization in Agriculture,* eds. D. Pimentel, 67–84. Boca Raton, Fla.: CRC Press, Inc.

Russo, V. M. 1995. Bedding, plant population, and spray-on mulch tested to increase dry bean yield. *HortScience* 30: 53–54.

Scott, M. L. 1986. *Nutrition of Humans and Selected Animal Species.* New York: John Wiley and Sons.

U.S. Dept. of Agriculture. 1936. *Agricultural Statistics 1936.* Washington, D.C.: U.S. Government Printing Office.

———. 1950. *Agricultural Statistics 1950.* Washington, D.C.: U.S. Government Printing Office.

———. 1966. *Agricultural Statistics 1966*. Washington, D.C.: U.S. Government Printing Office.
———. 1975. *Agricultural Statistics 1975*. Washington, D.C.: U.S. Government Printing Office.
———. 1988. *Agricultural Statistics 1988*. Washington, D.C.: U.S. Government Printing Office.
———. 1997. *Agricultural Statistics 1997*. Washington, D.C.: U.S. Government Printing Office.
Wilson, H., and J. Hansen. 1994. Global and hemispheric temperature anomalies from instrumental surface air temperature records. In *Trends '93: A Compendium of Data on Global Change,* eds. T. A. Boden, D. P. Kaiser, R. J. Sepanski, and F. W. Stoss, 609–614. Carbon Dioxide Information Analysis Center, Oak Ridge National Laboratory, Oak Ridge, Tenn.

Bibliography

Begon, M., J. L. Harper, and C. L. Townsend. 1990. *Ecology. Individuals, Populations and Communities*. Cambridge, Mass.: Blackwell Scientific Publications.
Beran, M. A., ed. 1995. *Carbon Sequestration in the Biosphere*. Berlin: Springer-Verlag.
Brady, N. C., and R. R. Weil. 1996. *The Nature and Properties of Soils*. Upper Saddle River, N.J.: Prentice-Hall.
Bright, C. 1997. Tracking the ecology of climate change. Pp. 78–94 in *State of the World 1997,* eds. L. R. Brown, C. Flavin, and H. F. French. New York: W. W. Norton.
Fluck, R. C. 1992. *Energy in Farm Production*. Amsterdam: Elsevier.
Fluck, R. C., and C. D. Baird. 1980. *Agricultural Energetics*. Westport, Conn.: AVI Publishing Company.
Gurney, R. J., F. L. Foster, and C. L. Parkinson. 1993. *Atlas of Satellite Observations Related to Global Change*. Cambridge, U.K.: Cambridge University Press.
Hammond, A. L., ed. 1990. *World Resources 1990–91*. New York: Oxford University Press.
Jackson, R. B. 1992. On estimating agriculture's net contribution to atmospheric carbon. In *Natural Sinks of CO_2,* eds. J. Wisniewski and A. E. Lugo, 121–137. Dordrecht, The Netherlands: Kluwer Academic Publishers.
Keeton, W. T., and C. H. McFadden. 1983. *Elements of Biological Science*. 3d ed. New York: W. W. Norton.
Schlesinger, W. H. 1991. *Biogeochemistry: An Analysis of Global Change*. San Diego, Calif.: Academic Press.
Stout, B. A. 1984. *Energy Use and Management in Agriculture*. North Scituate, Mass.: Breton Publishers.
———. 1990. *Handbook of Energy for World Agriculture*. London: Elsevier Applied Science.

Chapter 3

Community Ecology

Key Concepts

- Agroecosystems contain communities of organisms
- Soil organisms are vital in agroecosystems
- All ecosystems are interconnected

In the previous chapter, the concepts of the trophic pyramid and the food chain were introduced as convenient means for following the movement of material (carbon) and potential energy from producers through several levels of consumers. A food chain for a simple agroecosystem might consist of:

Corn → cattle → people

On the surface, this basic food chain might seem to cover the energy transfer within this system. This may actually be true for much *(but not all)* of the energy transfer in a particularly efficient system. But a closer examination will reveal that the agroecosystem is much more complicated than this simple food chain would indicate. Weeds may be present along with the crop plant, serving as an additional food source for the cattle. **Herbivorous** (plant-feeding) insects may remove some energy from corn or weeds, and they in turn may be eaten by various predators. People may obtain energy not only from cattle, but from many other systems as well, often as primary consumers of plant material.

In this chapter, we introduce some common features of food webs found in agricultural systems, including an introduction to the organisms found in complicated below-ground food webs. Since all agricultural soils contain various types of organisms, the simplest agricultural field, even a fallow site, can be considered an ecosystem with a characteristic community of organisms.

Above-Ground Food Webs

It is more realistic to replace the simple corn-cattle-people food chain with a more complicated diagram such as that shown in Figure 3-1. A **food web** shows the organisms present

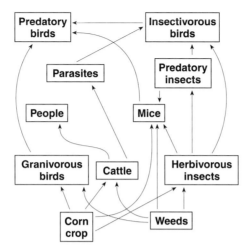

Figure 3-1 A generalized above-ground food web.

in an ecosystem, with arrows indicating direction of food (= energy) transfer. Many transfers and interactions may exist in an agroecosystem. The specific components and interactions depend on the particular agroecosystem, but many of these may be difficult to identify.

The basic corn-cattle-people system is expanded in Figure 3-1 to include primary consumers other than cattle. Herbivorous insects may feed on corn or on various weeds, and some consumers such as granivorous birds or mice may feed on the corn grain or on seeds of some weeds. Predatory insects may consume plant-feeding insects, and various parasites (flies, worms, etc.) may remove energy from the cattle population. Some groups of organisms may be placed clearly into particular trophic levels, but the assignment of other groups to trophic levels may be impossible. Insectivorous (insect-feeding) birds may not distinguish whether an insect food source was a primary consumer (herbivore) or a secondary consumer (predator). Predatory birds such as hawks or falcons may remove prey items from various levels. **Omnivores** feed on plant or animal matter. For example, mice may act as higher-level consumers of insects or as primary consumers when they feed on grain.

The arbitrary food web presented in Figure 3-1 is still greatly oversimplified, however. Many of the component boxes shown would probably consist of multiple species of weeds, insects, parasites, or birds. Separate boxes could be drawn for each of these species, and arrows added to show *all* energy transfers, from different species of weeds to different kinds of insects, etc. This would result in a highly complex food web that would be difficult to illustrate or interpret. In addition, many groups of organisms common in agroecosystems are not even shown in Figure 3-1, such as bacteria, fungi, and other parasites of plants, insects, or vertebrates.

How many different organisms are represented by each "box" in a food web depends on the particular crop species and environment. For example, although some 280 different insect species are reported to feed on sweet potato, only a few are responsible for most of the crop's problems (Jansson and Raman 1991). Presumably, these would account for most of the energy flow from plant to insect herbivores. During a three-year study, six to thirteen common insect herbivores and three to six common arthropod (insects and spiders) predators were regularly encountered in irrigated soybeans in Mississippi (Felland and Pitre 1991). The complexity of interactions among various groups depends on the habits of the individual organisms involved. **Generalists** feed on a variety of food sources and are said to be **polyphagous.** These include insects that may derive small amounts of energy from a range of crop plants and weeds, as well as nonspecific predators that choose prey items based simply on size or availability. **Specialists** are much more specific in their food habits. Some parasites may even be limited to a single food item.

Some agroecosystems may contain several different species that appear to have identical feeding habits. Ecologists use the term **guild** to refer to a group of species that use the same food resource or portion of the food resource. The term is not used so generally as to include all consumers of a particular plant together, but more specifically, as for a guild of mites feeding on a leaf surface, or a guild of fungi on plant roots. Of twenty-one species of ladybird beetles found in apple orchards in Hungary, several could be included in a guild of predatory insects feeding on aphids (Lövei, Sarospataki, and Radwan 1991).

The dynamics and even the structure of a food web change over time. Herbivores present during the early stages of crop growth may differ from those present in older crops. Availability of pollen or grain may determine the occurrence and abundance of some organisms. Near the end of the growing season, increasing amounts of dead material (leaves, insects, weed residues, etc.) may fall to the ground and enter soil food webs during the decomposition process.

Soil Organisms and Soil Biology

Soil is an excellent habitat for many different kinds of organisms. However, the rich biodiversity of soil is often overlooked because most of these organisms are microscopic. Nevertheless, some organisms are extremely abundant in soil. Typically, a liter of good agricultural soil may contain:

100,000,000,000	Bacteria
10,000,000,000	Actinomycetes
100,000,000	Fungi
10,000,000	Algae
10,000,000	Protozoa
10,000	Nematodes
500	Mites and insects
200	Enchytraeid worms
2	Earthworms

Many of these organisms are obscure and poorly known. Some of these groups contain species that are unknown and have not yet been described by science. It is likely that a handful of soil will contain undescribed species of bacteria, actinomycetes, fungi, algae, protozoa, nematodes, enchytraeids, and possibly mites. Therefore, at the present time, it is not possible to estimate the **biodiversity,** or number of species present, for agricultural soils. However, while it is impossible to identify most individuals to the species level, we can recognize some of the major groups of organisms that are present in most soils.

Representatives of all kingdoms of living organisms (see Chapter Endnote 3-1) occur in soil. Detailed descriptions and identification of many of these are presented elsewhere (see Dindal 1990). A general introduction to the most important groups common in most agricultural soils is presented here.

Bacteria

These members of the kingdom Monera are the most abundant soil organisms. However, because of their extremely small size, their total biomass in soil is lower than that of some other soil organisms. They are extremely important in the decomposition of organic material in soil (see Chapter 4), and many different kinds of soil bacteria are involved in this beneficial process. Others are beneficial in nitrogen fixation or in the mineralization of nitrogen and other nutrients. Bacteria are well-known as causative agents of disease, and different kinds of soil bacteria may cause diseases of plants, insects, and other soil animals. Note that different kinds of bacteria are responsible for these different functions. For instance, the bacteria that cause diseases in roots of plants are different from those that cause diseases in insects.

Many different types of bacteria live in the **rhizosphere,** the region of soil that is closely associated with plant roots. Much decomposition, exchange of nutrients, and other activity takes place in this region, and large numbers of beneficial bacteria are associated with these processes. As a group, these bacteria, which colonize the region around plant roots, are called **rhizobacteria.** Current research is examining whether rhizobacteria may interfere with pathogenic fungi or other pests that may attack plant roots.

Actinomycetes are a group of bacteria with a characteristic filamentous, or branched, appearance (Figure 3-2). They are abundant in soil, where they are very important as decomposers of organic material.

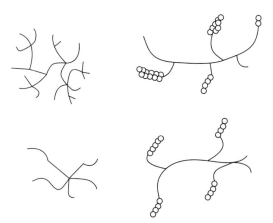

Figure 3-2 Various structures of actinomycetes.

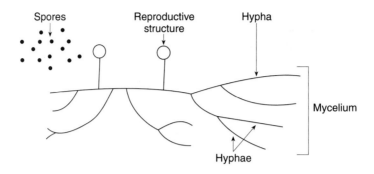

Figure 3-3 Generalized structure of a fungus.

Blue-green algae, or Cyanobacteria, are not true bacteria but are included in the kingdom Monera as close relatives of bacteria. They are filamentous and pigmented, and some kinds can fix nitrogen (N) in soils.

Fungi

Fungi comprise a separate kingdom in many recent classification systems. Many different kinds of fungi occur in soil in great numbers. Nutrients are obtained by filamentous strands, or **hyphae,** which collectively make up a mass called the **mycelium** (Figure 3-3). Fungi show a remarkable diversity of life cycles and reproductive habits. Various types of fruiting bodies and reproductive structures may be produced, depending on the type of fungus. The mycelia of many common fungi may go unnoticed in soil until mushrooms or other fruiting bodies are produced.

Fungi are important decomposers of organic matter, and this is their most important function in soil. **Mycorrhizae** are specialized fungi that live in close association with plant roots and aid plants in the uptake of nutrients (see Chapter 4). Some fungi are causative agents of plant diseases (see Chapter 10). Other fungi may infect insects or nematodes, and may benefit plants indirectly when they attack crop pests. **Endophytes** are fungi that live inside of plants. Their presence may make some plants less desirable to herbivores.

Plants

Plant roots are a major feature of agricultural soils. However, one group of plants and plant-like organisms that is common in soil and is often overlooked is the soil **algae.** Although favored by moisture in soils, algae are common even in arid regions, where they develop rapidly following rain. Algae are involved in the cycling of nutrients in soils, and photosynthetic forms can incorporate carbon and energy into soil food webs.

Protozoa

Although modern classifications place protozoa in the kingdom Protista, many zoologists study them as one-celled animals. They are frequent inhabitants of the moisture films in soil pores and are important consumers of soil bacteria. **Amoebae** are the most common protozoa in agricultural soils, and may consume fungal spores, nematodes, and other small organisms in addition to bacteria. **Slime molds,** fungus-like organisms related to protozoa only by the fact that they are also included in the kingdom Protista, may be involved in the decomposition of organic matter in some systems.

Animals

A remarkable diversity of animals inhabits agricultural soils. Most of these animals are **invertebrates,** a general term for animals without backbones. The sizes of soil animals vary greatly, and a specialized terminology may be used to describe these size ranges. Some authors have attempted to set measured limits to size the classes of soil invertebrates. However, definitions to these size classes vary, and some groups of animals such as ants or millipedes may occur in many different sizes. So, some simple rules of thumb are probably sufficient for recognizing these size classes. The **macrofauna** consists of animals large enough to see with the unaided eye. It includes larger animals as well as those as small as ants and termites. Members of the **microfauna** are too small to observe without a microscope. This group includes most nematodes and protozoans (Protista, but often included as animals in soil zoology). Members of the **mesofauna** are intermediate in size and require a hand lens for observation. A mite may be visible as a speck to your unaided eye, but you would probably need magnification to tell what it was. The mesofauna includes mites, Collembola, and enchytraeid worms (see the following section for descriptions of these groups). Since mites and Collembola are among the smallest members of the phylum Arthropoda, they are referred to collectively as **microarthropods.** A general introduction to classification and the relationships among the various animal phyla in soil is presented in Chapter Endnote 3-1 and Figure 3-13. Groups likely to be encountered in most agricultural soils are described in the following sections.

Nematodes. Nematodes, or roundworms, are microscopic, unsegmented worms (Figure 3-4). Most nematodes that inhabit soil range in length from 0.5 to 2.0 mm. From about twenty to as many as fifty different genera (some with undescribed species) may occur in a handful of typical agricultural soil. The different kinds of nematodes serve many different functions in the soil ecosystem. **Bacterivores,** which feed on soil bacteria, may be the most common nematodes in many agroecosystems (see Application 3-1 for clarification of terms). **Fungivores** feed only on hyphae and other fungal material in soils. The nematode herbivores in soil are also plant parasites, feeding on or in plant roots. Some of them are important agricultural pests (see Chapter 10). **Predators** feed on nematodes and other microscopic animals. Omnivores may eat other animals, but supplement their diet with fungi, algae, and other available material. Although present in most soils, omnivores and predators are usually less common than bacterivores, fungivores, and herbivores. Insect parasites may be beneficial when they feed on insect pests in the soil. Algivores and vertebrate parasites may be found occasionally in some soils. Although some are plant pests, soil nematodes as a group are mainly beneficial, since most kinds (bacterivores, fungivores, omnivores) are involved in decomposition and the recycling of nutrients.

Annelids. Earthworms, members of the phylum Annelida, are among the best-known soil inhabitants. Earthworms and other annelids are segmented worms that are considered more advanced than nematodes because they have a circulatory system and other more

40 ■ *Chapter 3*

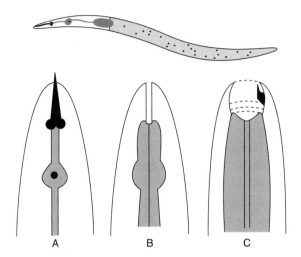

Figure 3-4 Nematodes are transparent under the light microscope, making many internal features visible. The top diagram shows a typical wormlike shape. The bottom diagram shows the head, mouthparts, and anterior end for a plant parasite (A), a bacterivore (B), and a predator (C).

Application 3-1 *Clarification of Feeding Terminology*

Some terms commonly used to describe the feeding habits of soil and aboveground invertebrates are:

Term as Noun	Adjective Form	Definition
Algivore	Algivorous	Organism that feeds on algae
Bacterivore	Bacterivorous	Organism that feeds on bacteria
Fungivore	Fungivorous	Organism that feeds on fungi
Herbivore	Herbivorous	Organism that feeds on plant material
Predator	Predatory	Organism that feeds on other animals or protozoa
Omnivore	Omnivorous	Organism that has more than one of the above feeding habits

The noun and adjective forms of these words are often confused. The phrases "bacterivorous nematodes" and "nematode bacterivores" mean the same thing even though different words are used as nouns and adjectives. All of these terms refer to different types of **consumers** (see Chapter 2). Many other terms are used as well, some of which are introduced later in this book. For instance, **phytophagous** is synonymous with "herbivorous." **Microbivore** is a general term used to describe general feeding on very small organisms, mainly bacteria and fungi. However, in some contexts, "microbivore" may be synonymous with "bacterivore."

complex features. More than 3000 different known species of earthworms are divided into a number of different families. Most European earthworms are members of the family Lumbricidae, but the diversity of earthworm species and families increases in tropical regions. The earthworm fauna of the United States show many important differences among the eastern, southern, and western regions of the country (see Dindal 1990). Also, some European lumbricids are **cosmopolitan,** since they have been introduced throughout the world with agriculture. The species of earthworms present in a particular site depends on the geographical location, soil type, and site history. Typically, a temperate agricultural soil contains four to six species (Edwards et al. 1995). Earthworms are particularly important in the ecology of agricultural soils (see Application 3-2).

Enchytraeids are small, white or gray-colored relatives of earthworms, classified in the family Enchytraeidae. Although larger enchytraeids are common in some habitats, most of those found in agricultural soils are about 5–20 mm long and are considered part of the mesofauna. Like earthworms, they are important in decomposition and nutrient cycling. Enchytraeids ingest a variety of materials including organic plant material, fungi, algae, and bacteria. They are generally considered omnivores, although fungi may provide a significant part of the diet for some species.

Arthropods. Invertebrates with jointed appendages are classified in the phylum Arthropoda. The classification of the arthropods discussed in this book is outlined in Table 3-1. We limit our discussion only to those arthropods important in agroecosystems. Many arthropod groups contain large numbers of species important in natural ecosystems. The most common and important arthropods in agricultural soils are discussed in the following sections.

Mites are small ($\leq$ 2.0 mm in length) members of the mesofauna that can be distinguished from insects by their major body features (Figure 3-5). Many different species of mites can occur in the soil environment. Their diversity and abundance are greatest in litter and the uppermost layers of soil. **Litter** is a layer of debris on the soil surface formed from fragments of plant and animal material (see Chapters 4 and 5). The feeding habits of different species of mites vary, even within a group (several different suborders of mites are common in agricultural soils). Some groups of soil mites contain mostly predators, which feed on other mites, nematodes, enchytraeids, and microscopic insects. The most abundant suborder of mites in soil, the oribatid mites, or "beetle mites" (descriptive of their characteristic shape), are mostly fungivores. However, some members of the various groups of soil mites may feed on algae, others on bacteria, and some on organic plant material, while others are parasites of vertebrates or invertebrates. As with other kinds of soil invertebrates, the feeding habits of many mite species are unknown. Most soil mites are involved in the decomposition process. Many aboveground species of mites are important animal parasites or plant pests (see Chapter 10).

Collembola, or **springtails,** are the most abundant soil insects. These small (mostly 1–5 mm) insects have an abdominal appendage that can be snapped against the ground to force the animal into the air (Figure 3-6). You may have noticed them as small specks appearing when a log or crop litter is turned over, and then rapidly disappearing from view as they "jumped" out of the way. Collembola are abundant in litter and upper layers of soil, where most of them feed on fungi. Some feed on bacteria, and nematode predation has been demonstrated in the laboratory for a few species.

Several other groups of insects are particularly important in soil. *Termites* (Figure 3-7) are large enough in size and numbers that they can have important effects on soil structure, particularly in the tropics, where their biodiversity is greatest. Their ability to use cellulose or

Application 3-2 Earthworms and Agriculture

Earthworms feed mainly on organic material originating from plants. However, their feeding habit is unusual since they ingest large amounts of soil along with organic debris. The great bulk of soil passes undigested through the digestive system, where it is enriched with nitrogen and other by-products of digestion and excretion, and is deposited as a **casting.** In addition to plant debris and soil, a variety of other substances are ingested (microscopic animals, fungi, bacteria, fecal material from larger animals, etc.). Because of this, earthworms are considered omnivores, although it is not clear how much of these various materials is actually digested. Different types of earthworms show important differences in their habitats and foraging habits. *Litter dwellers* inhabit and feed in layers of litter and other organic debris, rarely burrowing into mineral soil. *Vertical burrowers* live in typical burrows and emerge at night to feed at the soil surface. *Horizontal burrowers* remain in soil, burrowing horizontally to search for organic food sources.

Earthworms benefit *soil structure* by improving soil aggregation and porosity. Soil particles ingested by earthworms stick together to some extent in the castings, which form a basis for aggregates. The burrowing activities improve soil porosity and aeration. Drainage and water movement may improve as a result of the increased porosity (see Chapter 5 for further discussion of soil structure).

Earthworms benefit *soil fertility* in a variety of ways. They are decomposers of organic material and their activities accelerate nutrient cycling and the return of nitrogen, phosphorus, and other nutrients to the soil. Their castings are rich in nutrients and are sites of intense microbial activity. Movement of earthworms aids in incorporating organic matter into soil and in redistributing both organic matter and soil organisms to new locations. Even though earthworm numbers in soil are relatively low, their effects are significant because as a group, earthworms turn over (pass through their digestive system) relatively large quantities of soil.

A number of agricultural practices affect earthworm numbers in soil. Application of organic material (= earthworm food source) is the most important factor in increasing earthworm populations. For this reason, earthworms are usually more common in sites receiving organic fertilizer than in sites receiving inorganic fertilizer. Inorganic fertilizers may affect earthworms indirectly if they result in increases in plant biomass that could eventually serve as an organic food source. Likewise, crops that produce much residue are beneficial to earthworms. Plowing is detrimental to earthworms. Populations tend to be higher in systems with reduced tillage (see Chapter 14) than in systems with frequent cultivation. Earthworm numbers may also be reduced by pesticides, including herbicides, or by the burning of crop residues. Liming may increase numbers, especially in acidic soils.

Because of the various benefits they provide to soil structure and fertility, earthworms may improve the yields of agricultural crops. Therefore, they may be purposefully introduced into agricultural sites. Vermiculture, or "worm farming," provides a source of earthworms for this purpose.

Table 3-1 Outline of formal classification of arthropods discussed in this book.

Phylum Arthropoda
Class Crustacea (crustaceans: shrimp, lobsters, copepods, etc.)   Order Isopoda (isopods)
Class Diplopoda (millipedes)
Class Chilopoda (centipedes)
Class Symphyla (symphylans, symphylids)
Class Arachnida (arachnids: spiders, mites, scorpions, etc.)   Order Acarina (mites)   Order Araneae (spiders)
Class Insecta (insects)

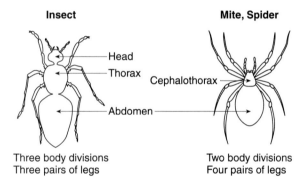

Figure 3-5 General drawing of most important differences between insects and spiders or mites.

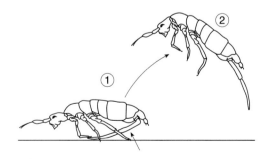

Figure 3-6 Side view of a springtail (order Collembola). The abdominal appendage is pushed against the ground to raise the animal into the air.

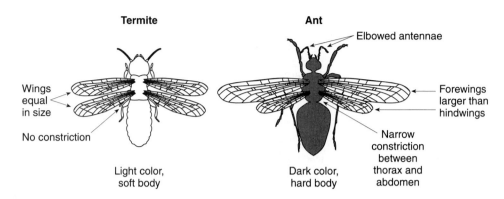

Figure 3-7 Basic distinction between termites and ants.

even wood as a food source, digested with the aid of protozoan or bacterial symbionts, makes them important in recycling of nutrients from these materials, as well as becoming pests when they feed on desirable wood or paper products. *Ants* (Figure 3-7) also affect soil structure and porosity. Ecologically, they are a very diverse group, with many different feeding and foraging habits. Most of the ants foraging in soil or litter feed as predators or scavengers. *Beetles* make up about 25% of all described animal species, and so it is not surprising that a variety of them inhabit soil, litter, and the soil surface. Many of the beetles found in this environment are predators. Others are decomposers, feeding on carrion (dead animals), dung, or fungi. Grubs or wireworms feed on roots or other plant material, often causing economic damage.

Other Soil Animals. The groups just described are usually the most common invertebrates in many agricultural soils. However, other animals are frequently present in agricultural soils, and may be locally abundant in some places. These include a variety of arthropods as well as members of other phyla.

In addition to those described previously, many other types of *insects* inhabit the soil environment. Larvae of many different *flies* are involved in decomposition. Many kinds of flies found in soil are nearly microscopic and feed on fungi.

Myriapod is a general term for a segmented arthropod with many pairs of legs. **Millipedes** (Figure 3-8) have two pairs of legs on most body segments and feed primarily as scavengers on dead plant material. **Centipedes** have one pair of legs per segment and are predators. **Symphylids** are small (mostly 5–15 mm long), whitish myriapods with up to twelve pairs of legs. They are omnivorous decomposers. Most symphylids feed on decaying plant material, but some may feed on and become pests of crop plants.

Spiders, which are closely related to mites (see Figure 3-5), are important predators in many ecosystems. They are common in litter, and some small species may enter the uppermost layers of loose soil.

Isopods are also called pill bugs, sow bugs, woodlice, or even "roly-polys" for their habit of rolling up into a ball when disturbed. These terrestrial crustaceans usually feed as scavengers on organic materials.

Snails and **slugs** are members of the phylum Mollusca. They feed on plant material, and some can become plant pests.

Figure 3-8 A millipede has two pairs of legs on most body segments. Most millipedes coil up when disturbed.

Flatworms (phylum Platyhelminthes) are important in agriculture as animal parasites. **Planarians** are free-living (nonparasitic) flatworms found in moist habitats. Some planarian species are predators on earthworms and other soil invertebrates.

Rotifers (phylum Rotifera) are microscopic predators or ingestors of organic debris. Most are aquatic, but some kinds occur in soil. **Hairworms** (phylum Nematomorpha), also called horsehair worms, are occasionally encountered as insect parasites. In some older texts, rotifers, hairworms, nematodes, and a few groups of marine organisms were combined into the phylum Aschelminthes. **Tardigrades** or "water bears," are a phylum of microscopic aquatic invertebrates that also occur in soil, feeding on algae, organic plant debris, bacteria, protozoans, and microscopic invertebrates.

Burrowing **vertebrates,** such as rodents, reptiles, and amphibians, may be important members of the agricultural soil fauna in some locations. The activities of some nonresident vertebrates, such as pigs or water buffaloes, can be disruptive to soil structure.

Although a number of different kinds of soil organisms have been discussed in this chapter, many others that have not been mentioned also occur in soil. More information on soil organisms can be obtained from texts on soil biology (e.g., Dindal 1990) or invertebrate zoology (e.g., Barnes 1980; Buchsbaum et al. 1987).

Benefits and Problems of Soil Organisms

Most soil organisms are beneficial to agroecosystems through their participation in the process of decomposition. Their activities speed up the decay of organic materials and the release and recycling of nutrients, which are discussed in more detail in the next chapter. Although a number of soil organisms feed on various types of organic materials, fungi and bacteria (including actinomycetes) are the most direct and abundant decomposers of organic matter.

Application 3-3 *Predators and Prey in Soil Ecosystems*

Since so many different kinds of organisms and predators live in soil, how do soil predators choose their prey? Size and opportunity are probably the most important factors affecting choice of prey by soil predators. A particular species of soil mite may be unable to attack a much larger beetle or ant. However, the same mite may be able to feed on other mites, various kinds of nematodes, or protozoa. Many soil predators show a similar lack of specificity. Finding prey items presents a special challenge in the soil environment, which offers a maze of pores and tunnels in which prey can hide. It may be most convenient simply to eat the most available prey item, which is often the most abundant species of potential prey. This nonspecific feeding by many different soil predators probably helps to stabilize soil food webs by keeping any one prey species from becoming too abundant.

Is predation by soil animals beneficial to agriculture? The answer varies, due to the lack of specificity of soil predators. A predatory mite may feed on plant-parasitic nematodes, which could indirectly benefit the plant. However, the same mite may feed on bacterivorous nematodes that are beneficial to decomposition. Nutrient release from decomposing organic material might be slowed as a result—a negative effect. One could devise many other scenarios in which a predator may feed indiscriminately on a beneficial or a harmful prey species. More information on predator-prey interactions and their implications in biological control is provided in Chapter 9.

The activities of many kinds of soil animals regulate the rate of decomposition of organic matter by bacteria and fungi. Larger invertebrates, such as earthworms, termites, and millipedes, and even small invertebrates, such as mites, springtails, and enchytraeids, break up pieces of organic matter of various sizes, exposing more surface area to decomposition. Grazing of bacteria and fungi by nematodes, mites, springtails, and protozoa exposes fresh surface area to decomposition. Populations of decomposer fungi and bacteria will grow at a steady rate if they continually have freshly exposed organic matter to colonize. Earthworm castings and other excretory products of soil invertebrates are often rich in nitrogen and other nutrients, causing them to become sites of high microbial activity. Movement of soil animals may redistribute organic matter and immobile decomposers, carrying them to new sites.

The benefits of earthworms and other macroinvertebrates in improving soil structure have already been mentioned. Some predators may feed on plant pests, although the extent to which this may benefit agricultural production is usually not clear (see Application 3-3). Other organisms, such as mycorrhizae or nitrogen-fixing bacteria, may provide special benefits to plants.

Although most soil organisms are beneficial, many cause disease or damage to agricultural crops. Many kinds of bacteria and fungi infect plant roots, causing premature root death and decay. A number of different insects and plant-parasitic nematodes feed directly on plant roots. More information on these pests and their management is provided in Chapters 10 and 11.

Below-Ground Food Webs

From the previous sections, it is evident that even the most simple agroecosystem may contain a complicated community of many different kinds of organisms. Using the most com-

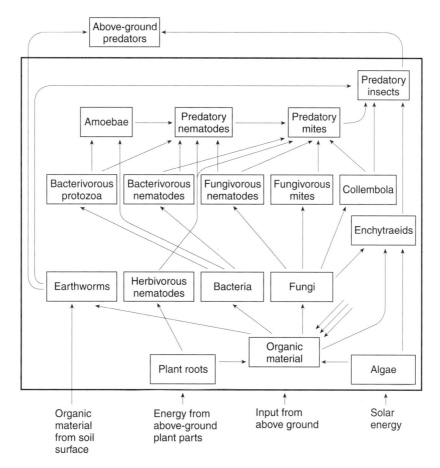

Figure 3-9 Simplified below-ground food web for soil organisms commonly found in agroecosystems.

mon soil organisms introduced in the previous sections, it is possible to construct a general below-ground food web that may be typical of an agricultural system (Figure 3-9). Arrows indicate the most important pathways of material and potential energy movement into the various consumers. For clarity, not all possible arrows are shown in Figure 3-9. For example, some mites or Collembola, which are primarily fungivores, may feed on some bacteria and algae as well. The set of three arrows entering the "organic material" component emphasizes that dead bodies and waste products of *all* types of soil organisms are recycled as organic material. Also, since the feeding habits of many soil organisms are not well known, it is possible that additional arrows could be drawn as new information becomes available.

Other authors (Moore and de Ruiter 1991) suggest dividing below-ground food webs into five trophic positions. This approach aids in the interpretation of Figure 3-9. Plants or organic material provide the basis for the rest of the food web. At the next position are organisms that are direct decomposers of organic material (bacteria, fungi, etc.) or primary consumers of plants. Herbivorous or plant-parasitic nematodes are present in most agro-ecosystems, but root-feeding insects or plant-pathogenic fungi and bacteria could be included as well. The various consumers of soil fungi and bacteria occupy the third trophic position. The next positions include intermediate predators and, finally, top predators.

Organization into these positions provides a convenient means for interpreting many key interactions in the soil system. Exceptions always occur, however, due to the complexity of soil ecosystems and the activities of some organisms at several positions. Although insects such as beetles or ants are shown as the top predators in this soil system (Figure 3-9), the position of predatory mites is not so clear. They feed on a variety of predators and non-predators, and are the top predators in some soil systems. Of course, dividing this group into different types of mites would be more informative.

A closer look at any component within the soil food web reveals additional levels of complexity. As an example, soil fungi could be divided into a number of different groups (Figure 3-10), among which the decomposers comprise the largest group in soil. Some of these groups of fungi use organic matter as their principal nutrient source while others depend on plant roots. This may vary with individual species within a group. Some plant parasites may feed only on roots of certain plants, while others may subsist on organic matter and act as **facultative parasites,** attacking plant roots only under certain conditions. The fungivorous mites, Collembola, and nematodes may feed on fungi of all kinds, as indicated in Figure 3-10 by the multiple arrows from the fungal groups. An interesting feature is that some fungi parasitize these consumers! So arrows could be drawn indicating the flow of energy and nutrients from higher-level consumers back to the fungi, further obscuring distinctions among levels or positions within the food web.

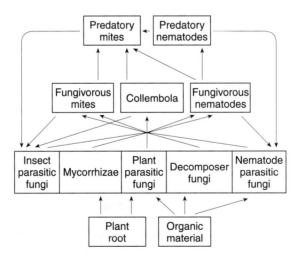

Figure 3-10 Detail of fungal components of a below-ground food web.

Cyclical Nature of Soil Food Webs

Any organism or part of an organism that is not completely consumed and digested will eventually be returned to the soil system as organic matter. Therefore, material and potential energy move from all soil organisms back into the pool of organic matter (Figure 3-11). Thus, the soil food web in Figure 3-9 should have downward arrows as well. When a top predator dies, it serves as an organic food source for bacteria and other organisms in the lower positions of the soil food web. Movement of materials and nutrients is much more cyclical in below-ground food webs than in above-ground food webs because of the great emphasis on decomposition.

Of course, the cycling of nutrients and potential energy cannot continue indefinitely. As organisms squeeze energy, carbon, and nutrients from a tiny piece of organic matter at every step, eventually all of the potential energy in the fragment will be converted and used, and all of the carbon (C) will be respired as CO_2. However, the soil system is not a closed system; it constantly receives fresh input of energy and material from above-ground systems (see Figure 3-9). Likewise, some energy and nutrients may be removed from soil systems by predators foraging in litter or the upper layers of soil, or by plants growing in the soil.

Diagrams such as Figure 3-9 provide a convenient map of the most important organisms and interactions present in an ecosystem. A deeper level of understanding could be achieved if information on the amounts and rates of transfer of energy or materials were known and provided. Measurement of transfers over time may reveal seasonal differences and would ensure that energy data represented realistic estimates of production. These transfers, as well as the composition of the food web itself, can change with soil depth or with increasing distance from the plant roots. Despite the difficulties in studying soil food webs, more detailed information on below-ground food webs in agroecosystems is available from a variety of sources (Hendrix et al. 1990; Lee and Pankhurst 1992; Moore and de Ruiter 1991). As the complex interactions of below-ground systems are better understood, it may be possible to manage these systems to provide improved maintenance of soil fertility and more efficient recycling of nutrients.

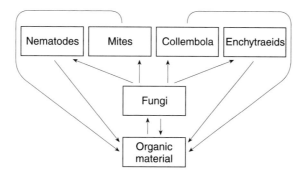

Figure 3-11 Portion of a below-ground food web emphasizing two-way interactions. After death, energy and material from all kinds of soil organisms are added to the other soil organic matter for decomposition.

Interconnection of Food Webs

Food webs are not isolated in nature or in agriculture, and a number of factors may connect different kinds of food webs. The connection between above-ground and below-ground food webs is particularly close. Many dead materials and varying amounts of crop residues and other organic materials pass from above ground into the below-ground system for decomposition. Recycled nutrients in soil are then available for use by plants. In fact, crop plants are the most important feature connecting the above- and below-ground systems since tops and roots are key producers for these systems. As energy is transferred from leaves into roots, it sustains the community of organisms that depend directly or indirectly on plant roots. So in practice, above- and below-ground systems are not really separated, but form two parts of the same agroecosystem.

The food webs of individual agroecosystems are interconnected to one another and to natural systems. Mobile animals such as birds or insects do not respect field boundaries, and may range over wide areas. In particular, top carnivores (such as predatory birds, shown in Figure 3-1) may be unable to sustain themselves from a single system. Instead, they must search for prey over a wide territory that may include many diverse habitats. Since supplies of grain, pollen, and nectar are seasonal in many crops, animals that depend on these food sources must migrate to locations where they are available. Insects from agroecosystems may fly into woodlands or over nearby streams or lakes, where they may be incorporated into different food webs. Erosion (see Chapter 5) may carry organisms as well as soil to other sites. One can think of many other ways in which food webs and ecosystems are interconnected.

Because agroecosystems are managed systems, the activities of people are one of the most important ways in which agroecosystems are connected to other kinds of ecosystems. Harvested crops and animals are often removed to urban ecosystems or other locations far from the original production site. People bring many materials into the agroecosystem that were produced elsewhere. These include many different agricultural inputs such as fertilizers, pesticides, seeds, transplants, etc., as well as unintentional introductions of crop pests and diseases.

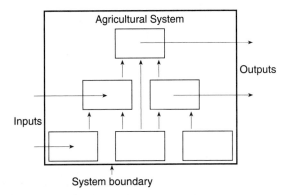

Figure 3-12 Because of interconnection, it is important to specify boundaries to the agroecosystem so that inputs and outputs can be defined.

The interconnection of food webs and ecosystems causes difficulty in tracking the movement of energy and materials. During the course of a summer, a bee may obtain energy-rich sugars from a variety of crops and native plants, and move a portion of this energy to a hive at a remote location. Under these conditions, it is critical to define clear system boundaries if one wishes to examine a particular agroecosystem (Figure 3-12). The energy or sugars removed by the bee could be treated as an output from the system. Otherwise, the tracking of such materials could continue indefinitely. On the other hand, awareness of the potential outputs from an agroecosystem can help in anticipating consequences that may result as these outputs enter other kinds of ecosystems (see Chapter 12).

Summary

The concept of the food chain is too simplistic for most agroecosystems. Complicated food webs of many different kinds of organisms make up the community of living organisms present in most agricultural fields. Although not readily observed and apparent, this abundant biodiversity is particularly evident in the complex communities of most agricultural soils. The many soil organisms perform a range of different functions in the soil environment. Through its various inputs and outputs, any agroecosystem is ultimately connected to other agricultural systems, natural ecosystems, and urban systems.

Topics for Review and Discussion

1. How are soil organisms beneficial in maintaining plant health?
2. Many people are not aware of the great biodiversity present in typical agricultural soils. Why do you think soil organisms have been so difficult to recognize and study?
3. What kinds of above-ground organisms are often present in crops grown in your area? Design a food web depicting their potential interactions on a typical crop.
4. Using a local field as an agroecosystem, list inputs and outputs to this ecosystem, and indicate how these are connected to other ecosystems.
5. In an ideal situation, measurements of transfer of energy or materials (like carbon) could be shown for every connection in a food web. Discuss the difficulties involved in obtaining such measurements. What information and experiments would be needed to obtain this level of resolution?

Literature Cited

Barnes, R. D. 1980. *Invertebrate Zoology*. Philadelphia: Saunders.
Buchsbaum, R., M. Buchsbaum, J. Pearse, and V. Pearse. 1987. *Animals Without Backbones*. 3d ed. Chicago: The University of Chicago Press.
Dindal, D. L. 1990. *Soil Biology Guide*. New York: John Wiley and Sons.
Edwards, C. A., P. J. Bohlen, D. R. Linden, and S. Subler. 1995. Earthworms in agroecosystems. In *Earthworm Ecology and Biogeography in North America,* ed. P. F. Hendrix, 185–213. Boca Raton, Fla.: Lewis Publishers.
Felland, C. M., and H. N. Pitre. 1991. Diversity and density of foliage-inhabiting arthropods in irrigated and dryland soybean in Mississippi. *Environmental Entomology* 20: 498–506.
Hendrix, P. F., D. A. Crossley, Jr., J. M. Blair, and D. C. Coleman. 1990. Soil biota as components of sustainable agroecosystems. In *Sustainable Agricultural Systems,* eds. C. A.

Edwards, R. Lal, P. Madden, R. H. Miller, and G. House, 637–654. Delray Beach, Fla.: St. Lucie Press.

Jansson, R. K., and K. V. Raman. 1991. *Sweet Potato Pest Management: A Global Perspective.* Boulder, Colo.: Westview Press.

Keeton, W. T., and C. H. McFadden. 1983. *Elements of Biological Science.* 3d ed. New York: W. W. Norton.

Lee, K. E., and C. E. Pankhurst. 1992. Soil organisms and sustainable productivity. *Australian Journal of Soil Research* 30: 855–892.

Lövei, G. L., M. Sarospataki, and Z. A. Radwan. 1991. Structure of the ladybird (Coleoptera: Coccinellidae) assemblages in apple: Changes through developmental stages. *Environmental Entomology* 20: 1301–1308.

Moore, J. C., and P. C. de Ruiter. 1991. Temporal and spatial heterogeneity of trophic interactions within below-ground food webs. *Agriculture, Ecosystems and Environment* 34: 371–397.

Bibliography

Angle, J. S., J. V. Gagliardi, M. S. McIntosh, and M. A. Levin. 1996. Enumeration and expression of bacterial counts in the rhizosphere. In *Soil Biochemistry,* vol. 9., eds. G. Stotzky and J. M. Bollag, 233–251. New York: Marcel Dekker.

Coleman, D. C., and D. A. Crossley, Jr. 1996. *Fundamentals of Soil Ecology.* San Diego, Calif.: Academic Press.

Crossley, D. A., Jr., D. C. Coleman, P. F. Hendrix, W. Cheng, D. H. Wright, M. H. Beare, and C. A. Edwards, eds. 1991. *Modern Techniques in Soil Ecology.* Amsterdam: Elsevier.

Didden, W. A. M. 1993. Ecology of terrestrial Enchytraeidae. *Pedobiologia* 37: 2–29.

Edwards, C. A., and P. J. Bohlen. 1996. *Biology and Ecology of Earthworms.* 3d ed. London: Chapman and Hall.

Fenchel, T. 1987. *Ecology of Protozoa.* Madison, Wisc.: Science Tech Publishers.

Hendrix, P. F., ed. 1995. *Earthworm Ecology and Biogeography in North America.* Boca Raton, Fla.: Lewis Publishers.

Hendrix, P. F., R. W. Parmelee, D. A. Crossley, Jr., D. C. Coleman, E. P. Odum, and P. M. Groffman. 1986. Detritus food webs in conventional and nontillage agroecosystems. *BioScience* 36: 374–380.

Lobry de Bruyn, L. A., and A. J. Conacher. 1990. The role of termites and ants in soil modification: A review. *Australian Journal of Soil Research* 28: 55–93.

Schuster, R., and P. W. Murphy. 1991. *The Acari.* London: Chapman and Hall.

South, A. 1992. *Terrestrial Slugs: Biology, Ecology and Control.* London: Chapman and Hall.

Stork, N. E., and P. Eggleton. 1992. Invertebrates as determinants and indicators of soil quality. *American Journal of Alternative Agriculture* 7: 23–32.

Yeates, G. W., T. Bongers, R. G. M. de Goede, D. W. Freckman, and S. Georgieva. 1993. Feeding habits in soil nematode families and genera—an outline for soil ecologists. *Journal of Nematology* 25: 315–331.

Chapter Endnote 3-1 *Classification of Living Organisms*

Every different species of organism known to science has a unique **scientific name.** This system of Latin names was developed by the Swedish naturalist Linnaeus in 1753 for plants, and in 1758 for animals. The system provides us with a universal terminology for referring to specific organisms without confusion or translation. A scientific name consists of two parts: the **genus** (capitalized), which includes closely-related species, and the **species,** which is assigned to the particular organism. Thus, humans are classified as *Homo sapiens*.

The many genera and species of organisms are organized into higher levels of classification as well. Related genera are grouped into a **family.** Related families comprise an **order.** Orders are grouped into a **class,** and related classes into a **phylum** (for animals) or **division** (for plants). Finally, all related phyla or divisions make up a **kingdom.** The complete classification of humans is:

Kingdom = Animalia
Phylum = Chordata
Class = Mammalia
Order = Primates
Family = Hominidae
Genus = *Homo*
Species = *sapiens*

Intermediate classifications (e.g., suborder, superfamily, subfamily, etc.) may be used for some groups of organisms, but the preceding scheme is the most basic and consistent for all groups.

The organization of living things into kingdoms has been the source of some confusion, as new and unusual kinds of organisms were discovered. For example, how would you classify the genus *Euglena*, which has chlorophyll like a plant, but can move like an animal? Historically, living organisms were divided into animals or plants. Today, many authorities and authors of biology texts, such as Keeton and McFadden (1983), recognize five kingdoms:

- *Monera* includes bacteria and blue-green algae.
- *Fungi* were formerly classified as plants, but are not photosynthetic and have several other characteristics that are different from plants.
- *Protista* includes a variety of unicellular and primitive multicellular organisms that traditionally have been difficult to classify. This kingdom includes protozoa (still classified as animals by some zoologists), slime molds, euglenoids, dinoflagellates (responsible for red tides), and those algae classified in the division Chrysophyta (diatoms, yellow-green algae, golden-brown algae).
- *Plantae* still includes green algae (Chlorophyta), brown algae (Phaeophyta), and red algae (Rhodophyta), as well as the more traditional plant groups.
- *Animalia* includes all groups formerly classified as animals, except for the protozoa.

Viruses are not included in the classification schemes for living organisms, although they have some characteristics of living organisms.

(continued)

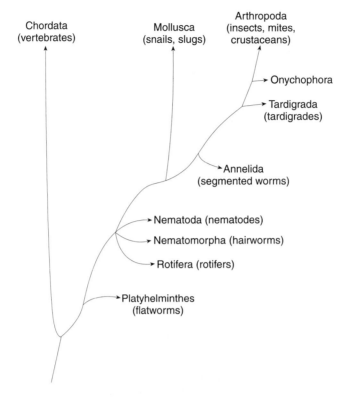

Figure 3-13 Simplified phylogenetic tree showing relationships among animal phyla found in terrestrial habitats.

Many different kinds of living organisms, particularly animals, are discussed in this chapter. Zoologists often use the **phylogenetic tree** to express relationships among different animal phyla (Barnes 1980). Many of the animal phyla are marine, and are unlikely to be encountered in agricultural habitats. A simplified phylogenetic tree of terrestrial animal phyla is presented in Figure 3-13. Onychophora (not discussed in this text) is an obscure group, intermediate between Annelida and Arthropoda.

Taxonomy is the branch of biology that deals with the classification of organisms. Despite vast efforts by scientists to identify, describe, and name species, *many organisms remain unidentified*. Although some scientists search for new species in tropical rain forests and other remote habitats, *it is likely that unidentified species will be present in most agricultural soils*. In these cases, it is still possible to identify the organism at some higher level of classification (phylum, class, order, family, etc.), even if it is not a named species. Dindal (1990) provides a guide to the higher-level classification of soil organisms.

Chapter 4

Nutrient Cycling and Decomposition

Key Concepts

- The macronutrients and micronutrients essential to plants
- Chemistry and cycling of nitrogen
- Recycling of nutrients through decomposition

The subject of biogeochemistry was introduced in Chapter 2 with the carbon cycle. In fact, any chemical element that is taken up by living organisms could be the subject of a biogeochemical cycle. Such elements are eventually returned to the physical environment following the death and decomposition of the organism itself or some member of the food web that has consumed it. **Nutrient cycling** refers to the movement and recycling of elements that are essential for life.

Essential Nutrients for Plant Growth

A number of different elements are required as essential nutrients for plant growth (Table 4-1). **Macronutrients** are required in relatively large quantities, and **micronutrients** are needed only in small, or trace, amounts. Therefore, the only difference between a macronutrient and a micronutrient is the *amount* of nutrient that is needed by the plant. In addition to the micronutrients listed in Table 4-1, silicon (Si), sodium (Na), cobalt (Co), and vanadium (V) are essential to some plants, but maybe not to all. Several other elements may have minor, nonessential functions in some plants.

Plant growth varies greatly depending on the amount or concentration of each nutrient supplied (Figure 4-1). Figure 4-1 is presented without units because the actual plant response would depend on the particular nutrient, crop plant, and range of environmental conditions. In general, there is some range in levels of each nutrient (from L1 to L2 in Figure 4-1) that results in optimum plant growth. Note that similar plant performance would be

Table 4-1 Essential elements for plant growth.

Macronutrients			Micronutrients[a]		
Symbol	Element	Common Form[b]	Symbol	Element	Common Form[b]
C	Carbon	CO_2	B	Boron	BO_3^{2-}
H	Hydrogen	H_2O	Cl	Chlorine	Cl^-
O	Oxygen	O_2	Cu	Copper	Cu^{2+}
N	Nitrogen	NH_4^+, NO_3^-	Fe	Iron	Fe^{2+}
P	Phosphorus	$H_2PO_4^-, HPO_4^{2-}$	Mn	Manganese	Mn^{2+}
K	Potassium	K^+	Mo	Molybdenum	MoO_4^{2-}
Ca	Calcium	Ca^{2+}	Zn	Zinc	Zn^{2+}
Mg	Magnesium	Mg^{2+}			
S	Sulfur	SO_4^{2-}			

[a] Silicon (Si), sodium (Na), cobalt (Co), and vanadium (V) are essential for some plants.
[b] Most common form(s) for uptake and use by plants.

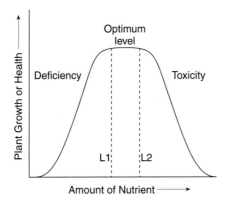

Figure 4-1 Generalized relationship between plant and growth and amount or concentration of an essential nutrient.

achieved at both levels L1 and L2, but that more fertilizer would be used (at more expense) at L2 to achieve the same result. Below L1, nutrient deficiency has an adverse effect on plant performance, while above L2, excess levels of the nutrient may be toxic. Nutrient deficiencies and toxicities result in very characteristic damage symptoms on many crop plants (Bennett 1993). Toxicity may result from excessive amounts of some essential nutrients, from nonessential elements such as aluminum (Al) or arsenic (As), or from heavy metals such as chromium (Cr), cadmium (Cd), nickel (Ni), or lead (Pb). Excess Na or Cl from saline soils or irrigation water is a common source of toxicity for many crop plants (see Chapter 15).

The essential elements are available to plants in many forms. The most commonly used forms are summarized in Table 4-1. Photosynthesis is the means by which carbon (C), hydrogen (H), and oxygen (O) are incorporated into plants, as discussed in Chapter 2. Note that CO_2 is the actual source of O that is incorporated into organic molecules; O from water is taken into plants during photosynthesis but is given off as O_2. However, plants use some O_2 for their own respiration as well. With the exception of C, H, and O, the principal source of the other essential elements is ions dissolved in water in the soil. These are taken up by plant roots. In addition to the most common ions (Table 4-1), many other sources of essential elements are possible. For example, some C, H, and O may enter plants as bicarbonate (HCO_3^-) ions. Some elements may even be absorbed through leaves as atmospheric gases or nutritional sprays.

Nitrogen in Agroecosystems

Nitrogen is an essential element for plant and animal growth since it is a constituent of many different kinds of organic compounds, including proteins and nucleic acids. There is an abundant supply of nitrogen (N) in the air (78% is N_2 gas), but since plants cannot use N_2 gas directly, the forms available to them are found primarily in the soil. This available N can be in short supply in agroecosystems and often must be supplemented by N-containing fertilizers.

Forms of N in Soil

The N in soil is divided into organic forms and inorganic forms. **Organic N** originates from living organisms and is a part of the organic compounds remaining after the death and decomposition of organisms. **Inorganic N** refers to all forms of N that have been freed by mineralization from organic (C-containing) compounds, including:

Ammonium ion	NH_4^+
Nitrate ion	NO_3^-
Nitrite ion	NO_2^-
N_2 gas	

as well as other forms such as nitrous oxide (N_2O) or N that is trapped in some clays and other minerals.

The NH_4^+ and NO_3^- ions are readily soluble in water and can be taken up by plant roots, but these and other forms of N also may be used by soil bacteria during the decomposition process. These inorganic forms of N are converted into organic N compounds forming the tissues and other structures of the bacteria. This tie-up of N as organic N in microbial tissues is called **immobilization,** and the immobilized N is unavailable to roots of crop plants. The reverse of this process is **mineralization,** in which the organic N compounds are broken down and converted into inorganic N, often by decomposition. Thus,

Immobilization:	Inorganic N → organic N
Mineralization:	Organic N → inorganic N

In the soils of agroecosystems, the amount of N tied up in the tissues of living or dead plants, animals, and bacteria is usually much greater than the amount of inorganic N. As a result, the amount of inorganic N is limited and quickly immobilized by plants, bacteria, or other organisms.

Recycling of N in Soil

Mineralization and related processes result in the recycling of organic N into forms that can be used to meet the nutritional needs of plants. The decomposition of organic N compounds provides an energy source to the bacteria and fungi that carry out the decomposition. The basic reaction by which the N from an amine group (NH_2) is removed from a larger organic molecule (represented here as "Org") is:

$$Org-NH_2 + H_2O \rightarrow org-OH + NH_3 + energy$$

The liberated ammonia (NH_3) reacts further with water or acids present in the soil solution to form the ammonium ion (NH_4^+). The hydrolysis of NH_3 is given by:

$$NH_3 + H_2O \rightarrow NH_4^+ + OH^-$$

The NH_4^+ released by mineralization can follow several pathways. This form of N can be used by plants, or it may be further transformed by the process of nitrification (see next paragraph). Depending on the soil type, some of the NH_4^+ may be tied up by certain types of clays. Since NH_3 is a volatile gas, there is also the possibility of some **volatilization** losses of this intermediate product.

Nitrification is a two-step conversion of ammonium (NH_4^+) to nitrate (NO_3^-). These reactions are carried out by *nitrifying bacteria*. The rate of these reactions depends not only on the presence of the bacteria, but also on soil aeration (need adequate O_2 supply), temperature, and a number of other factors. The first step is the oxidation of NH_4^+ to nitrite (NO_2^-):

$$2NH_4^+ + 3O_2 \rightarrow 2NO_2^- + 2H_2O + 4H^+ + energy$$

Nitrite can be toxic to plants, but it is a short-lived intermediate product that is quickly converted to nitrate:

$$2NO_2^- + O_2 \rightarrow 2NO_3^- + energy$$

A number of different things may happen to the NO_3^- produced, the most important of which is that this form of inorganic N can be readily used by plants and other organisms such as bacteria and fungi. Since NO_3^- is readily soluble in water, some of it may be lost during drainage in the process known as **leaching.** Depending on field conditions, some NO_3^- may be lost to denitrification.

Denitrification is the loss of N from soil in gaseous form (usually as N_2O or N_2). It is a particular problem in soils that have poor drainage and aeration. The N_2O produced from NO_3^- may be lost by volatilization or may be further reduced to N_2 gas. Both reactions are conducted by *denitrifying bacteria:*

$$2HNO_3 \rightarrow N_2O + H_2O + 2O_2$$
$$2N_2O \rightarrow 2N_2 + O_2$$

Adding N to Soil

The recycling of organic N is one means by which inorganic N sources to plants may be increased. Additional N may be brought into the agroecosystem primarily by fixation of N_2 from the atmosphere or by addition of N fertilizer. Some N also may be fixed by lightning and added from NH_4^+ and NO_3^- dissolved in rain or snow.

Nitrogen fixation is the process by which N_2 gas from the atmosphere is converted into NH_3. The NH_3 serves as a substrate that can be converted into NH_4^+ or organic N compounds that can be used by plants. The process is carried out by soil organisms, especially certain bacteria and blue-green algae, which may be in close association with the roots of plants. The atmospheric N_2 can reach these organisms quite easily through the numerous air spaces present in soil. The concentration of N_2 in the air spaces of most well-aerated soils is similar to that in the atmosphere (about 78%).

Symbiotic N Fixation. Many plant species in the family Leguminosae are colonized by bacteria in the genus *Rhizobium*, providing the best-known examples of symbiotic N fixation (see Application 4-1). The bacteria modify root hairs and other root tissues to form the nodules in which they reside (Figure 4-2). These nodules can be distinguished from other swellings on the root system by their liquid-filled centers, which are apparent when the nodules are cut or squeezed. Also, the nodules can usually be removed or scraped off of the roots quite easily. The bacteria use carbohydrates from the plant as an energy source and, in turn, provide the plant with N. Since both organisms benefit from the association, it can be considered a **symbiosis.**

Although N fixation by symbiotic bacteria is common in many legumes, it occurs in some nonlegume species as well. The most common examples are found in some genera of trees or shrubs that colonize cleared sites, such as *Alnus* in temperate regions or *Casuarina* in tropical or subtropical climates. The bacterial symbionts differ from those found on legumes and are usually not *Rhizobium* spp.

Nonsymbiotic (asymbiotic) N Fixation. A number of organisms may fix N directly, without an association with roots of higher plants. These include some bacteria such as *Azotobacter* spp. and various blue-green algae, which can be important in N fixation in some wet soils. When these organisms die and decompose, their N-containing compounds may be mineralized into forms that can be taken up by plant roots.

Nonsymbiotic N fixation is particularly important in some rice production systems. Nitrogen fixation by blue-green algae in the water or at the soil surface is significant in flooded rice. *Azolla* spp., which are common water weeds in some locations, contain a blue-green algal species that fixes N. The contribution of N fixation from these and other sources exceeds the losses from denitrification expected under flooded conditions.

Fertilizer. Large amounts of supplementary N are added to many agroecosystems in the form of fertilizers. Although there are many different forms and kinds of N fertilizers, their use and uptake by plants follow similar principles. As examined earlier, the form in which N is present changes as an organic N source proceeds through mineralization and nitrification:

$$\text{Org-N} \rightarrow NH_3 \rightarrow NH_4^+ \rightarrow NO_3^-$$

Application 4-1 *Legumes and N Fixation*

The benefits of including legumes in crop rotations have long been recognized for improving soil fertility and N management. Many commonly grown food crops like bean, pea, peanut, or soybean, and forage crops such as alfalfa or clover are particularly useful for this purpose. A wide variety of tropical legumes may be included in appropriate cropping systems (see Chapters 13 and 14). The amount of fixed N varies depending on the legume species and environmental conditions, but generally ranges from about 50–200 kg/ha per year, although higher amounts are reported in some cases. Alfalfa is one of the most productive crops for N fixation, with levels near the top of this range, but this may be expected of a long-lived perennial crop that can sustain N fixation for most of the year. Short-lived annual crops typically fix amounts in the range of 50–100 kg/ha.

Nitrogen-fixing bacteria supply substantial amounts of N to legumes, but this may not meet all the N needs of the plant. Allen (1975) reports that fixation supplies about 75% of the N required for plant growth. The remainder comes from nitrates and other forms in the soil mentioned in this chapter. The fulfillment of the plant's requirement for N depends on the growth stage of the plant. For example, N fixed by *Rhizobium* spp. is particularly important in the pod-filling stages of grain legumes like soybean. Because of N fixation, the tissues of legumes are particularly rich in N, making them especially useful as feed sources for livestock. The breakdown of legume crop residues over time can supply some N to a subsequent crop grown in rotation.

A number of conditions favor development of N-fixing bacteria, their nodules, and the rate of N fixation. Proper temperature for the bacteria and the legumes is important, as are good soil aeration, healthy plants, adequate water for nutrient uptake, favorable soil pH, and other factors. Adequate levels of certain nutrients like Ca, P, and K are critical, but excess levels of N from other sources may inhibit nodule formation. Of course, it is essential that the *Rhizobium* bacteria are present in the soil!

This last requirement should not be taken for granted, since different species of legumes may require different *Rhizobium* species, or even strains within species, as their symbionts. This is a complex area, since some bacteria may associate with different types of related plants, while others are highly specific. If a particular crop has not been grown in a site recently, the required *Rhizobium* species or strain may not be present or may occur only at very low levels. To ensure the presence of the proper N-fixing bacteria, crops may be *inoculated* with the appropriate *Rhizobium* type. Seed may be either sprayed with inoculant or mixed with a peat-based material containing *Rhizobium*. Inoculation usually increases both crop yield and the %N in plant tissues.

As any fertilizer or other N source is added to the system, the forms of N change over time following this sequence. A fertilizer consisting mainly of NH_4^+ or NO_3^- can be used rapidly by plants, whereas a fertilizer consisting primarily of organic N must first be mineralized and converted into NH_4^+ or NO_3^- before the N can be taken up by plant roots. Mineralization may take some time, depending on the organic N source and the environment, and so the release and uptake of N from organic N fertilizers tends to occur much more slowly and steadily through a crop season than uptake from fertilizers containing NH_4^+ or NO_3^-. The N ultimately used by the plant is in the same form (mainly NH_4^+ or NO_3^-) whether it originated

Figure 4-2 Nodules formed by nitrogen-fixing bacteria on the roots of a peanut plant.

from an inorganic or an organic fertilizer. Some important differences between organic and inorganic N fertilizers are introduced in Application 4-2.

The Nitrogen Cycle in Agriculture

Fertilizer application and N fixation are the two major means by which N from the abundant N_2 pool in the atmosphere can be added to an agroecosystem. Although addition of supplementary N may not be of particular concern with natural ecosystems, it is of critical importance in agricultural systems because these systems are usually characterized by a harvest that removes a substantial amount of N.

Application 4-2 *Organic and Inorganic N Fertilizers*

Organic nitrogen fertilizers are carbon compounds that contain organic N as the N source. They originate from living organisms or their products and release their N to plants following decomposition and mineralization. **Inorganic nitrogen fertilizers** do not contain carbon. **Synthetic fertilizers** are made by industrial processes. Examples of some common synthetic fertilizers are given in Table 4-2. Most of these also are inorganic fertilizers, with one important exception. Urea is an organic compound that can be made synthetically. It is also an excretory product produced by many different kinds of organisms. There are natural sources of some of the synthetically produced inorganic N fertilizers as well, the most important of which is probably Chilean nitrate, a source of $NaNO_3$.

Table 4-2 Some commonly used synthetic N fertilizers.

Fertilizer	Chemical Formula
Ammonium hydroxide[a]	NH_4OH
Ammonium nitrate	NH_4NO_3
Ammonium phosphate	$NH_4H_2PO_4$
Ammonium sulfate	$(NH_4)_2SO_4$
Anhydrous ammonia	NH_3
Diammonium phosphate	$(NH_4)_2HPO_4$
Potassium nitrate	KNO_3
Sodium nitrate	$NaNO_3$
Urea	H_2NCONH_2

[a] Ammonium hydroxide is a common base used to make N solutions of various other materials.

The availability of N from any fertilizer depends on its use and environmental conditions. In general, N is released more slowly from organic fertilizers (time is needed for decomposition and mineralization), although some inorganic fertilizers may be formulated for slow release as well. Since many inorganic fertilizers produce readily soluble forms of N, the N needs and plant uptake may be satisfied soon after application. However, these fertil-
(continued)

izers may be more subject to loss by leaching, since the N forms are more soluble than those contained in organic fertilizers. On the other hand, some manures may contain high concentrations of NH_3 gas that can be lost by volatilization. Some forms of fertilizer may cause salt accumulation or pH effects. A typical "burn" from too much fertilizer often results from excess salts such as Na^+ or Cl^- rather than from excess N. Soil pH may be decreased by some fertilizers, particularly those containing NH_4^+ as the N source. The NH_4^+ taken up by plants is balanced by the release of H^+, producing an acidic soil reaction. This acidification process occurs naturally to a lesser extent because H^+ ions are also released during the mineralization of NH_4^+ from any source.

Although synthetic fertilizers are convenient for ready and predictable use, their manufacture requires large amounts of energy. The N for most synthetic fertilizers originates from N_2 in the atmosphere, which is combined with H_2 in the Haber process to form ammonia:

$$N_2 + 3H_2 \rightarrow 2NH_3$$

The high energy requirements for this combination are met under natural conditions by lightning or energy used by N-fixing organisms, but in the synthetic process, fossil fuels are used. The NH_3 produced from this reaction serves as a starting point for the synthesis of many different N fertilizer products. These subsequent reactions require additional energy as well. For example, estimates of the amount of energy needed to produce ammonium nitrate range from 51.5–76.7 MJ per kg of N produced (Fluck and Baird 1980), equivalent to 12,300–18,300 kcal per kg of N. Therefore, 1.23–1.83 million kcal of energy would be required to produce the ammonium nitrate needed for an application of 100 kg/ha of this synthetic fertilizer. This is equivalent to the amount of potential energy contained in 150–223 L of gasoline, assuming that a L of gasoline contains 8200 kcal of potential energy (figure calculated from data in Pimentel and Dazhong 1990). Of course, many more L of fuel would actually be required to produce the fertilizer, since the use of any fuel is not 100% energy efficient.

Examples such as this demonstrate the energy requirements for fertilizer production. Energy availability is a major limitation in the production of synthetic N fertilizers, more so than availability of material, since N_2 is abundant in the atmosphere. This energy is not a direct input into the agroecosystem, unlike the solar energy used by plants in photosynthesis (see Chapter 2). But it must be considered as an indirect energy cost when examining energy use or energy efficiency in agricultural systems.

The major events in the N cycle of a typical agroecosystem are summarized in Figure 4-3. The major inputs from atmospheric N_2 are shown as N fixation, synthesis into inorganic fertilizers, and rainwater. The crop plants contain a substantial amount of immobilized N in their tissues, which is removed through harvest to supply N to consumers such as animals or people. Cattle and other grazing animals in turn may return a portion of the N to the system as manure, but other animals, such as birds or insects that feed on the crop, also harvest some N and may return it to the system in various forms.

The dynamic process of N recycling takes place in the soil, where dead plant material and other forms of organic matter accumulate. As this material breaks down, the N may be made available to plants through mineralization and nitrification or it may be immobilized as it

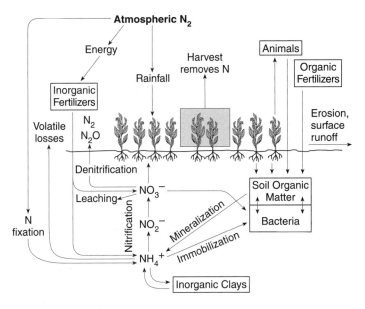

Figure 4-3 The nitrogen cycle in the agroecosystem.

is incorporated into bacteria and other soil organisms. These events have been described in detail earlier in this chapter, but the net result is that in most agricultural systems, much more N is tied up in plants, soil organisms, and organic matter than is available in inorganic forms in the soil. This limited pool of inorganic N is the subject of intense competition among plants and various soil organisms. Although plants compete with bacteria and other decomposer organisms for N, they also depend on these organisms for the recycling and release of N, a paradox that will be examined in more detail later in this chapter. Losses of N from the agroecosystem through leaching into groundwater, erosion, and runoff in surface water are examined in more detail in Chapter 12.

Macronutrients and Micronutrients in Agroecosystems

As with N, the other nutrient elements follow characteristic cycles as they are supplied to and used by agricultural systems, and as they are recycled from plants and animals back to soil.

Soil Chemistry of Phosphorus

The soil chemistry of inorganic forms of P is complex, but some of the key features of critical importance to agriculture are summarized here. Four different forms of P are in equilibrium in aqueous solution, and the predominant form of P depends on the soil pH (see Chapter 5 for more information on soil pH):

$$H_3PO_4 \leftrightarrow H_2PO_4^- \leftrightarrow HPO_4^{2-} \leftrightarrow PO_4^{3-}$$
$$pH<2 \quad pH>2 \quad pH>7 \quad pH>12$$
$$\phantom{pH<2 \quad} pH<7 \quad pH<12$$

Several forms of P may occur simultaneously in the soil solution, especially when pH is near neutrality (pH = 7.0), but $H_2PO_4^-$ becomes more common in acidic soils (pH < 7.0), and HPO_4^{2-} in alkaline soils (pH > 7.0). Both of these ions can be absorbed readily by plants *if* they are in solution.

The problem with P is that while many inorganic forms occur in soil, many of these are insoluble, so that the P is not available for uptake by plants. In acidic soils, phosphates react with Fe, Al, and Mn ions and their hydroxides to form insoluble precipitates. This problem increases as soil acidity increases. In alkaline soils, soluble phosphate compounds react with Ca^{2+} and calcium carbonate ($CaCO_3$) to form insoluble compounds. Although PO_4^{3-} is theoretically soluble at pH of 12 or above, soil pH doesn't reach these levels, and insoluble PO_4^{3-} compounds such as $Ca_3(PO_4)_2$ generally precipitate out at relatively low levels of soil alkalinity. No matter what the soil pH is, phosphates will always have a tendency to form insoluble compounds with some element (e.g., Ca, Al, Fe, Mn) that is very common in soil. Brady and Weil (1996) examine the interactions of soil pH and the occurrence and solubility of various important P compounds, and conclude that P availability to plants, while usually limited, is probably greatest when soil pH is between 6.0 and 7.0. Superphosphates [$Ca(H_2PO_4)_2$ and $CaHPO_4$] are commonly used fertilizer sources of P since they are the most soluble of the Ca phosphates.

The Phosphorus Cycle in Agriculture

Although P is essential for many biochemical functions in animals and plants, it is probably best known for its importance in energy storage and transfer. As was the case with N, P is present in the agroecosystem in both organic and inorganic forms. Organic P is not available to plants until the organic materials have been decomposed and the P is mineralized to soluble inorganic forms (Figure 4-4). Phosphorus is immobilized as it is taken up by plants or soil organisms and incorporated into their organic compounds. The many chemical reactions of inorganic P compounds in soil are summarized by the equilibrium between soluble and insoluble inorganic P compounds shown in Figure 4-4. This "equilibrium" is very uneven, and the strong movement toward insoluble precipitates represents the major sink of P in the agroecosystem. These reactions deplete the available (= soluble) soil P and limit its uptake by plants.

A principal source of P for agriculture is mined phosphate rock, which is acidified by industrial processes and converted to more soluble forms of P. Organic sources of P fertilizer from plants or animals also may be used, but release of the immobilized P is slower. Losses of P through leaching occur but can be limited by its low solubility. Losses through erosion of surface soil and in runoff water tend to be more important in many agroecosystems. The ability of plants to uptake the limited soluble P from soil can be improved by mycorrhizae (see Application 4-3 and Figure 4-5).

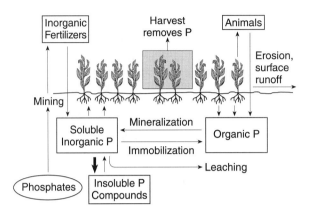

Figure 4-4 The phosphorus cycle in the agroecosystem.

Application 4-3 *Mycorrhizae*

Mycorrhizae are fungi that are common on the roots of most plants, where they improve nutrient uptake by the plant. The relationship is a **symbiosis** since it benefits both plant and fungus. The fungus growing on or within plant roots obtains and uses carbohydrates from the plant as a carbon and energy source. The fungus improves the uptake of nutrients such as P, N, or Ca to the plant since these nutrients can be absorbed from fungal hyphae by the root cells. In a sense, the network of fungal hyphae acts as an extension of the root system, increasing the distance over which nutrients can be absorbed as they move through the hyphae and eventually into the root system. This increased range is important in reaching an immobile nutrient such as P, and mycorrhizae are particularly effective in improving plant uptake of this critical element. The increased nutrient uptake resulting when mycorrhizae are present generally leads to improved plant growth and performance. Although intermediate forms occur, two general types of mycorrhizae are particularly common (Figure 4-5):

Ectomycorrhizae. The hyphae of these fungi grow around and between root cells but do not actually enter individual cells. The mycelium typically forms a thick mat on root surfaces. Morphological changes to the root system can result, such as swollen roots or extreme branching. These mycorrhizae are especially important on trees, and aid in the colonization of cleared sites by woody plants and forest trees.

Endomycorrhizae. The hyphae from these fungi grow directly into cells, where they form specialized feeding hyphae. Some of these hyphae may be modified into swollen, rounded structures called *vesicles,* and so these fungi also may be called **vesicular-arbuscular mycorrhizae,** abbreviated as VAM. They do not cause extensive morphological changes in roots, and are common on the roots of most crops and other plant species, with the exception of plants in the family Cruciferae (cabbage, broccoli, etc.). Although at present it is impractical to inoculate crops with mycorrhizae, future research is aimed at the development of methods and technologies for this application (Sylvia 1994).

(continued)

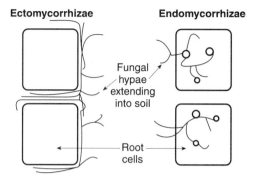

Figure 4-5 Generalized diagrams of two kinds of mycorrhizae on cells of plant roots.

Potassium in Agriculture

Potassium is a critical element in osmotic regulation and many other biological functions. Many forms of K are highly soluble and the element is abundant in most soils, yet limitations in the K that is available to plants can occur for several reasons. Most of the K in soil is bound up in insoluble compounds in minerals, and a portion of the K may be tied up by various kinds of clays. The soluble forms of K are subject to heavy losses through leaching, especially in areas with high rainfall and soils with poor water-holding capacity. Under such adverse conditions, the amount of inorganic fertilizer K lost by leaching may approach the amount taken up by crop plants. Finally, any K taken up by plants or other organisms will be immobilized until decomposition and mineralization eventually occur. This is particularly important because plants have a tendency to take up excessive amounts of K, which doesn't harm the plant, but may reduce its availability. In some cases, soluble K may be removed from soil and tied up in early-season weeds, making it unavailable at critical stages of crop growth. The main sources of inorganic K fertilizers are mined deposits of K salts, especially KCl and K_2SO_4. N, P, and K are the three main macronutrients supplied in inorganic fertilizers (see Application 4-4).

Other Macronutrients and Micronutrients

As was the case with N, P, and K, the availability of each of the other macro- and micronutrients depends on the relative amount that is present in soluble inorganic form(s) in the soil solution, which can be absorbed by roots. All of these elements can be tied up in soil minerals and other insoluble forms, or immobilized while they are part of plant or animal tissues or other organic materials. Availability in the soil solution depends on soil pH, the nature of the soil, and related environmental conditions, which will determine whether the concentrations are excessive or deficient. Many of these elements are included in commercial fertilizers along with N, P, and K. A few of these elements also may be applied during the management of soil pH. Fields with acidic pH may be limed with materials such as CaO, $Ca(OH)_2$, or limestone (mostly $CaCO_3$). **Dolomitic limestone** contains Mg^{2+} as well as Ca^{2+}. Gypsum ($CaSO_4$) sometimes is used in the management of alkaline soils. Sulfur originating as industrial pollutants may be added to ecosystems as SO_2 or other gases absorbed

Application 4-4 NPK Fertilizers

Many different types and mixtures of NPK fertilizers are available commercially. Traditionally, analysis of N, P, and K is reported as a series of three numbers, such as 4-8-8 or 10-10-10. From these, the total %N by weight is given directly by the first figure (4% or 10% in these examples). The second and third numbers represent P and K, but not directly, which can be a source of confusion. It is typical to report P as an equivalent amount of "available phosphoric acid," or P_2O_5, whether P_2O_5 is included in the fertilizer mixture or not. Since P_2O_5 is only about 44% P by weight, the second figure must be adjusted accordingly. K is reported as an equivalent amount of "soluble potash," K_2O, which is 83% K by weight. Therefore, a 4-8-8 fertilizer actually contains 4% N, 3.5% P (8% × 0.44), and 6.6% K (8% × 0.83) by weight.

---

by plants, or may enter soil as one of the components of acid rain ($SO_2 + H_2O \rightarrow H_2SO_3$). More information on the soil chemistry of macro- and micronutrients and their effects on crop plants is provided by Bennett (1993), Brady and Weil (1996), and several other references at the end of this chapter.

Decomposition

The subject of decomposition has already been introduced since this process is so important in the recycling of nutrients. Decomposition is the breakdown of organic matter, involving a number of interconnected stages. **Organic matter** consists of living organisms and various substances and organic compounds that originated from living organisms.

Types of Decomposition

Abiotic decomposition processes do not require living organisms to carry out the decomposition of organic matter. Chemical oxidation reactions result in the breakdown of some organic materials over time, but the most dramatic abiotic decomposition process is the burning of organic matter such as crop residues. Much of the C from the organic matter is converted to CO_2 gas by burning:

$$\text{Organic matter} + O_2 \rightarrow CO_2, \text{etc.}$$

The "etc." indicates that this is not a formal chemical equation and that other products are produced, depending on the nature of the organic matter. But the major reactants and product are shown. If the O_2 supply is adequate, most of the organic C will be lost as CO_2.

Biotic decomposition processes require living organisms to carry out the decomposition. Usually, the organisms most directly responsible for decomposition are bacteria and fungi. These organisms use the organic matter as an energy source during their respiration. In aerobic respiration, O_2 is used by the decomposer organisms:

$$\text{Organic matter} + O_2 \rightarrow CO_2 + \text{other organic products}$$

Anaerobic respiration occurs in situations when O_2 is limited, and methane or other gases are produced instead of CO_2:

$$\text{Organic matter} \to CH_4 + \text{other organic products}$$

Note that in both biotic decomposition reactions, organic matter is decomposed, but other organic products are formed or left over. This emphasizes the fact that biological decomposition is a multistep process, unlike the fairly complete decomposition achieved when organic matter is burned in fire. Biological decomposition of some substances may take years, with a succession of different bacterial and fungal decomposers working on the substrate over time. Each of these may change the original organic material into other form(s) that then serve as suitable substrates for further decomposition by the next group of organisms.

Most decomposition activity, particularly in agriculture, takes place in the soil or at the soil surface, although decomposition is certainly not limited to the soil environment (e.g., dead leaves on plants, rotten fruit, etc.). Decomposition occurring in the soil is of particular importance since nutrients released from organic matter there can be recycled easily into plant roots. An entire community of soil organisms is associated with the decomposition process in most soils. Each of these extracts some of the energy and C that was stored in the C compounds present in the original organic matter.

Sequence of Events in Decomposition

As implied from previous sections, organic matter has many forms. Decomposition of most materials follows a fairly general sequence as organic matter passes though several forms:

Living organisms → dead organisms → litter → detritus → humus → inorganic compounds

Litter consists of fragments of organisms that are large enough that the original source can still be recognized upon close examination. For instance, in litter removed from the soil surface of a field or forest, it is possible to identify fragments of different kinds of leaves, sticks, straw, insect parts, or other distinct materials. **Detritus** consists of small fragments and crumbs of decomposed litter that are no longer recognizable. **Humus** is a dark, brown mixture of various substances and compounds, including organic compounds freed during decomposition and by-products produced by soil organisms. It represents the final stages of organic matter decomposition. These forms of organic matter are most common in the uppermost layers of the soil profile (see Chapter 5). Up to this point, all nutrient elements are present in large organic compounds in fragments arising during the decomposition process. Inorganic compounds are freed from the organic compounds in humus through the process of mineralization, which already has been discussed in detail for some elements such as N. In the preceding sequence, the double arrows at several points indicate that conversion of organic matter from one form to the next is not a simple conversion, but is a multistep process.

Role of Soil Organisms in Decomposition

In the soil, most of the direct decomposition of the various forms of organic matter is carried out by various forms of bacteria and fungi. However, many other soil organisms are essential for regulating the sequence of decomposition. During their activities, they may move the relatively immobile bacteria and fungi to new locations. Several groups of invertebrates break up larger pieces of litter into smaller pieces. Earthworms break up organic matter, in addition to feeding on plant material in the soil organic matter. Microarthropods, including many species of mites and Collembola, feed on fungi and break up pieces of organic matter during their feeding. Enchytraeid worms also feed on fungi and perform a similar function. The activities of these and other microinvertebrates constantly expose a fresh surface area of litter and detritus to direct decomposition by bacteria and fungi. Nematodes, protozoa, and some other types of small animals may be too small to break up pieces of organic matter, but their feeding still stimulates decomposition. As these animals remove dead bacteria, fungi, and microbial by-products from the surface of organic matter particles, decomposers can colonize the freshly exposed surfaces, increasing the rate of decomposition. However, too much feeding on bacteria and fungi could reduce their numbers and slow decomposition, so a balance between the decomposers and their consumers is critical. Many kinds of soil animals are known to benefit mineralization, including earthworms, nematodes that feed directly on bacteria or fungi, and many of the predators that are present in the soil environment. By feeding on and digesting the organic tissue of bacteria and other organisms, these animals may free inorganic nutrients and return them to the soil environment. Their waste products may be rich in N or other nutrients, which may stimulate growth of additional decomposers.

In summary, soil animals may benefit decomposition in several ways:

- Moving decomposers to new locations;
- Breaking litter into small pieces;
- Exposing fresh surface area to decomposition, increasing bacterial and fungal growth rates; and
- Increasing rate of mineralization of nutrients.

Factors Affecting Decomposition Rate

Decomposition rates of various substances can be measured experimentally by burying them in litter bags, which are open-mesh bags allowing access by decomposers and microinvertebrates. The rate of decomposition is affected by many factors, among which the most important are:

- *Soil organisms.* The right kinds of bacteria, fungi, and other organisms must be present for decomposition.
- *Temperature.* The activity and growth rate of many bacterial and fungal decomposers increase as temperature increases. This is one reason why organic matter breaks down more quickly in tropical than in temperate environments. Composting (see Application 4-5) is a method of taking advantage of the increased decomposition rates at higher temperatures.
- *Climate and soil.* Aeration, water-holding capacity, and other characteristics of the soil environment combine to affect decomposition rate. In the midwestern United States, for example, soil organic matter content generally increases as we move north (cooler temperatures, less decomposition) and as we move east (increased soil moisture, less decomposition).

Application 4-5 *Composting*

Composting is the natural breakdown of organic materials to produce a humus-like substance. The process can be accelerated in the "compost piles" of home gardeners or in industrial-scale composting facilities. The process is affected by aeration (usually done under aerobic conditions), moisture, and, of course, the composition of the materials included in the system. Although there are many different types of recipes for proportions of materials added to a compost pile, the principles by which decomposition is accelerated are similar. Initially, bacteria break down some of the more easily decomposed material, releasing heat during the process. However, since this occurs within a pile of material, the heat is trapped and the temperature of the pile rises over a few days. Decomposition then proceeds at elevated temperatures, sometimes reaching 60° C (140° F) to 80° C (175° F), depending on conditions. Although these temperature are too high for most organisms, thermophilic bacteria, which are adapted to high temperatures, colonize the compost pile and proceed with decomposition. Since the rates of bacterial activity and decomposition increase with temperature, decomposition proceeds much more quickly in a compost pile than it does in natural soils. As decomposition continues, the C:N ratio of the material in the pile decreases. As decomposition eventually diminishes, the temperature stabilizes and more typical bacteria and fungi recolonize the compost. Of course, **compost** (technically, any material resulting from decomposition) can be removed for use at any stage of the process, which is why the nature of this material is highly variable.

Almost any biodegradable material can be used as a source for compost, including by-products from kitchens or food-processing industries, yard trimmings, crop residues, etc. Municipal solid wastes typically include many yard trimmings and other biodegradable material that could be composted.

Several problems may occur with composting. Nitrogen may be lost by volatilization as NH_3 or N_2, or by leaching as NO_3^- if the compost pile is too loose or not sealed well. If compost is removed too early in the decomposition process while the C:N ratio is still relatively high, it is considered "immature" or unstable, and its use can tie up available soil N when it is applied to soil. Composts derived from urban wastes may be contaminated by traces of metals and other nonbiodegradable materials that can gradually build up over time if composts from these sources are reapplied to the same site. Also, a compost made from materials low in a particular nutrient will remain low in that nutrient. Finally, shipping large volumes of compost from source to field can be a major cost. However, several of these potential problems may be relatively easy to avoid in some situations, making the use of compost in agriculture particularly attractive.

- *Season.* Fresh, succulent plant material produced in spring breaks down more quickly than does older material. Crop residues at the end of the season may contain higher levels of lignin and other woody materials that are more difficult to break down.
- *Tillage.* Plowing and cultivation usually hasten the breakdown of crop residues and decrease the amount of organic matter in soil (see Chapter 14 for more detail on tillage practices).
- *Composition of the material (C:N ratio).* The most important factor affecting decomposition rate is the nature of the material itself. Organic materials can contain

various levels of lignins and other compounds that are slow to break down, or alkaloids that may be detrimental to soil organisms. The ratio of the amount of C to the amount of N in a substance (**C:N ratio**) is an important indicator of decomposition rate and N availability.

C:N Ratio, Decomposition, and N Availability

In general, the higher the C:N ratio of a material, the longer it will take to decompose. The C:N ratios of common organic materials vary widely (Table 4-3). Young crops and their residues have lower C:N ratios than older crops. Note that materials that we would expect to break down very slowly (like wood or straw) have much higher C:N ratios than materials that decompose quickly (like microorganisms). This is because a larger number of different types of organisms can decompose materials with low C:N ratios. Also, as decomposition proceeds, the C:N ratio decreases (compare the C:N ratios of straw and organic matter in Table 4-3), so more organisms can participate as decomposition proceeds. Initial decomposition of wood and straw is particularly slow, since only a few kinds of organisms participate in the decomposition. Usually these are specially adapted organisms such as wood-rotting fungi or those bacteria that compete well for the N that is so limited in materials with high C:N ratios.

The C:N ratio in organic crop residues has two important consequences in the availability of N to plants, if organic materials are used as a N source or crop amendment. First, the

Table 4-3 C:N ratios for some selected materials.

Material	C:N Ratio	Reference
Sawdust	400:1 to 600:1	Brady and Weil 1996
Wood	157:1	Wasilewska and Bienkowski 1985
Wheat straw	80:1	Brady and Weil 1996
Cornstalks	48:1	McSorley and Frederick 1999
Maize crop residue	43:1	Tian, Kang, and Brussaard 1992
Rice straw	42:1	Tian, Kang, and Brussaard 1992
Partially composted yard waste	36:1 to 46:1	McSorley and Gallaher 1996
Grasses (weeds at end of season)	33:1	McSorley and Frederick 1999
Soybean crop residue	30:1	Cavigelli 1998
Prunings from woody tropical legumes	13:1 to 28:1	Tian, Kang, and Brussaard 1992
Velvetbean (legume) crop residue	19:1	McSorley and Frederick 1999
Farmyard manure	16:1	Iakimenko et al. 1996
Soil organic matter	14:1	Iakimenko et al. 1996
Soil organic matter (upper soil)	12:1 to 17:1	Bravard and Righi 1991
Young alfalfa hay	13:1	Brady and Weil 1996
Nematodes	8:1 to 12:1	Wasilewska and Bienkowski 1985
Fungi	10:1	Brady and Weil 1996
Sewage sludge	8:1 to 9:1	Iakimenko et al. 1996
Bacteria	6:1	Freckman 1988
Bacteria	3:1 to 4:1	Wasilewska and Bienkowski 1985
Urea	0.4:1	Calculated

rate at which N becomes available through decomposition depends on the C:N ratio. So straw, woodchips, and other materials with C:N ratios much greater than 30:1 would be ineffective as organic N fertilizers because it takes them so long to break down, that little significant N would be released during the life of most crops. The second problem is more subtle and depends on the degree of competition between plants and decomposers (bacteria, fungi) for limited amounts of available N in soil. Recall that some of the soil bacteria are more efficient than plants at competing for and immobilizing inorganic N supplies. If C-rich materials like wood or straw are added to soil, the bacteria may use the available N in soil for the decomposition of these C sources, and so the soil N would no longer be available to plants. This phenomenon is called "**N rob**" because the bacteria and other microorganisms rob the plants of this limited N source. The implications of this problem in agriculture and its management are discussed in more detail in Chapter 14. Adding a mixture of woodchips and inorganic N fertilizer to a crop at the same time is especially wasteful, since much of the inorganic N would go not to the plant, but to the decomposition of the woodchips!

Summary

A number of chemical elements are essential nutrients for plant growth. Nitrogen is much in demand by plants, but often in short supply in agroecosystems. Much N is immobilized in organic compounds in living or dead organisms and must be freed through mineralization before it can be used by plants. Examination of cycles of major nutrients like N or P reveals important inputs and outputs of these materials to agroecosystems. Many nutrients are recycled in soil following decomposition of organic materials. This multistep biological process involves bacteria, fungi, and a sequence of other soil organisms.

Topics for Review and Discussion

1. Why might plants have difficulty in obtaining P even when this element is abundant in soil?
2. Suppose a soil was sterilized to eliminate most soil organisms and biological activity. What limitations would this impose on fertilizer use and N uptake by crop plants?
3. Why is breakdown and loss of soil organic matter a more serious problem in tropical than in temperate regions?
4. Discuss differences in the use of organic and inorganic N fertilizers. How would the availability of N and its use by plants differ over time for these two types of fertilizers?
5. Explain the significance of the C:N ratio in decomposition and nutrient release. Suppose the C:N ratio of a compost was too high for sufficient N release. What methods could be used to lower the C:N ratio of a compost or other organic N source?

Literature Cited

Allen, O. N. 1975. Symbiosis: Rhizobia and leguminous plants. In *Forages,* eds. M. E. Heath, D. S. Metcalfe, and R. F. Barnes, 98–104. Ames, Ia.: The Iowa State University Press.
Bennett, W. F. 1993. *Nutrient Deficiencies and Toxicities in Crop Plants.* St. Paul, Minn.: The American Phytopathological Society.
Brady, N. C., and R. R. Weil. 1996. *The Nature and Properties of Soils.* Upper Saddle River, N.J.: Prentice-Hall.

Bravard, S., and D. Righi. 1991. The dynamics of organic matter in a latosol-podzol toposequence in Amazonia (Brazil). In *Diversity of Environmental Biogeochemistry*, ed., J. Berthelin, 407–417. Amsterdam: Elsevier.
Cavigelli, M. A. 1998. Carbon. In *Michigan Field Crop Ecology*, eds., M. A. Cavigelli, S. R. Deming, L. K. Probyn, and R. R. Harwood, 17–27. Michigan State University Extension Bulletin E-2646. East Lansing, Mich.: Michigan State University.
Fluck, R. C., and C. D. Baird. 1980. *Agricultural Energetics*. Westport, Conn.: Avi Publishing Company.
Freckman, D. W. 1988. Bacterivorous nematodes and organic matter decomposition. *Agriculture, Ecosystems and Environment* 24:195–217.
Iakimenko, O., E. Otabbong, L. Sadovnikova, J. Persson, I. Nilsson, D. Orlov, and Y. Ammosova. 1996. Dynamic transformation of sewage sludge and farmyard manure components. 1. Content of humic substances and mineralisation of organic carbon and nitrogen in incubated soils. *Agriculture, Ecosystems and Environment* 58:121–126.
McSorley, R., and J. J. Frederick. 1999. Nematode population fluctuations following decomposition of specific organic amendments. *Journal of Nematology* 31:37–44.
McSorley, R., and R. N. Gallaher. 1996. Effect of yard waste compost on nematode densities and maize yield. *Supplement to Journal of Nematology* 28:655–660.
Pimentel, D., and W. Dazhong. 1990. Technological changes in energy use in U.S. agricultural production. In *Agroecology*, eds., C. R. Carroll, J. H. Vandermeer, and P. M. Rosset, 147–164. New York: McGraw-Hill.
Sylvia, D. M. 1994. Role of mycorrhizae in sustainable agriculture. In *Environmentally Sound Agriculture*, eds., K. L. Campbell, W. D. Graham, and A. B. Bottcher, 559–566. St. Joseph, Mich.: American Society of Agricultural Engineers.
Tian, G., B. T. Kang, and L. Brussaard. 1992. Biological effects of plant residues with contrasting chemical compositions under humid tropical conditions—decomposition and nutrient release. *Soil Biology and Biochemistry* 24:1051–1060.
Wasilewska, L., and P. Bienkowski. 1985. Experimental study on the occurrence and activity of soil nematodes in decomposition of plant material. *Pedobiologia* 28:41–57.

Bibliography

Berthelin, J. 1991. *Diversity of Environmental Biogeochemistry*. Amsterdam: Elsevier.
Butcher, S. S., R. J. Charlson, G. H. Orians, and G. V. Wolfe, eds. 1992. *Global Biogeochemical Cycles*. London: Academic Press Limited.
Cavigelli, M. A. 1998. Nitrogen. In *Michigan Field Crop Ecology*, eds., M. A. Cavigelli, S. R. Deming, L. K. Probyn, and R. R. Harwood, 28–43. Michigan State University Extension Bulletin E-2646. East Lansing, Mich.: Michigan State University.
Chameides, W. L., and E. M. Perdue. 1997. *Biogeochemical Cycles*. New York: Oxford University Press.
Coleman, D. C., and D. A. Crossley, Jr. 1996. *Fundamentals of Soil Ecology*. San Diego, Calif.: Academic Press.
Harrison, A. F., P. Ineson, and O. W. Heal, eds. 1990. *Nutrient Cycling in Terrestrial Ecosystems*. London: Elsevier Applied Science.
Hendrix, P. F., R. W. Parmelee, D. A. Crossley, Jr., D. C. Coleman, E. P. Odum, and P. M. Groffman. 1986. Detritus food webs in conventional and no-tillage agroecosystems. *BioScience* 36:374–380.

Jarrell, W. M. 1990. Nitrogen in agroecosystems. In *Agroecology,* eds., C. R. Carroll, J. H. Vandermeer, and P. M. Rosset, 385–411. New York: McGraw-Hill.

King, L. D. 1990. Soil nutrient management in the United States. In *Sustainable Agricultural Systems,* eds., C. A. Edwards, R. Lal, P. Madden, R. H. Miller, and G. House, 89–106. Delray Beach, Fla.: St. Lucie Press.

Schlesinger, W. H. 1991. *Biochemistry: An Analysis of Global Change.* San Diego, Calif.: Academic Press.

Sprent, J. I., and P. Sprent. 1990. *Nitrogen Fixing Organisms.* London: Chapman and Hall.

Chapter 5

Physical Factors and the Agroecosystem

Key Concepts

- The influence of climate on agroecosystems
- Limiting factors
- Hydrological cycle
- Soil environment and plant growth

Climate

Most crops are grown where they produce a reasonably good yield with minimal inputs. While it is certainly possible to grow bananas in Alaska, the inputs needed to produce even one good hand of the fruit in Anchorage would cost much more than the price of importing them from a much warmer and more suitable environment. Climate, therefore, plays a large role in determining what crops can be grown in which systems, and when. Indeed, it is the extremes in climate that limit agricultural production in most areas. Climate, in turn, strongly depends on location, and is defined by an interaction of many factors, including day length, temperature patterns, and precipitation patterns.

Agricultural Climate Types

Seven different agricultural production zones can be recognized on a global scale (Rice and Vandermeer 1990). These zones differ from the biomes described in most ecology texts, and do not include regions where agricultural production is naturally impossible, such as high alpine or arctic zones. These zones are described on the basis of temperature and moisture regimes, and are generally delineated by latitude:

1. *Wet tropical.* This zone is located between about 5° N and 5° S latitude, and generally receives rainfall in excess of 15 cm per month for at least eleven months out of the year. A number of locations near the equator fall into this zone.

2. *Wet-dry tropical.* This region is located between about 5° and 25° N and S latitudes. The zone has distinct wet and dry seasons, includes much of the world's tropical agriculture, and covers large areas of Africa, India, southeast Asia, and Central and South America.
3. *Cool tropical.* This agricultural climate type occurs at high altitudes; generally at 1000 m elevation or greater. The Andean region of South America is the most extensive area of this climate type.
4. *Moist midlatitude.* Located in selected regions between about 25° and 55° N and S latitudes, most of the eastern United States and a large part of Europe lie within the moist midlatitude zone, as well as a number of important locations in the Southern Hemisphere.
5. *Dry midlatitude.* Also located between 25° and 55° N and S latitudes, this region receives much less rainfall than the moist midlatitude zone. Central North America and central Asia fall into this agricultural climate zone.
6. *Mediterranean.* These regions are located on the coasts of some continents, between 30° and 40° N and S latitude. They tend to have moist winters and dry summers. As the name implies, many countries around the Mediterranean Sea have this climate type, as well as agriculturally important regions of California.
7. *Arid.* While these regions are not delineated solely according to latitude, they are found primarily around 30° N and S latitudes for reasons described in more detail later. Although important in the southwestern United States, even larger expanses of arid lands occur in the Middle East, Africa, and Australia.

These climatic types are approximate, of course, and gradual changes and transitions occur between zones. **Subtropical** climates have seasonal differences in rainfall, as in the wet-dry tropical zone, and in temperature, as in the moist midlatitude region. Killing frosts may occur occasionally. Subtropical climate effects are most evident near the northern and southern extremes of the wet-dry tropical zone. Around the northern limits of this zone in the United States, for example, some subtropical features are evident in the coastal region of South Carolina. Subtropical climate is characteristic of central and south Florida, but subtropical features gradually tend to give way to more tropical conditions in Cuba.

Solar Radiation, Seasons, and Day Length

The tilt of the earth is on an angle of approximately 23° 22' from a line drawn perpendicular to the plane of the earth's orbit (Figure 5-1). Both this tilted axis and the rotation of the earth around the sun are responsible for seasonal fluctuations in the amount of incoming solar radiation to all areas of the world. In the Northern Hemisphere, December 21 is the winter solstice. At this point in the earth's orbit around the sun, the Northern Hemisphere is tilted away from the sun, and the Southern Hemisphere is tilted toward the sun, so areas south of the equator receive the greatest amount of solar radiation. During this period of time, as we move northward from the equator, days get shorter and nights get longer. In fact, all areas north of about 66° N latitude are in continual darkness at this time because the tilt of the earth keeps the northern latitudes hidden from the sun's light at all times.

As spring approaches in the Northern Hemisphere, days slowly lengthen as winter turns to spring, while days begin to get shorter in the Southern Hemisphere as summer turns to autumn. On March 20 (the vernal equinox), both hemispheres receive an equal amount of

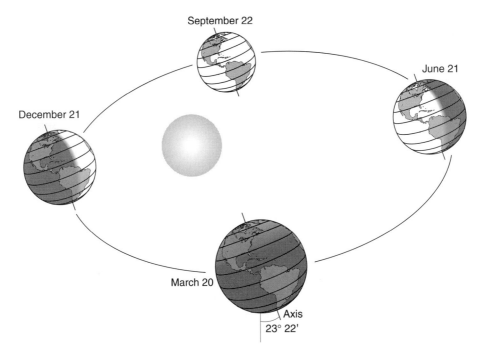

Figure 5-1 Annual movement of the earth around the sun.

the sun's energy, and the length of a day at any given latitude in the Northern Hemisphere should correspond exactly to the length of a day at the corresponding southern latitude.

By June 21, or the summer solstice in the Northern Hemisphere, the situation just described for the winter solstice is reversed. At this point, the Southern Hemisphere is tilted away from the sun, and areas north of the equator receive the majority of the incoming solar radiation. During this time, as we move northward from the equator, days get longer and nights get shorter. All areas south of about 66° S latitude are in continual darkness at this time.

As the earth continues its rotation around the sun, day length in the Northern Hemisphere gradually decreases, while days lengthen in the Southern Hemisphere as summer returns to that area. On September 22 (the autumnal equinox), both hemispheres again receive an equal amount of incoming radiation from the sun, and the length of days at opposing latitudes is again the same. It is interesting to note that throughout the path of the earth's rotation, the equatorial zone of the planet receives nearly constant amounts of solar radiation, and day length in this region shows very little variation.

Day length has very important implications for agricultural systems. Some plants flower and fruit in response to length of night (**photoperiodism**). When this occurs, plants cease vegetative growth, and begin to move into the reproductive stage of their development. Different plants respond in various ways to day length. **Short-day** (long-night) **plants** such as soybean flower only when the night is longer than a certain critical value. **Long-day** (short-night) **plants,** such as many varieties of onion, flower or bulb only when the night is shorter than some critical value. Some plants are **day neutral,** and are not affected by day length. For most plants, length of night either accelerates or slows down flowering, but neither immediately triggers reproduction nor prevents it entirely. In many cases, plant growth and timing of plant processes are related to the accumulation of temperature or heat units over time. Heat units and their influence are discussed in greater detail in Chapter 6.

Different plants may show various adaptations to light intensity as well. The photosynthetic pathways used by certain plants enable them to adapt to lower light intensities (see Application 2-3). For instance, C3 plants are well adapted to various light levels and can exhibit maximum growth under cooler conditions and lower light intensities. On the other hand, C4 plants are adapted to higher light intensities and are generally more efficient in water use and photosynthesis. Plants using the CAM photosynthetic pathway are well adapted to high light intensities and regions with very low moisture. Plant adaptations to light intensity are important to consider when planning farming systems, and will be considered further in Chapter 13.

Temperature Patterns

To illustrate why temperatures decrease with increasing latitude, Figure 5-2 shows how sunlight strikes the earth at three different latitudes on the same date. On June 21, when the Northern Hemisphere is tilted toward the sun, parallel rays of sunlight strike the earth at any given time. Since the sun is much larger than the earth, the rays are parallel; if the sun was smaller or farther away, the rays might appear in the shape of a fan as they reach the earth's surface.

The parallel rays do not strike the earth at the same angle, however. Since the surface of the earth is curved, the incoming rays hit the earth at more oblique angles with increased latitude. While the angle at which the sun hits the earth's surface is perpendicular at 23° N latitude on June 21 (due to the curvature of the earth), that angle decreases as we move farther north. Therefore, the amount of energy that strikes any given area of earth will decrease with increased latitude. The equator of the planet tends to be warmest since it receives the largest number of photons per unit area (e.g., square meter). As you move away from the equator and toward the pole, the light from the sun hits a larger unit of land, so the number of photons hitting a square meter closer to the poles will decrease and temperatures will decline.

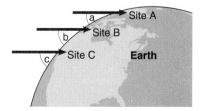

Figure 5-2 Angle of sunlight striking the earth with changing latitude. Angle c > angle b > angle a, so the energy received by site C is greater than that received at sites B or A.

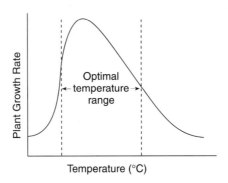

Figure 5-3 The relationship between plant growth rate and temperature.

These patterns again have important implications for agricultural productivity. As with nearly all physical factors, extremes in temperature have adverse effects on plant growth and development. Below a certain minimum, or above a certain maximum, temperature plant growth is slow or nonexistent (Figure 5-3). There is a range of temperatures between those extremes that is considered optimal for the plant, and that range varies with each crop. For instance, some tree crops die if they are ever exposed to freezing temperatures (coffee); others (cherry) need a period of time below freezing temperatures in order for the tree to produce. Most vegetables also have a range of temperatures in which they are most productive. For example, cucumbers need nights above 10° C (50° F) in order to produce fruit, while lettuce must be grown in cool temperatures since it will go to seed if it is grown in warm to hot temperatures. Many grains are available in spring or winter varieties, and yield varies according to the region in which the grain is grown and the variety used.

Precipitation Patterns

We know that warm air holds more moisture than does cool air, so as warm air cools, it will release much of its moisture as precipitation. Understanding the planet's precipitation patterns is easier if we first look at how air moves across the earth's surface because global air circulation patterns are responsible for many of the precipitation patterns that we observe on earth.

As air over the equator is warmed, it picks up moisture and rises (Figure 5-4). When the air reaches cooler altitudes, it releases its moisture as precipitation, causing a very wet belt stretching from about 10° N latitude to 10° S latitude. The cooler and drier air then moves away from the equator and descends at northern and southern latitudes of about 30°. These latitudes tend to contain many of the world's deserts, since as the descending air warms up, it absorbs any available moisture from the land and plant surfaces. At about 60° N and S, air again warms over the continental landmasses, picks up moisture, rises, and releases that moisture as precipitation. The air then descends again in the polar regions, giving rise to cold, dry polar deserts.

The air circulation patterns on earth are strongly affected by the rotation of the earth on its axis. If the earth didn't rotate, air would simply rise as it warmed, cool as it moved north

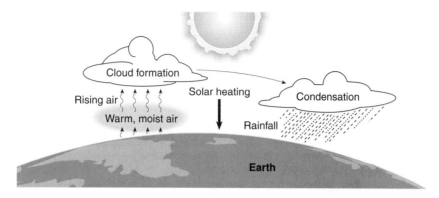

Figure 5-4 Cloud formation by convection and rainfall produced by condensation.

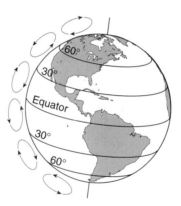

Figure 5-5 Movement of surface air as a result of the spin and rotation of the earth.

or south, and descend again in a constant, circular pattern. Wind movement, then, would always be to the north or to the south. However, the rotation of the earth puts a spin on the winds, as shown in Figure 5-5. Because of this rotation, the air patterns are deflected to the east or to the west.

While air movement is important in determining precipitation patterns, resulting winds can be important in agriculture as well. Direct damage to crops from wind occurs in many situations (Figure 5-6). In other cases, water loss from plants may be accelerated by wind.

The Potential Effects of Climate Change

There have been numerous predictions on how future climate change could affect plant growth and productivity. In 1988, L. M. Thompson reported that corn and soybean grown in the midwestern region of the United States showed great variation in yield, and described

Figure 5-6 Banana leaves damaged by wind. Note fraying around the edges of the leaves.

how this variation could be linked to weather patterns associated with climate change. While increases in air temperature may actually increase productivity in some regions of the world, climate change would likely result in unpredictable rainfall, which could lead to decreased productivity if drought or flood conditions occur. In addition, an increase in temperature would lead to higher evapotranspiration rates, which may reduce the amount of available water held by soil (Hatfield 1990). Climate change could also lead to changes in the atmospheric carbon dioxide (CO_2) concentration, which could modify temperatures and further affect climate (see Chapter 2).

Limiting Factors

Plants in an agroecosystem will grow until a constraining factor prevents further growth and production. Any resource essential to the growth of an organism that is in short supply is considered to be a **limiting factor.** The concept of a limiting factor was first presented by the German chemist Justus von Liebig. He simply stated that plant production is limited by that factor that is in shortest supply, relative to the need of the plant for that factor. In other words, if a plant is supplied with all essential nutrients, water, light, heat, oxygen, etc., but the soils in which it is grown are limited in phosphorus, it is the concentration of phosphorus in the soil that will determine when productivity ceases. The plant cannot continue to produce in the absence of an essential nutrient. The same is true for humans. If we are supplied with the right temperature, nutrients (food), and water, but we are in a closed system with only a limited amount of oxygen, we will no longer be able to survive once the oxygen level drops below a certain level, even though we still have plenty of food and water.

Once the limiting factor in an agroecosystem has been identified, increasing that factor may result in yield and quality increases until the next limiting factor is reached. Increasing a nonlimiting factor will do very little to improve plant productivity, and may even decrease it by throwing off the balance of the system or approaching toxicity levels of that factor.

Water and the Water Cycle

Water is one of the most important factors in agricultural production. Water makes up approximately 70% of the live mass of nonwoody plants, and is directly involved in critical metabolic processes such as respiration, photosynthesis, and nutrient transport. Because of this, it is crucial to ensure that an agricultural crop has sufficient quantities of water for metabolism and growth for the production of a high yielding, good-quality harvest.

When we look at pictures of the earth taken from space, we can easily wonder how any areas on the planet can be deficient in water. Approximately two-thirds of the earth's surface is covered by water, and countless lakes, streams, and rivers penetrate the large landmasses that make up the remaining one-third of the earth. However, much of this water is unavailable for human, animal, or agricultural use. About 97% of the water on earth is saline, and cannot be readily used without either expensive desalination procedures, or through the natural processes of water cycling (see the next section). Of the remaining 3%, about 77% is tied up in glaciers or ice fields in the polar regions of the earth. This leaves a total of only about 0.65% of the earth's water available for use by humans, animals, and plants. In addition, some of this available water has been polluted by humans, leaving it unfit for drinking or agricultural use.

Conservation of water is often an important part of agricultural production. In many parts of the United States, water is readily available and is therefore often used without restraint. However, in other sections of the United States, such as in the arid southwest, water is scarce, and farmers pay close attention to how much water is necessary for good yields. Outside of the United States, many farmers are unable to irrigate their crops at all, or must find innovative ways to provide fruits and vegetables with water while minimizing evaporative loss.

Water is a physical factor in plant growth. As such, plant growth in response to water shows a pattern similar to that of most other physical factors (Figure 5-7). Unlike other physical factors, however, there is a fairly wide range of soil water levels within which a plant will show maximum growth. In this case, adding water by irrigation will not affect yield

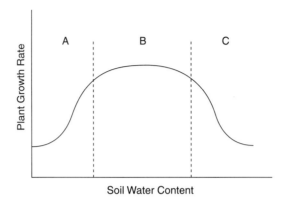

Figure 5-7 The relationship between plant growth rate and soil water content. (A) Insufficient soil water to support high rates of growth, irrigation may be necessary. (B) Optimal moisture levels for high rates of growth. (C) Too much moisture to support high rates of growth; drainage systems may need improvement.

much (if at all), and may even decrease yield. The challenge for agricultural scientists and ecologists is in determining how much water a crop needs, and in meeting those needs without exceeding them. In areas where farmers are paying substantial amounts for their water use, the challenge becomes economic as well as ecological.

The Hydrological Cycle

The movement of water through and between the ecosystems and the environments of the earth defines the **hydrological cycle.** The majority of the water that moves through agroecosystems around the world originates in the world's oceans (Figure 5-8). As sunlight warms the ocean, water evaporates from the surface and condenses to form clouds. The clouds can then move over land, dropping moisture as precipitation as temperatures cool. A large amount of this water actually falls back into the ocean, while about 10% of it falls on land. Water can also be added to the air by evaporation from lakes, rivers, streams, and soil. In addition, plants require water for photosynthesis and other metabolic activities, and release much of it back to the atmosphere through the process of **transpiration.**

When water falls on an agroecosystem, three main events may occur. Much of the water is simply retained by the soil, although how much water the soil holds depends on the soil structure and texture, which is discussed in more detail later in this chapter. Water may also be taken up immediately by plants in the system, particularly if the soils are relatively dry and the plants have been water-limited. Water that is not taken up by plants and that cannot be held by the soil either evaporates back into the air, or percolates through or runs off of the land on which it has fallen. From there, it may serve to refill groundwater stores or replenish aquifers. One of the major concerns, in terms of water quality, in this cycle is that water that moves through the soil may carry with it pollutants or agricultural chemicals, and these chemicals can eventually end up in aquifers, water supplies, or in lakes and rivers

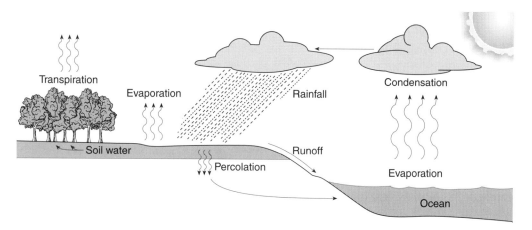

Figure 5-8 The earth's hydrological cycle.

(see Chapter 12). In addition, if water simply runs off the land over the soil surface, it may carry much of the topsoil if that soil is not firmly anchored by roots from vegetation. This results in both soil erosion and silting, which occurs when topsoil settles into lakes and rivers (see Application 5-1).

The amount of water that is lost to the atmosphere as water vapor from an agricultural system depends both on transpiration and evaporation rates, collectively referred to as **evapotranspiration** (Figure 5-9). Evapotranspiration rates increase in response to an increased potential of the environment to evaporate water. We know that water vapor moves from regions of high concentration to low concentration, so anything that causes a larger gradient in water concentration between the soil and the environment, or between plants and the environment, will increase water loss to evapotranspiration. Increased solar radiation or increased temperature will lead to increased transpiration in plants as well as a greater capacity of the atmosphere to absorb water vapor, increasing the water vapor gradient and the movement of water vapor from agroecosystem to environment. If the relative humidity of the air is lowered, the atmosphere may become dryer than the soil or the plant, increasing the gradient and therefore increasing evapotranspiration rates. Wind can also serve to blow vapor-saturated air away from the surfaces of the soil or the plant, replacing it with less saturated air, which will then pull more moisture from the soil or plants.

Soils and the Soil Environment

In all agroecosystems, soils are an essential component contributing to plant growth and development. Except in the unique case of hydroponic gardening, plants cannot grow without the soil medium to provide them with anchorage and support, aeration, water, and essential nutrients. Soils also serve as a recycling center for nutrients and organic waste, and provide a stable environment for soil organisms.

Application 5-1 *The Problem of Soil Erosion*

Soil erosion consists of two very different components: the on-site degradation of one location, and the off-site settlement of eroded particles on another. Ecologically, each is as potentially destructive as the other, although the productivity of most agricultural systems is more profoundly affected by degradation than by deposition.

What Is Erosion?

Erosion is simply the movement of soil, including its removal from one point and its deposition at some other point. Both the removal and deposition of soil can cause serious problems. Productive topsoil may be moved away from a particular agricultural site or field. Of course, this soil that is removed must go somewhere, and this can cause pollution and sedimentation in other parts of the ecosystem. A certain degree of erosion occurs in all natural systems—some soil is moved by wind and some by water. A small amount of erosion can be tolerated in most systems, since at the same time that soil is being removed from a site, new soil is being formed there from parent material, although very slowly. Unfortunately, agroecosystems are sometimes most heavily subject to soil erosion since the intensive management and cultivation of these soils affects many of the factors that lead directly to soil loss.

Why Is Erosion a Problem in Agroecosystems?

- The degradation of soil by erosion inevitably leads to a loss of soil productivity. Most of the organic matter and nutrients in a soil are located in topsoil, and it also provides the best structure and aeration, and the largest population of soil organisms. When topsoil is moved at a rate faster than it can be created, the entire function of the agroecosystem can be affected. The loss of nutrients and organic matter alone will usually cause the need for replacement of the lost nutrients by fertilizers.
- The removal of topsoil from the profile may lead to a thinning of the profile, providing less depth for the development of plant roots and a decrease in the water-holding capacity of the soil.
- As erosion carves gullies and rills, water runoff from agricultural fields is facilitated. This means that less water actually penetrates the soil surface, so the amount of water available to plants for uptake may be decreased.

What Happens Off-Site?

- Eroded soil sediments may end up in rivers and streams. As these sediments settle into bodies of water, they take up space in these rivers, streams, and lakes, leaving less space available to absorb heavy rains. Therefore, flooding is much more likely to occur unless dredging is undertaken to remove the deposited sediment.
- Eroded soil often contains nutrients or chemicals that can end up in water supplies or natural systems.

(continued)

How Does Erosion Occur?

Water Erosion

Water erosion occurs when the impact of a raindrop detaches soil particles from the rest of the soil, and the detached particles are then carried by water to a different location where they are deposited. This is a very common method of erosion, and it occurs naturally at a slow pace, although the pace is dependent on the amount and intensity of rainfall. Soils particularly susceptible to natural erosion have low infiltration capacity (water doesn't penetrate the soil easily) or poor structural stability (water detaches soil particles easily), which are often affected by such factors as the amount of organic matter present, soil texture, kinds of clays present, or the layering of soils.

The most destructive erosion, however, is often helped along by human activity. Although we can't change the amount of water that falls or the rainfall intensity, our management of agricultural systems can make them more or less vulnerable to erosive forces. For example, cultivation of a steeply sloped hill will lead to more soil loss by erosion than if it had been left in natural vegetation. When plants are present in a system, the roots serve to hold the soil together, thus preventing runoff. When we clear land for cultivation, we remove that vegetation, causing the land to become quite susceptible to erosion. Even after agricultural plants have been established, erosion is still more likely since many agricultural systems tend to leave large portions of the land bare to prevent competition for resources. Generally, runoff is more likely on soils that are steeply sloped, rather than on soils with a long and low gradient.

In general, soils that are covered by natural vegetation and grass are least susceptible to erosion, followed by dense plantings of forage crops (which still serve to keep the ground covered with vegetation). The planting of small grains may lead to erosion in certain conditions. The planting of row crops is particularly bad for soil protection during early stages of growth, before the plants provide much soil cover. In this situation, erosion can be reduced by applying plant residues or mulch to the exposed ground between rows. Clean fallow is particularly susceptible to water erosion since the ground is kept bare for extended periods of time.

Most management practices associated with reducing water erosion have three major objectives: improving water infiltration and preventing runoff, reducing sediment load of runoff water, and directing runoff water into predetermined channels to minimize the formation of rills and gullies in the fields (which could make cultivation difficult). Examples of excellent management practices to reduce erosion include the use of a dense forage crop in rotation with row crops (improves soil structure so water infiltration improves), conservation tillage (leaves crop residues on the soil surface), contour tillage (cultivation across the slope), strip cropping (strips of close-growing crops alternated with other crops to catch runoff), terraces (ridges or benches line each row perpendicular to the slope to prevent runoff and improve infiltration), and vegetative barriers (natural vegetation or grasses or shrubs planted along the slope to catch runoff).

Wind Erosion

Wind erosion occurs when soil particles detach from larger clods or soil aggregates, and are carried by wind to a different location where they are deposited. Wind erosion can be as

(continued)

destructive to agricultural systems as water erosion, but it is perhaps less common since it occurs primarily in areas where the soil surfaces are dry, such as in arid and semiarid regions. Wind not only can erode away the fertile topsoil of a system, but it may even blow shallowly rooted crops away, expose the roots of crops that are more deeply rooted, or cover plants in drifts when the soil is deposited back at another location. Windblown soil particles can cause abrasions on fruits and vegetables, affecting market quality.

Several factors affect the susceptibility of a soil to wind erosion. Covering the soil surface with plants or crop residues can provide the best protection against wind erosion. Soil moisture is also helpful in deterring wind erosion since water holds soil particles together and makes them less likely to detach from each other in the wind. For this reason, any agricultural practices that serve to maintain soil moisture also will help to prevent wind erosion. Intuitively, it is rather obvious that more erosion will occur in areas of high wind velocity and air turbulence, so planting windbreaks (tall trees, shrubs, sugarcane, elephant grass) to serve as natural barriers to the wind can be essential in these areas. Soil conditions can affect wind erosion: A rough surface causes a still layer of air at the soil surface, reducing erosion; and stable clods and aggregates in the soil tend to be quite resistant to the abrasive effects of wind. For this reason, management practices that improve soil structure can be important in reducing wind erosion. Conservation tillage (see Chapter 14), which leaves crop residues on the soil surface, helps to roughen the surface of the soil, and also provides a protective cover to the soil particles beneath it. The incorporation of organic matter into the soil can also dramatically improve structure. Planting rows perpendicular to the direction of the prevailing winds can also reduce erosion.

Soil Conservation Efforts

In the United States, soil conservation practices on lands susceptible to erosion include a wide variety of methods. The most frequently used methods are those that provide some cover to the soil surface. These can include the retention of crop residues and groundcovers or the establishment of native or introduced grasses and legumes. The planting of trees or other windbreaks is a common practice. Less frequently used methods include planting of contour grass strips, establishment of field windbreaks, and alley cropping (United States Department of Agriculture 1998). Financial incentives provided by the United States government encourage farmers to adopt soil conservation methods. Through the use of these various methods, the average erosion across the United States between 1986 and 1997 was approximately 15.7 kg soil/ha, with a total of over 16 million hectares of land (almost 15 million hectares of which is cropland, pasture, or rangeland) protected from erosion by the end of 1995 (United States Department of Agriculture 1998).

The Many Purposes of Soils
Soil as a Medium for Plant Growth

Anchorage and Support. Roots grow down into the soil, anchoring plants so they won't topple as they grow upward and increase their aboveground biomass. Without soil (as in hydroponic gardening), growers need to provide artificial support for plants, usually in the form of wire mesh or bamboo rods.

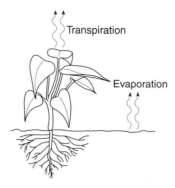

Figure 5-9 The components of evapotranspiration.

Oxygen/Aeration of Roots. Plant roots need oxygen to survive and grow. Because soil particles are unevenly shaped, they do not fit together perfectly, but instead leave air spaces between the particles. This provides a system of ventilation for oxygen to reach from the soil surface to the roots of plants. Soil that is highly compacted or waterlogged will lose many of these air spaces, which seriously decreases the amount of oxygen that can reach root systems.

Water. Different kinds of soils hold water with varying degrees of tenacity, but soil is still the primary water storage site for agricultural plants. Plants need an enormous amount of water for normal growth and productivity (between 200–1000 g of H_2O for each gram of plant dry matter is needed for photosynthesis, nutrient transport, and other metabolic processes). Since plants need water continuously, but rainfall only occurs sporadically, the water-holding capacity of the soil directly affects plant growth and development.

Nutrients. Plants obtain most of their needed nutrients from the soil. Only three essential nutrients (carbon, hydrogen, and oxygen) are obtained from gas exchange with the atmosphere. All other nutrients must be taken up from the soil environment. Soils act as a nutrient "bank," holding important ions until they can be taken up by plant roots.

Soil as a Recycling System for Nutrients and Organic Wastes. Within the soil system, organic wastes accumulate along with the dead bodies of fungi, plants, animals, and even humans. The microorganisms that live in the soil are involved in the decomposition, recycling, and subsequent assimilation of the nutrients and elements from these organisms and substances so they may cycle through the system and be reused by subsequent generations of plants and animals (see Chapter 4).

Soil as a Habitat for Soil Organisms. The soil is home to literally billions of organisms that are essential in the function of ecosystems in general, and agroecosystems in particular (see Chapter 3). Without the action of certain bacteria that live in the soil, plants would be unable to assimilate nitrogen into their systems. Without the decomposing action of a collection of soil microorganisms, many nutrients would be forever tied up in forms that are unusable to both plants and animals. In short, without these organisms, most agricultural systems would be unable to function, much less to produce enormous, or even adequate, yields.

Soil Components

Organic Versus Mineral Soils. Soils are composed of both an organic ("organic matter" was defined in Chapter 4) and an inorganic (mineral) fraction. The proportions of these two fractions depend on a variety of factors, including climate, activity of soil organisms, topography, time, and weathering. As a rule, mineral soils contain less than 20% organic matter, except for a surface layer of organic debris. Most agricultural soils are mineral soils, and most of these actually contain less than 10% organic matter; 3–5% is probably a typical average for fertile mineral soils in the midwestern United States. Soil organic matter content generally decreases in warmer and tropical climates. Organic soils contain more than 20% organic matter, and form when the rate of organic accumulation exceeds the rate of decomposition. They often form underwater as aquatic plants die and settle, but do not rapidly decompose. Some organic soils are highly prized as rich agricultural soils, while others tend to be acidic and are not as useful for agroecosystems. Organic soils can occur almost anywhere in the world, but because decomposition rates are greater in warmer climates, the largest concentrations of organic soils are found in Russia, Canada, and northern Europe.

Major Components of Mineral Soils. In general, all soils are made up of solids (mineral matter and organic matter) and pore spaces. Mineral matter is composed of sand, silt, and clay, which are different sized classes of soil particles. The proportions of these particles in a soil determines the **soil texture.** Mineral matter is formed from the breakdown and weathering of parent materials (rocks, minerals, etc.). The organic matter content of a soil can change over time, and can be successfully managed in an agroecosystem. Organic matter is important for cultivation and plant growth, and serves to increase both the water-holding capacity and the fertility level of soils (Figure 5-10).

Soil Structure and Texture

As indicated previously, mineral soils are formed of three major components: sand, silt, and clay (Table 5-1). Soil texture is defined on the basis of percent composition of these components, and is determined by the use of a soil textural triangle (Figure 5-11). Each corner of the triangle is represented by one of the major classes of separates (particles), and the addition of other particle sizes modifies that texture. Almost all soils consist of more than one, and usually all three, separates.

Soil texture is an important determinant of the potential inputs needed in an agroecosystem. It affects such functions as water-holding capacity, movement of water through soil, and the nutrient-holding capacity of the soil. Soil texture can indicate irrigation needs, potential fertility problems, and drainage difficulties that may be encountered.

Water Movement Through Soil. Pore space affects the rate of movement of air and water through soil. The size of the pore spaces is largely determined by soil texture (sand particles result in larger pore spaces than clay or silt alone) and by aggregate size. Clay particles tend to clump together, forming large aggregates that increase water flow and the size of pore space. In this case, we can speak of **macropores,** the space that is formed

Figure 5-10 Note the dark color of a muck soil (center) containing 80% organic matter, compared to other soils. A silty clay loam (left) retains some crumbs and aggregates compared to a sand (right).

Table 5-1 The United States Department of Agriculture System of Soil Separates.

Separate	Diameter (mm)	Feel
Very coarse sand	2.00–1.00	Grains easily seen; sharp, gritty
Coarse sand	1.00–0.50	
Medium sand	0.50–0.25	
Fine sand	0.25–0.10	Gritty; each grain barely visible
Very fine sand	0.10–0.05	
Silt	0.05–0.002	Grains invisible to eye; silky to touch
Clay	< 0.002	Sticky when wet; dry pellets hard, harsh

between soil aggregates, and **micropores,** the space found within soil aggregates (between clay particles) (Figure 5-12).

Both macropore and micropore spaces may be filled with water or air, and the percent filled fluctuates rapidly over time. For instance, a heavy rainfall can fill up pore spaces with water, decreasing the amount of air in the soil. On the other hand, the evaporation of water from the soil back into the atmosphere can increase the amount of air in the pore spaces.

The soil solution is the water that is held in the pore spaces of a soil, and this solution contains the various nutrients and compounds that are necessary for the growth and development of a plant. Water moves through a soil in response to water potential gradients (from high concentrations to low concentrations), so it can move in any direction, and can move in either a vapor or a liquid state. Because of the large pores found between large sand grains, water moves through sandy soils much more quickly than it moves through the tiny soil pores found in clay soils.

If the pore spaces are not water-filled, they contain air. The composition of this air varies greatly depending on plant use, soil type, and microbial activity. Generally, soil air has a

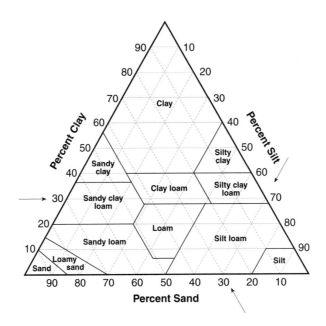

Figure 5-11 The soil texture triangle. The numbers are the percentages of sand, silt, and clay in a soil of a given type.

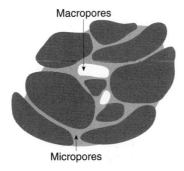

Figure 5-12 Soil macropores and micropores. Dark areas represent sand grains or other large soil particles that do not contian pores. Lighter gray areas represent clay aggregates with micropores contained within them. At field capacity, macropores are well-drained and micropores are filled with water available for plant uptake.

higher relative humidity than the atmosphere, and also has a higher CO_2 concentration, since respiration of plant roots and soil organisms removes O_2 from the soil pores and releases CO_2. In fairly dry soil, gradients in CO_2 cause the gas to diffuse out of the soil, and O_2 to diffuse into the soil rapidly enough to prevent CO_2 toxicity or O_2 deficiency in the soil. However, when soils have a high water content, diffusion of gases is extremely slow, so soils that are not well-drained may have a very high CO_2 level and a low O_2 level, which can inhibit plant growth and microbial activity.

The water-holding capacity of a soil depends largely on soil texture, which determines the size of the pores (both micropores and macropores). Water that falls on pure sand tends to pass immediately through it, since the pores between the large sand grains are large, and very little water is held between the grains. Clay tends to hold a lot of water since the pores are much smaller, and water is held with more tension by the particles and within small micropores, so it does not readily move through. Silt is intermediate in water-holding capacity. In all soil types, the addition of organic matter serves to increase the water-holding capacity.

Rapid movement of water through soil can have serious consequences for agroecosystems. Since most nutrients become available in the soil solution before plants take them up, rapid movement of water through soil can cause these available nutrients to be leached out of the soil if rainfall occurs before uptake. In addition, movement of water out of the soil means that less water is available for plant uptake and use, resulting in rapid drying of the soil and fairly rapid wilting of the plants if more water is not applied.

In aggregated soil (soil in which particles clump together to form clods), water movement speeds up since the pores are larger and water is not held very tightly. Drainage is not likely to be a problem in these soils. However, if aggregation is destroyed by excessive tillage, water movement may be directly affected since water will no longer pass quickly through large pores, but instead must make its way through smaller pores. In these situations, drainage may slow down substantially, perhaps leading to flooded conditions if rainfall or irrigation is heavy.

When all of the macropores have been drained of their water by gravity, but all of the micropores are filled, the soil is said to be at **field capacity.** If you imagine dipping a sponge into a bucket of water, then simply allowing all of the water in the sponge to run out (without squeezing the sponge), the water left in the sponge is analogous to the water in a field at field capacity. As the field dries out and the micropores empty, less water is available for plant uptake. Eventually, any water left in the field is held with such tenacity by the soil particles that the plants are unable to take up any water, and the plants will wilt. This is known as the **wilting point.** Plants do not grow well in soils with water levels below the wilting point (no water available) or above the field capacity (no oxygen available to the roots).

The Soil Profile

Vegetation, climate, weathering processes, topography, and other natural processes can modify soils over time. Materials may be blown in or deposited, while other materials may be lost from the soil by erosion or leaching. Still other materials may be carried upward with water along an evaporation gradient, while others move downward in leachate. In addition, weathering processes continually convert parent material to mineral soil, organic matter decays, and chemical reactions occur within the soil to cause changes in the soil composition over time. Changes in soils that have occurred over time can often be seen by examining a soil profile.

A **soil profile** is a vertical section in the soil that extends from the soil surface down to the parent material, which may be igneous, sedimentary, or metamorphic rock. As we cut through the soil, we often see distinct layers, called **soil horizons,** which become more obvious as soil ages. Each horizon differs in some fundamental way from the horizons surrounding it. These horizons are commonly identified by letters, which may vary among different sources, but identify similar characteristics. The most common horizons, or **master horizons,** may be further subdivided, and soils may or may not have all of these master horizons and subdivisions present.

The surface layer of a soil that contains organic matter is the *O horizon* (Figure 5-13), which occurs above the underlying mineral soils in the profile. Since organic matter accumulates only when the soil is not readily disturbed, the O horizon is often absent from agricultural fields. Plowing and tilling of soil mix the organic matter into the underlying mineral layers, unless conscious efforts are made to keep an O horizon present. The O master horizon can be further differentiated into three different layers:

Oi: This layer is formed by the dead remains of plants and animals that are either slightly decomposed, or have not yet started to decompose.

Oe: Partly decomposed organic matter makes up this subdivision. Although some decomposition has occurred, it is still possible to determine what the source of this organic matter was.

Oa: This layer is made of up very well-decomposed organic matter. The amorphous appearance of this organic matter makes it impossible to determine its source.

The *A horizon* is generally found immediately beneath the O horizon, or on the soil surface when no organic matter is present (Figure 5-13). The A horizon is commonly referred

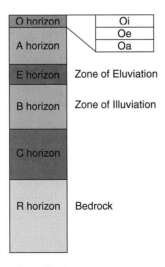

Figure 5-13 Typical horizons of a soil profile.

to as topsoil, and is the most fertile mineral layer of an agricultural soil. Organic matter tends to accumulate in this zone, and the amount of organic matter present tends to affect the color of the soil. Over time, eluvial horizons may develop in this zone as some materials like clay and iron are lost through leaching. Eluvial horizons are lighter in color than the rest of the A horizon and tend to be located near the bottom of this horizon. The A horizon is where the majority of crop roots develop in agroecosystems, and it may be seriously affected by tillage.

An *E horizon* may occur in some profiles. This horizon, or zone of maximum leaching (**eluviation**), sometimes forms underneath the A horizon in some soils. Leaching in this region has led to the loss of clay particles, minerals, and organic matter, leading to a higher concentration of larger sized particles. This zone is most common in high rainfall areas.

The *B horizon* is also known as the zone of accumulation (**illuviation**). Materials that have leached from the A and E horizons tend to accumulate in this region, as do materials resulting from the weathering of parent material below. As a result, this layer tends to be high in clay, but is deep enough that it is not normally disturbed by common tillage practices. If these B horizons are too high in clay, they may not let water through easily and may present drainage problems in agricultural systems if rainfall is high.

The *C horizon* is very similar in appearance to the parent material that lies beneath it. Soil-forming processes have not yet had much time to affect these soils. They differ only slightly from the *R horizon,* or bedrock, beneath them, which has not weathered at all.

In addition to all of these horizons, some soils show *transition horizons,* which are (as the name implies) transitional between the master horizons. These may show properties that are intermediate between two horizons, and are thus designated by two letters (AE, EB, BC). To complicate matters, some soils do not show all of the master horizons. Very young soils will not have been exposed to much leaching or organic matter accumulation, and may only exhibit an A and a C layer. As soils age, more horizons generally appear in the profile.

Agricultural Modifications to Soil Profile. Certain soils are more favorable to agricultural productivity than others (Application 5-2). In addition, agricultural practices can, over time, cause changes in the soil profile that may be helpful or detrimental to future productivity. In soils that are mixed together by plowing, the top layers of soil lose the layered structure that is commonly seen in natural systems, forming a **plow layer.** This can allow the incorporation of organic material into the mineral soils below, thus resulting in increased fertility of the field.

On the other hand, frequent plowing can lead to **hard pans,** or compacted layers that resist root penetration, and may also prevent water movement through the soil. If roots are unable to penetrate this layer, they may not be able to grow deeply enough to either anchor the plant or to provide it with the water and nutrients it needs. Hard pans may result from the leaching of surface minerals to deeper zones (mineral hard pan), or from frequent plowing to a given depth, also referred to as a plow pan, which can also restrict root penetration and water movement.

In some tropical agroecosystems, plowing can cause irreversible damage to a field. In some areas, a soil type called **plinthite** forms as iron and aluminum accumulate in the soil. If this plinthite is exposed to the air and dries, it forms a substance that is as hard as brick, and subsequently cannot be cultivated.

Application 5-2 Soil Classification and Suitability for Agriculture

Soil classification is based on such factors as soil profiles, texture, origin, vegetation, minerals, climate, and age. The various types of soils are divided into ten soil orders:

Histosols: organic soils (bogs, peat, muck). These soils contain greater than 20% organic matter; some contain as high as 50–80% organic matter. Bogs are too acidic to be of much use agriculturally, except for specially adapted crops like cranberry. Peat soils are often harvested and sold commercially for home gardens, potted plants, and greenhouse production. The high organic content of some peat soils may also provide a convenient fuel source. Muck soils are used extensively to produce high-value crops because they have excellent water-holding capacity and are high in nutrients.

Vertisols: expandable clay soils. These soils crack under alternating wet/dry conditions. They are difficult to plow (too sticky in the rainy season and too hard in the dry season), so any tillage must be done in a very short time period when the soil is intermediate in moisture levels. For this reason, Vertisols tend to be better suited to mechanized tillage rather than use of animals or human power, which may not be able to plow the entire field during the critical period of time available for cultivation. If irrigation water is available, flooded rice production is a good agricultural option for Vertisols. The heavy clay base of these soils tends to make root penetration difficult.

Young Soils (Characteristics Depend Strongly on Parent Material)

Entisols: new soils (e.g., volcanic deposits or sand deposits). These recently formed soils (by geologic standards) lack significant profile development. Agricultural productivity of these soils varies with location and soil properties. Very sandy Entisols are difficult for agricultural production, but Entisols that have been laid down in floodplains are highly productive. With proper fertilization and irrigation, many Entisols can be cropped successfully, depending on such factors as depth of profile and drainage.

Inceptisols: some profile development is evident (e.g., volcanic soils, etc., after some vegetation has altered the soils). These are as variable in use as the Entisols, and their productivity varies with location and management. Inceptisols formed on slopes should be left to the native vegetation to prevent erosion caused by cultivation.

Middle-Age Soils (Characteristics Depend on Climate and Vegetation, as Well as on Parent Material)

Aridisols: desert soils. Not much leaching into mineral layers has occurred over time, since rainfall is scarce. Aridisols are not suitable for agricultural use unless they are

(continued)

irrigated. Productivity of animal-based systems is also low since natural vegetative growth is extremely slow. Some of these soils can be extremely productive if well irrigated and fertilized, but accumulation of salts through irrigation can be a problem in some areas of the world.

Mollisols: grassland soils. Mollisols are very fertile and are some of the most productive soils in the world. Many of these soils are very intensively cultivated, so fertilizer applications are sometimes necessary to maintain soil fertility. To prevent degradation from continual production, careful management of these soils is recommended.

Alfisols: found in hardwood forests in cool to warm, humid areas. Alfisols are good soils that tend to be quite fertile, and generally have texture and structure favorable for good agricultural production. Because they are normally found in relatively humid areas, soil moisture is usually sufficient for growth. However, Alfisols can tend toward acidity, and may need to be limed to improve production.

Spodosols: found in coniferous forests in cold, humid regions. Some leaching of nutrients and organic matter into the subsurface horizon has occurred, forming a "spodic" horizon. These soils are not naturally fertile, and need to be substantially fertilized for optimum agricultural crop production. In addition, Spodosols are acidic, so liming may be necessary for production.

Old Soils (Much Weathering Has Occurred; Nutrient Poor)

Oxisols: highly weathered and low in nutrients. Most nutrients have been leached out to great depths from these tropical soils. In some cases, accumulation of Fe and Al is evident. A mineral hard pan may be present. Oxisols may be used effectively in agroforestry systems if deep-rooted trees can tap into the deeper soil layers to remove nutrients and recycle them through the system. If used for other agroecological systems, substantial fertilization is important, and irrigation may be needed as well.

Ultisols: weathered soils that are acidic and high in Fe. Under good management, Ultisols can actually be quite productive, particularly if provided with adequate fertilizer and lime. The reddish soils (colored by iron oxides) of the southeastern United States fall into this order.

Summary of Agricultural Potential

Good for agriculture: Mollisols, Alfisols, and (in some cases) Histosols

Respond to management (fertilization or irrigation): Entisols, Inceptisols, Aridisols, Spodosols, and Ultisols

Poor for agriculture: Vertisols and Oxisols (these require a *lot* of inputs to be productive)

Cation Exchange Capacity and Soil Fertility

The **cation exchange capacity** (CEC) of a soil is related to how many positively charged particles are held by the soil and are therefore available to be "exchanged" for other charged particles. In practice, the CEC refers to the ability of soil particles to hold nutrients. If a soil has a good CEC, ions that are bound to soil particles will be readily released into the soil solution and made available for plant uptake. Ions that are released from soil particles are considered to be available for uptake by a plant since they are dissolved in the soil water.

Clay and humus have the highest cation exchange capacities of all soil particles. They are fine particles that have a fairly large surface area (high surface-to-volume ratio), and have many charges that attract positive and negative ions, as well as water. Clay soils commonly undergo **isomorphous substitution,** a process that replaces a charged atom with an atom of similar size, but often with varying charge. This process leads to the high number of positive and negative charges associated with clay particles, and is responsible for the high cation exchange capacity of these soils. Cation exchange occurs when ions that are attached to soil particles by attraction to the surface charges are released into the soil solution in exchange for another charged ion (Figure 5-14).

In the example in Figure 5-14, Ca^{2+} was released into the soil solution in exchange for two H^+ ions formed by the dissociation of water. The Ca^{2+} is now available for plant uptake, and the negative surface charges on the clay particles have been satisfied by two positively charged H^+ ions. While it is true that the higher the CEC of the soil, the more potentially fertile the soil will be, it is also true that since many of the positively charged ions that are held by soil particles are exchanged for hydrogen ions, the pH of the soil will strongly affect its cation exchange capacity.

Soil pH

The pH of the soil refers to the relative concentrations of H^+ and OH^- ions in the soil solution. Since pH is based on a log-10 scale, each change in pH represents a tenfold change in the concentration of the H^+ ions:

$$pH = -\log[H^+]$$

The pH scale ranges from 0 to 14, or from H^+ ion concentrations ranging from 10^0 (= 1.0) to 10^{-14}M. When a soil solution has a neutral pH (7.0), it has a balanced concentration of H^+ and OH^- ions; there is one H^+ ion for every OH^- ion found in the soil and each ion is present at a molar concentration of 10^{-7}. Soil pH refers to the H^+ ion concentration only; a corresponding term representing the OH^- ion concentration, called pOH, can be determined by simple subtraction of pH from 14.0. So if pH = 3.0, then pOH = 11.0, $[H^+]$ = 10^{-3} M, and $[OH^-]$ = 10^{-11}M. If the pH of the soil solution is less than 7.0, the solution is acidic; if it is greater than 7.0, the reaction of the soil solution is basic or alkaline.

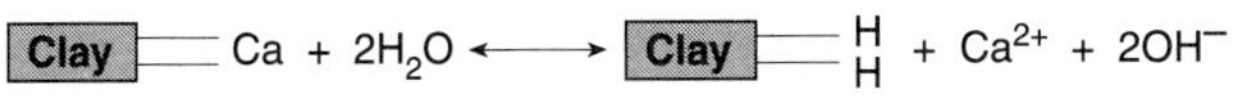

$$\boxed{Clay} — Ca + 2H_2O \longleftrightarrow \boxed{Clay} \genfrac{}{}{0pt}{}{—H}{—H} + Ca^{2+} + 2OH^-$$

Figure 5-14 Cation exchange in soil. The Ca^{2+} ion is exchanged for two H^+ ions from water molecules.

Soil pH is one of the most important properties involved in plant growth. Certain crops grow very well in acidic soils, and others in alkaline soils. Most cultivated crops do best at a pH of between 6.0 and 7.0 since most plant nutrients are readily available in this range (see Chapter 4). In humid regions, soil pH may dip as low as 4.5 because many bases are leached from the soil by frequent rainfall as they become available. By contrast, soils in arid regions may reach a pH as high as 8.5 because very little leaching of bases occurs.

Soil acidity (low pH) results from many natural processes within an agroecosystem. Since pH of soil is often a direct reflection of the pH of the parent material, the development of acidic soils from acidic parent materials is common. Respiration of plant roots and soil organisms releases carbon dioxide, which reacts with water to produce carbonic acid (H_2CO_3). The carbonic acid breaks down further to release hydrogen ions, contributing to soil acidity. In addition, plants tend to take up many base-producing ions, such as Mg^{2+} and Ca^{2+}, leaving some of the more acidic ions (Al^{3+} and H^+) behind in the soil. Those bases that are left in the soil, or made available in the soil solution by the cation exchange process, may be leached from the soil by rainfall, especially in highly humid areas. Soil alkalinity (high pH) usually occurs in more arid regions. In these cases, bases are released as parent materials weather, but since rainfall is limited, leaching seldom occurs and the bases accumulate in the soil. This accumulation of basic salts can lead to serious salinity problems in some areas of the world (see Chapter 15).

Changes in soil pH can also occur as an indirect response to human activity. Acid rain, which commonly has a pH between 4.0 and 4.5, can acidify the soil on which it falls over time. Acid rain originates when nitrogen and sulfur-containing gases emitted by factories, automobiles, or burning forest and crop residues react with water in the atmosphere to form H_2SO_4 and HNO_3.

In agroecosystems, cultivation and fertilization practices can also affect soil pH. Many commercial fertilizers lower soil pH because nitrification of ammonium ions (NH_4^+) results in the release of two hydrogen ions for every ammonium ion oxidized. No-till systems can increase surface acidity of soils because acids from fertilizer applications or organic matter decomposition tend to localize in the upper layers. The addition of sewage sludge to a system can also serve to acidify soils. Conversely, soil pH may be increased as a result of irrigation, since salts present in the irrigation water are deposited into the soils and left behind when the irrigation water evaporates.

Fortunately, soil pH can be modified by a variety of management practices. If soil pH is too high, organic matter can be added. As this organic matter decomposes, it forms acids that will decrease soil pH. Farm manures, however, should be used with caution because they may contain salts and may actually create osmotic problems. Inorganic chemicals such as ferrous sulfate, elemental sulfur, or sulfuric acid (in irrigation systems) can also be added to the soil; hydrolysis of these chemicals serves to acidify the soil. If pH is too low, the most common solution is to lime the soil. The application of carbonates, oxides, or hydroxides of Ca^{2+} and Mg^{2+} will supply bases to replace H^+ and Al^{3+} on the colloids, thus increasing soil pH.

Summary

Growth of agricultural crops in any given region depends strongly on physical factors. Climate, including temperature and precipitation regimes, often determines the potential range of any given crop since climate itself cannot be modified by management. Soil texture and fertility can also limit the distribution of some crops, particularly if inputs of fertilizer or organic matter are not available or are available only in limited quantities. Some common methods used in agricultural production cause changes to the soil profile, which may lead to a decrease in productivity over the long term. Soil fertility can be directly modified by application of organic or inorganic fertilizers, and acidity can be adjusted with management practices or by adding lime.

Topics for Review and Discussion

1. How does the tilt of the earth cause changes in seasons? Why don't the tropics show seasonality?
2. Describe a typical soil profile from surface to unconsolidated parent material. Explain why soil profiles are important diagnostic tools in agriculture.
3. Explain how soil pH affects fertility. Do you think that most plants do better in soils that are slightly basic or slightly acidic? Why?
4. Sand, silt, and clay soils all differ in their ability to hold water and to allow good drainage. Explain how water moves through each of these soils, and discuss the implications of this movement for agricultural systems. Which soils are best suited for cultivation, and why? What inputs are necessary to make each soil better adapted for cultivation?
5. If water cycles through the system according to the hydrological cycle, and if water takes up over two-thirds of the global surface, why are scientists becoming increasingly concerned about a global water shortage in the future?

Literature Cited

Hatfield, J. L. 1990. Climate change and the potential impact on the soil resource. *Journal of the Iowa Academy of Science* 97(3):82–83.

Rice, R. A., and J. Vandermeer. 1990. Climate and the geography of agriculture. In *Agroecology*, eds. C. R. Carroll, J. H. Vandermeer, and P. M. Rossett, 21–63. New York: McGraw-Hill.

Thompson, L. M. 1988. Effects of changes in climate and weather variability on the yield of corn and soybeans. *Journal of Production Agriculture* 1:20–27.

United States Department of Agriculture. 1998. *Agricultural Statistics 1998*. Washington, D.C.: U.S. Government Printing Office.

Bibliography

Brady, N. C., and R. R. Weil. 1996. *The Nature and Property of Soils*. Upper Saddle River, N.J.: Prentice-Hall.

Demillo, R. 1994. *How Weather Works*. Emeryville, Calif.: Ziff-Davis Press.

Gibbons, B. 1984. Do we treat our soil like dirt? *National Geographic* 166:350–388.

Klages, K. H. W. 1942. *Ecological Crop Geography*. New York: Macmillan.

Pimentel, D., C. Harvey, P. Resosudarmo, K. Sinclair, D. Kurz, M. McNair, S. Crist, L. Shpritz, L. Fitton, R. Saffouri, and R. Blair. 1995. Environmental and economic costs of soil erosion and conservation benefits. *Science* 267:1117–1123.

Plaster, E. J. 1997. *Soil Science and Management*. Albany, N.Y.: Delmar.

Schlesinger, W. H. 1991. *Biogeochemistry: An Analysis of Global Change*. San Diego, Calif.: Academic Press.

Smith, R. L. 1996. *Ecology and Field Biology*. 5th ed. New York: HarperCollins.

Tivy, J. 1992. *Agricultural Ecology*. Essex, U.K.: Longman Scientific and Technical.

Chapter 6

Growth and Dynamics of Populations

Key Concepts

- Dry matter accumulation and phenology of crop plants
- Dynamics of pest population growth
- Predictions of population growth

Biomass and production of a species (crop, herbivore, or pest) in an ecosystem can increase over time in two different ways: through growth in the size of individuals or through increases in the numbers of individuals. Growth and development of an individual may follow a somewhat predictable sequence of events that is dependent on the physiology of the individual organism. Increases in population size, or the numbers of a particular species, may also follow predictable patterns. **Population dynamics** is the study of the changes in the sizes of populations over time. The speed and magnitude of these changes can be very impressive for small organisms with short life cycles and high reproductive rates, but the principles of population dynamics apply to humans and other long-lived organisms as well (see Chapter 19). Fluctuations and dynamic changes of populations of different kinds of organisms are key features of all ecosystems. Over several seasons or many years, they may result in successional changes in the species composition of the ecosystem (see Chapter 7).

Although dynamic change is expected in all ecosystems, an important, constant feature separates many agroecosystems from natural systems. In agricultural systems, the population size of the crop plant producer is usually controlled by management (planting density, thinning, etc.). The main production feature during the season is an increase in the size of individual plants, rather than an increase in the number (population size) of crop plants. Therefore, we preface our study of population dynamics with an examination of some key features of plant growth.

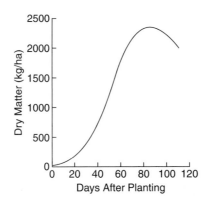

Figure 6-1 Typical pattern of total plant dry matter accumulation during the growing season.

Dynamics of Plant Growth

Many different physical, chemical, and biological factors affect plant growth. Some of these, in particular temperature, photoperiod, and light intensity, are so seasonal in nature that they are often good predictors of plant growth, assuming that other factors are adequate.

Plant Dry Matter (Biomass) Accumulation

Many crop species follow a fairly typical pattern in their buildup of dry matter (biomass) during the growing season (Figure 6-1). Although Figure 6-1 is generalized, it illustrates some typical features of such curves. Dry matter accumulation is gradual early in the season, but increases at a more rapid rate as leaf area (and photosynthetic activity) increases. Dry matter accumulation does not continue indefinitely during the season, but peaks and finally declines. This decline is more noticeable if plants are left in the field beyond the optimum harvest date, and results from the dropping of older leaves and other senescent plant parts.

A rough approximation of a typical curved pattern of dry matter accumulation for some plants may be obtained from a cubic equation of the form:

$$Y = a_0 + a_1 x + a_2 x^2 + a_3 x^3 \qquad \text{(Equation 6-1)}$$

where Y = dry matter yield and x = number of days after planting. The coefficients a_0, a_1, a_2, and a_3 depend on the plant species and environmental conditions, and could be positive, negative, or zero. A quadratic equation results if $a_3 = 0$.

The **partitioning,** or allocation of new dry matter to various plant parts, changes as the season progresses. Early in the season, most plants partition much of their production energy into rapid growth of leaves, which will be used to capture even more energy. Later in the season, allocation shifts toward production of fruit or grain, the harvested products of many crops (Figure 6-2). The shaded areas in Figure 6-2 represent the relative allocation of energy to the various parts of the crop plant. The various plant parts show peaks and declines in their dry matter accumulation at different times. Typically, leaf biomass declines first, as lower leaves become less functional in photosynthesis and fall from the plant.

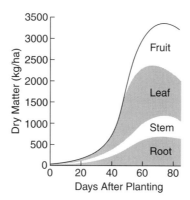

Figure 6-2 General pattern for partitioning of dry matter during plant growth. Uppermost line (bold) represents total plant dry matter.

Figures 6-1 and 6-2 are generalized figures showing typical plant growth patterns. Growth curves of this type have been developed for many different crops in a great variety of situations. A few examples illustrate growth of sugarcane in Hawaii (Evensen et al. 1997), winter wheat in Oklahoma (Rao and Dao 1996), *Medicago* species in Minnesota (Zhu et al. 1998), and corn roots in a clay loam soil (Nickel, Crookston, and Russelle 1995).

Heat Units and Degree-Days

Knowing the number of days since planting may be useful in providing an approximation of plant growth during the season. But seed planted in a "cold" soil may take several days longer to emerge than seed planted under warmer conditions. Enzymatic activities and growth rates are directly affected by temperature. In fact, except for animals like birds and mammals, which maintain a constant body temperature, the activity and growth rates of most organisms are related to temperature. The length of crop cycles, as well as the life cycles of many organisms, are longer at lower temperatures. As a result, life cycles can often be described more accurately in terms of time and temperature, rather than in terms of time alone.

Heat units, measured in **degree-days** or *degree-hours,* are the number of days or hours of accumulated temperature above some threshold. For instance, a temperature of 20° C maintained for four days results in a heat unit accumulation of forty degree-days above a 10° C threshold ([20° – 10°] × 4 days). The threshold or base temperature is the minimum temperature required for growth, and varies with crop species. For corn, a lower threshold of 12.8° C (55° F) is used (Aldrich and Leng 1972), while for potato systems, 5° C (41° F) may be more suitable (Pinkerton, Santo, and Mojtahedi 1991). For many crops, the minimum threshold temperature may be near 10° C (50° F). An upper limit to heat unit accumulation may be used as well, since growth may also be limited if temperatures are too hot.

Degree-day accumulation can be approximated by:

$$DD = \sum_{i=1}^{d} T_{mean} - T_{base} \qquad \text{(Equation 6-2)}$$

Application 6-1 *Sample Calculation of Heat Units*

Using the methods described in the text, we calculate the number of degree-days (DD) accumulated for one week with daily temperatures as shown here, if plant growth limits are 10° and 30° C.

Day	T_{max}	T_{min}	T_{mean}	Heat Units
Sunday	22	18	20	10 DD
Monday	28	18	23	13 DD
Tuesday	34	30	32→30	20 DD
Wednesday	25	17	21	11 DD
Thursday	20	8	14	4 DD
Friday	10	8	9	0 DD
Saturday	12	6	9	0 DD
				$DD_{10} = 58 \, DD_{10}$

Degree-days summed over the entire week are listed with a subscript as DD_{10} to indicate that 10° C was the base temperature. Since extra heat units are not accumulated above the maximum limit, the T_{mean} for Tuesday, which is 32°, is rounded down to 30°, the upper limit. No heat units are accumulated when T_{mean} is below the 10° threshold (Friday and Saturday). Degree-days cannot be negative. Since this method uses T_{mean} in the calculation to provide an *approximation* of DD, a slight error is included in the calculation for Saturday. With a T_{max} of 12° C, a few hours above 10° C may have been reached. However, this may be only a fraction of one degree-day, compared to fifty-eight degree-days for the entire week. More accurate estimates of heat unit accumulation can be obtained if hourly accumulation is measured (degree-hours). Errors in estimating degree-day accumulation can be minimized by integration or approximation of sine functions (see Beasley and Adams 1996 and references cited therein). If extreme accuracy is required, the position of temperature measurement must be chosen carefully. Temperatures may be very different if measured in the soil, under a plant canopy, or on the surface of an upper leaf.

---

where T_{mean} is daily mean temperature, T_{base} is the minimum temperature for development, and d is the number of days in the time period. The temperature differences (heat units per day) are summed over the number of days (see Application 6-1 for sample calculation). T_{mean} can be approximated as the average of T_{max} = daily maximum temperature and T_{min} = daily minimum temperature (Figure 6-3):

$$T_{mean} = \frac{(T_{max} + T_{min})}{2} \quad \text{(Equation 6-3)}$$

T_{max} and T_{min} are often available from weather records, and provide a convenient starting point for estimating heat unit accumulation.

Degree-days can provide a better prediction of many plant growth processes (especially vegetative growth) or insect population growth, and could be substituted for days after planting in Figure 6-1 and several other figures in this chapter. An opposite effect, which is also very important in agriculture, is the accumulation of *chilling days* or *chilling hours* below

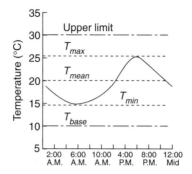

Figure 6-3 Diagram of daily temperature fluctuation, indicating temperatures used in heat unit calculation.

a certain temperature. Plants such as some temperate fruit crops must accumulate a required number of days or hours below the threshold before they can set fruit. This limits the distribution of crops such as apples, pears, or cherries in some tropical locations. Although the trees could be grown in these climates, they would not produce fruit if the chilling requirements were not met.

Phenology and Plant Growth Stages

Phenology is the study of the sequence in which certain biological and developmental events occur. A number of distinct stages can be recognized in the growth of a crop plant, such as a stage from planting to emergence, a vegetative growth stage, flowering, etc. Some of these events, such as emergence and vegetative growth, depend heavily on heat units or time. One of the most significant events in plant development is the change from the vegetative growth stage to the reproductive growth stage following flowering. The flowering response of many plant species depends on photoperiod, and the various responses of different plants to photoperiod were presented in Chapter 5. Response to photoperiod is particularly important in the development of soybean, and usually determines where a particular variety can be grown successfully (see Application 6-2).

Once flowering occurs, the pattern of plant growth may change. Vegetative growth of **determinate plants** stops or is greatly reduced at flowering. Future growth is devoted to development of reproductive structures and seed production. **Indeterminate plants** continue with some vegetative growth after the first flowering. Multiple cycles of flowers and fruit may occur. Cultivars of the same plant species may have different degrees of indeterminate and determinate characteristics. More than 50% of the yield of four of the cowpea cultivars shown in Table 6-3 was obtained on the first harvest date. Yield of the other cultivars was divided more evenly over several harvests. A strong determinate response can be particularly important in obtaining fruit of uniform age and size for mechanical harvest.

The phenology of major crop plants is well known, and their development can often be divided into recognized growth stages. For example, several stages can be recognized for potato, such as planting to emergence, vegetative growth, tuber initiation, tuber bulking, and tuber maturation (Rowe and Secor 1993). The recognized growth stages of soybean have been referred to often since their definition in the 1970s (Fehr and Caviness 1977). These include an indefinite number of vegetative stages (based on the number of leaf nodes) and eight reproductive stages, from beginning bloom (R1) to full maturity (R8). References to defined growth stages are increasing in other crops as well. The different plant growth stages present distinct biological environments and habitats as they change through the season, affecting other organisms in the agroecosystem to varying degrees. For example, the plant canopy, shading, air movement, and insects present in an emerging corn or soybean crop are very different from those present in the later stages of crop maturity.

Application 6-2 *Soybean Maturity Groups*

The sensitivity of soybean to photoperiod is well known. Soybean is a short-day plant, and most cultivars begin flowering soon after the day length shortens (= night length increases) below some critical value. Soybean cultivars have been developed to mature at different times so that the crop can be grown over a wide range of latitudes. Soybean cultivars are grouped into *maturity groups* based on photoperiod response and maturity date. There are twelve maturity groups, from 00 (earliest) to X (latest), and eleven are applicable to soybeans grown in the United States. Maturity group X is adapted to tropical latitudes. Table 6-1 shows the maturity groups for just a few of the many soybean cultivars developed for use in the United States and Canada. Approximate zones in North America to which different maturity groups are best adapted are also indicated, although cultivars in a maturity group can usually be grown a few zones away with acceptable results.

Table 6-1 Examples of soybean cultivars in various maturity groups adapted for growth in the United States and Canada.

Maturity Groups	Cultivars	Approximate Locations for Best Adaptation
00	Maple Arrow, McCall	N. Minn., Manitoba
0	Dawson, Evans	Mid-Min., Ontario
I	Hardin, Hodgson	Wisc., Mich., Ontario
II	Beeson, Century, Corsoy	Ia., N. Ill.
III	Pella, Williams	Ill., Ind., Oh.
IV	Crawford, Kent, Stafford	Mid-Mo., S. Ill. Ky.
V	Essex, Forrest, Hill	Ky., Tenn.
VI	Centennial, Davis, Lee	Ark., N. Miss., N. Ala., N.C.
VII	Bragg, Braxton, Ransom	Miss., Ala., Ga.
VIII	Cobb, Hutton, Kirby	La., S. Ga., S. Ala.
IX	Jupiter, Santa Rosa	Fla.

Why are maturity groups and photoperiod responses so important when planting soybean? Consider the data in Table 6-2, which indicate an approximate date on which daylength falls below a certain number of hours, for three different locations. In Springfield, Illinois, at approximately 40° N latitude, the longest day of the year is June 21, with just over fifteen hours of daylight. Beyond June 21, daylength shortens, dropping below fifteen hours on June 29 and below fourteen hours on August 9. A soybean cultivar that flowers when

(continued)

daylength drops below fourteen hours would flower on about August 9 and would therefore mature earlier than a cultivar that responds to thirteen-hour daylength. Cultivars can be chosen so that they will mature at the most convenient time for a particular location.

Table 6-2 Approximate dates at which daylength falls below a specified interval at three different U.S. locations.

Location	Approximate Latitude	First Date When Daylength is Less Than:		
		15 hrs	14 hrs	13 hrs
Lake City, Fla.	30° N	—	July 9	August 25
Springfield, Ill.	40° N	June 29	August 9	September 5
Grand Forks, N.D.	48° N	August 1	August 25	September 9

Dates approximated from data presented by List (1945)

Problems can occur when using cultivars from maturity groups poorly adapted to a particular latitude. From Table 6-2, it is evident that a cultivar that responds to fifteen-hour daylength is more suitable for Grand Forks, North Dakota, while one that responds to thirteen-hour daylength is better adapted to Lake City, Florida. What if seeds from these two cultivars were mixed up and planted at the opposite location? A cultivar responding to a daylength of less than thirteen hours would not flower in Grand Forks until September 9, when the long, summer daylengths finally fall below this critical value. The crop would mature, but may be in danger from frost before harvest. At the other extreme, a cultivar responding to a daylength of less than fifteen hours may flower prematurely in Lake City, producing a low yield of beans on very small plants. For many years, it was difficult to produce good soybean yields in tropical countries where daylength does not vary much during the year. More recent development of cultivars in maturity groups IX and X has provided productive options for these locations.

Plant Population Dynamics

Total plant biomass accumulated per hectare during the growing season depends on both the size of individual plants and the number of plants present. The plant population size can be counted as the number of individual plants present per hectare, per square meter, per measured length of row, or per some other unit of space.

Crop Plant Population

In most agroecosystems, the population density of a crop is controlled by planting or thinning operations. Of course, a number of uncontrolled factors may affect the final plant stand, such as seed quality, weed competition, or damage from pests or weather. But unlike natural ecosystems, the initial (before any mortality) population of seeds or transplants per unit of land is fixed. This initial plant density can be manipulated by changing the row spacing (distance between rows of plants), the distance between plants in rows, or the broadcast seeding rates for plants not grown in rows.

Crop yield is related to plant density, and an optimum density to achieve maximum yield can be determined (Figure 6-4). The seeding rate needed to achieve this optimum density is determined by research and experience within a particular geographic area. Achieving this optimum density requires a balance:

- If plant population density is too low, the maximum usage of incoming solar energy will not be achieved (see Chapter 2).
- If plant population density is too high, intraspecific competition and related problems may result (see Chapter 7).

Table 6-3 Yield concentration of cowpea cultivars showing various levels of determinate and indeterminate characteristics.

Cultivar	Percent of Total Yield Harvested Each Day[a]						
	July 25	August 2	August 12	August 30	September 12	September 27	October 17
Whippoorwill	42	44	8	6	0	0	0
Pinkeye Purplehull	66	24	2	7	0	0	0
Texas Purplehull	52	40	7	0	0	0	0
Purple Knuckle	61	31	5	3	0	0	0
California Blackeye #5	34	23	38	5	0	0	0
Mississippi Silver	86	11	2	1	0	0	0
Tennessee Brown	28	23	20	12	6	6	5

[a] Calculated from data presented by Gallaher and McSorley (1993)

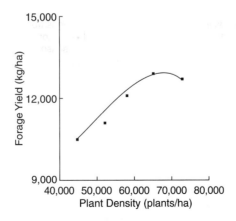

Figure 6-4 Relationship between tropical corn forage biomass and plant density, based on three-year average of data from Cuomo, Redfearn, and Blouin 1998.

Population Dynamics of Weeds

Weeds colonize and build up high populations readily on agricultural land (see Chapter 7). Many weed species are well-adapted to sites disturbed by agriculture. Two features of weed biology are particularly important in understanding the dynamics of weed populations. First, weeds are highly prolific, with many kinds producing extremely large numbers of seeds. For example, *Amaranthus spinosus* (spiny amaranth) was reported to produce up to 235,000 seeds per plant (Holm et al. 1977). Second, weed seeds often survive in soil for a long time. Seeds of most of the worst weeds typically survive in soil for several years, and survival over *decades* is not unusual for some species. In addition, weed seeds exhibit various degrees of **dormancy,** or interruption in their development, so that not all seeds germinate at the same time. Instead, a strong supply of germinating weeds can be available each year for a number of years. The **seed bank,** the reservoir of weed seeds in the soil, is available to germinate over a long period of time. If weeds are allowed to go to seed in a site, levels in the seed bank increase quickly, resulting in problems for many years.

Some key stages in the development of a weed population are shown in Figure 6-5. Various environmental and nutritional factors determine the rate of weed production at each stage. Some of the more important mortality factors affecting various stages are also shown,

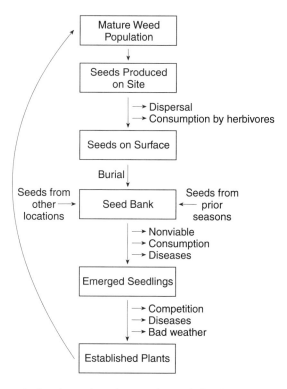

Figure 6-5 Key stages in the dynamics of a weed population in a site, including some important mortality factors.

although many other factors could be added. Of course, any mortality factor affects different weed species to varying degrees. The overall sequence of events (Figure 6-5) determines the change in the weed population in a site over time. As we gain more detailed knowledge about seed production, growth, and mortality at each stage, our predictions about future weed populations will improve, as will our understanding of which stages of a particular weed are most susceptible to management. Numbers of emerging and established weeds are critical in the end, since their density determines the degree of competition with crop plants (see Chapter 7) and the severity of their effects on yield.

The seed bank is an integral part of the development of a weed population (Figure 6-5). Seeds produced on-site in a season are added to those already present in the seed bank from past seasons. Also, because seeds are dispersed efficiently by wind, animals, water, and other means, seeds from other locations (such as a weedy neighboring field) may be added to the seed bank of a site. Since dormancy is a key feature of seed survival, not all seeds will move along the path from seed bank to emerged seedlings in a season. Some seeds may bypass the seed bank altogether, germinating to produce multiple crops of some weed species in the same season. But for most weed species in the United States and other temperate locations, the seed bank is the site for overwintering survival.

Two features of weed population dynamics cause particular problems in agriculture (Figure 6-6). The height of the population curve in Figure 6-6 (how many weeds there are) is of particular concern, as already mentioned. But the timing of weed emergence is an important and troublesome feature as well, particularly in the establishment of the first generation of weeds in a season. Following winter, seeds may come out of dormancy for a variety of reasons. Depending on species, the breaking of dormancy and seed germination may be stimulated by moisture, temperature, heat units, cultivation, exposure to light or various chemical compounds, or combinations of various factors. However, the environmental cues to break dormancy tend to be rather specific for a particular weed species. As a result, the breaking of dormancy (for those seeds programmed to break dormancy in that season) often tends to be fairly synchronized. This means that large numbers of weeds emerge at about the same time. This population of emerged weeds of similar age is known as a **cohort.** The timing of emergence of a cohort may be quite short (sometimes only a few days). Such a

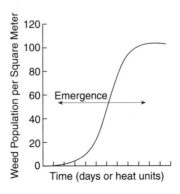

Figure 6-6 Typical population curve for emergence of a weed species during the early part of a season. The arrow indicates the time over which emergence occurs.

rapid appearance of a large weed population can cause obvious problems to crops. Prediction of weed emergence is critical in anticipating problems and timing control procedures. Weather data, analysis of the content of the seed bank, previous field history, and scouting or mapping weed patch locations within the field can be important in predicting the timing of weed emergence. In Quebec, the timing of emergence of lambsquarter *(Chenopodium album)* in the spring can be predicted from heat unit accumulation (Leblanc and Cloutier 1995).

Exponential Growth

The simplest type of population growth is **exponential growth,** also called *density-independent population growth,* which applies to populations growing in an unlimited, favorable environment. If we assume that conditions for growth are optimal, that food supply is unlimited, and that birthrates and death rates per individual are not affected by (are independent of) population size, then the growth potential of the population is unlimited, and it can rapidly build to very large numbers.

Model for Predicting Exponential Growth

Bacteria are noted for rapid population growth, dividing and forming new individuals in generation times ranging from only 15–20 minutes up to a number of hours or days, depending on the bacterial species and temperature. Figure 6-7 shows a bacterial population that divides (doubles in size) every thirty minutes. The figure charts the growth of the population over a ten-hour period from an initial population size of ten individuals. Note the especially great increase in numbers at eight to ten hours. What was a low population initially has exploded under the conditions of unlimited growth.

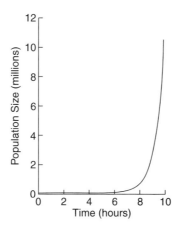

Figure 6-7 Buildup of a bacterial population that doubles every 0.5 hours, from an initial population size of ten individuals.

Growth of this bacterial population, or any other population that grows exponentially, can be described by the formula:

$$\frac{dN}{dt} = rN \qquad \text{(Equation 6-4)}$$

If N is population size and t is time, the expression dN/dt is technically (from calculus) the rate of change of population size with time, or the growth rate of the population. Therefore, the growth rate of the population is related to the population size (N) times a constant (r). This constant, **r,** is the **intrinsic rate of increase,** a characteristic of the species representing the maximum rate of population growth that the species is biologically able to support. This net growth rate is the difference between birthrate (b) and death rate (d):

$$r = b - d \qquad \text{(Equation 6-5)}$$

Note that from Equation 6-4, if $r = 0$, population size does not change (births = deaths), while $r < 1$ would indicate the rate at which a population is dying out.

Using methods from calculus and differential equations, Equation 6-4 can be solved, yielding the following solution over the time interval from time = 0 to time = t:

$$N_t = N_0 e^{rt} \qquad \text{(Equation 6-6)}$$

N_0 = Initial population size, at time = 0
N_t = Population size at time = t
r = Intrinsic rate of increase
e = 2.71828 . . ., a mathematical constant (base of natural logarithms)

Once r is known, population size can be estimated for any time in the future (see Application 6-3). The graph in Figure 6-7 was made from Equation 6-6, with $N_0 = 10$ and $r = 1.386$.

Estimating r from Experimental Data

We must know r in order to use Equation 6-6 to make predictions about population growth. The constant r can be estimated from life table data (discussed later in the chapter) or graphically. If we take the natural logarithm of each side of Equation 6-6, the following expression results:

$$\ln N_t = \ln N_0 + rt \qquad \text{(Equation 6-7)}$$

If data on population size are collected at various times, as in Figure 6-8A, r can be estimated graphically. If $\ln N_t$ is plotted against time, the graph should approximate a straight line (Figure 6-8B). According to Equation 6-7, the slope of this line is r and the y-intercept is $\ln N_0$.

Assumptions for Exponential Growth

To apply the equations predicting exponential growth, we must make several important assumptions:

1. Unlimited favorable environment (including unlimited food supply).
2. Constant environmental conditions, as defined in assumption #1.

Application 6-3 *Sample Calculations of Exponential Growth*

Example 1. Calculate the size of a population after seven time units if it grows exponentially, with $r = 0.80$ from an initial population size of twenty individuals.

$$N_t = N_0 e^{rt}$$
$$N_7 = 20e^{(0.80)7} = 20e^{5.6}$$
$$= 20(270) = 5400$$

Example 2. Calculate r for a bacterial population that doubles in size every thirty minutes. (This example is based on data presented in Figure 6-7.) Caution: We need to be careful about time units since r is based on the time unit. Growth rates per hour, per minute, or per day will be different. For consistency of comparison with Figure 6-7, we use a doubling time of 0.5 hours.

$$N_t = N_0 e^{rt}$$
$$N_{0.5} = N_0 e^{r(0.5)}$$

After a time of 0.5, N_0 will double in size to $2N_0$, so $N_{0.5} = 2N_0$:

$$2N_0 = N_0 e^{r(0.5)}$$
$$2 = e^{0.5r}$$
$$\ln(2) = \ln e^{0.5r}$$
$$0.693 = 0.5r$$
$$r = 1.386$$

Example 3. An insect produces 150 offspring per female during a period of four weeks. If we start with two females, what is the population size expected after twelve weeks if the population grows exponentially?

$$N_t = N_0 e^{rt}$$

Based on information in the first sentence, we must first find r:

$$150 = (1)e^{r(4)}$$
$$\ln(150) = \ln e^{4r}$$
$$5.01 = 4r$$
$$r = 1.25$$

Now for the population level after twelve weeks:

$$N_{12} = (2)e^{(1.25)(12)}$$
$$N_{12} = 2e^{15}$$
$$N_{12} = 6{,}540{,}000$$

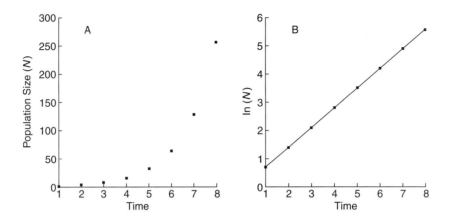

Figure 6-8 (A) Typical form of exponential growth data with population size *(N)* on *y*-axis. (B) A straight line results when ln *N* is plotted on *y*-axis.

3. Stable age distribution. In a **stable age distribution,** the proportions of all age groups in the population are always the same from generation to generation. (Of course, a population with 30% adults would grow differently than a population with 70% adults.)

These assumptions are impossible! We can only approximate them in real situations. Real data will not be as clean as the examples shown in Figures 6-7 and 6-8, but still can be used to obtain *estimates* of r and make forecasts of population growth under a *particular set of conditions*. Two common deviations from exponential growth are shown in Figure 6-9. Variation in N_t over time can result in the pattern shown in Figure 6-9A. This variation can result from error in measuring population size or from slight changes in environmental conditions (such as temperature) over time, violating assumption #2. If conditions are less than optimum, a pattern such as that shown in Figure 6-9B will occur. The *r* value estimated from the dotted line will be less than the true intrinsic rate of increase in an unlimited environment, but it still may be useful in making predictions under particular environmental conditions.

Applications to Agriculture

If exponential growth depends on an unlimited environment and food supply, why should we be concerned about it in real agricultural situations? Unlimited food is not available in real situations. Even if an organism was able to grow exponentially for a while, it would eventually deplete its food supply and the population would crash.

Unfortunately, populations of some agricultural pests may grow exponentially for a while, resulting in serious problems. Several hectares of corn may seem like an unlimited food supply to the first few insect pests arriving in a cornfield. Populations may then grow exponentially, resulting in massive numbers causing widespread destruction. The insect population may crash, but the field could be destroyed in the process. A common concern is the population growth of certain bacteria and fungi that cause diseases in plants or animals (see Chapter 10). The remarkable growth rates of bacteria have already been examined

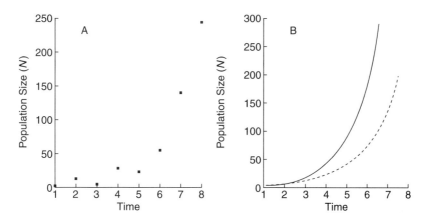

Figure 6-9 Deviations from exponential growth. (A) Variation in population size *(N)* over time. (B) Actual growth (dotted line) is less than exponential growth (solid line).

(see Figure 6-7), but obvious symptoms might not be noticed until the exponential growth phase is well underway. Early detection of low numbers of these pathogens can be difficult, but is important in predicting buildup of large populations of disease-causing agents.

Life Tables

Life tables provide a convenient means for organizing population data and studying age structure and change in population size over time. This approach works best when a cohort, or population of individuals born at the same time, is available for study. A **survivorship curve** shows the survival of members of the cohort at various ages, and can take various forms (Figure 6-10), depending on the species involved.

An example of a typical life table is shown in Table 6-4. The first column, x, represents the age of the cohort (in weeks in this example). Only data for females are included in life tables such as this, to avoid complications of different sex ratios, etc. The second column, l_x, is survivorship, the estimated proportion of the cohort alive at the midpoint of the time interval starting at time $x = 0$. So, if we started with a cohort of 1000 females at time $x = 0$, and 900 were still alive at time $x = 5$, the survivorship for the interval $x = 0$ to 5 would be $(1000 + 900)/2 = 950$ = average number alive in the interval, divided by 1000 (starting number for cohort) = 0.95. Caution: This l_x is only an approximation and is calculated differently by some sources. The third column, m_x, is the **fecundity** of a female in that time interval. Thus, a female that is alive in the interval from $x = 20$ to $x = 25$ produces an average of 1.0 female offspring. However, a proportion of only 0.40 of the original cohort is still alive during this interval, so the net production for the cohort per female during this interval is only $l_x m_x = 0.40$. For the early ages, the $m_x = 0$ values represent immature stages with no reproduction.

Of particular interest is the sum of all of the $l_x m_x$ values for all age classes:

$$R_0 = \sum l_x m_x \qquad \text{(Equation 6-8)}$$

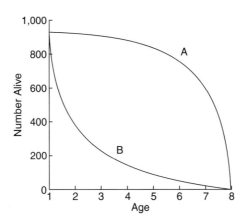

Figure 6-10 Survivorship curves show numbers of individuals alive at various ages. Curve A shows low infant mortality, with most individuals dying as older adults. Curve B shows high mortality of seeds or young offspring.

Table 6-4 Example of life table, with hypothetical data.

Age in weeks, x	Survivorship, l_x	Fecundity, m_x	$l_x m_x$
0	0.95	0	0
5	0.85	0	0
10	0.75	0	0
15	0.60	0	0
20	0.40	1.0	0.40
25	0.30	2.5	0.75
30	0.10	3.0	0.30
35	0.01	0	0
			$\Sigma l_x m_x = 1.45$

R_o is the **net reproductive rate,** representing the total number of female offspring produced per female during one generation. If:

$R_o = 1$ Stationary population, replaces itself
$R_o > 1$ Increasing population
$R_o < 1$ Decreasing population

R_o indicates how many times the population multiplies in a generation.

The intrinsic rate of increase, *r*, can be found from life table data. Methods for obtaining various approximations of *r* from life tables can be found in more specialized texts (see Carey 1993).

If life tables can be constructed, they provide important insight into which stages have the highest natural mortality or the highest fecundity rates. Since many pests and weeds produce many offspring, it is important to control them before they reach their most productive stages for reproduction. If the feeding rates of two different insect pests were similar (e.g., equal amount of leaf consumed per individual), the one with the higher R_o has the potential to cause more damage and problems over time due to more rapid increases in numbers.

Logistic Growth

Although populations may grow exponentially for a time, the assumptions of unlimited food supply and optimal environment are eventually violated, and the world does not fill up with bacteria, insects, or other rapidly growing organisms. There is an upper limit to growth as an increasing population competes with itself for food, space, and other necessities of life. **Logistic growth,** or *density-dependent population growth*, describes population growth in a limited environment. This is a much more realistic situation than exponential growth.

In logistic growth, there is a **carrying capacity** (given the symbol **K**), or upper limit to the population size that a given environment will support. A typical pattern of logistic growth in a population over time is shown in Figure 6-11. The size of the population building up over time is limited by, and levels off to, the carrying capacity. The growth rate of a population under these conditions is described by:

$$\frac{dN}{dt} = rN\left(\frac{K-N}{K}\right) = rN\left(1 - \frac{N}{K}\right) \quad \text{(Equation 6-9)}$$

The growth rate of the population, *dN/dt*, is related to *K* as well as to *r* and *N*. The growth rate is not constant, but depends on how near the current population size *N* is to the carrying capacity. Equation 6-9 shows some interesting responses at very low and very high population densities (values of *N*):

- If the population size *(N)* is very small ($N \rightarrow 0$), competition within the species will be very low, so the growth rate approaches *rN* = exponential growth. Under these conditions ($N \rightarrow 0$), the term of Equation 6-9 in parentheses approaches 1.0. So Equation 6-9 then resembles Equation 6-4, and exponential growth is predicted. Note similarities between the early part of the growth curve in Figure 6-11 and the exponential growth curve of Figure 6-7.
- If *N* is large ($N \rightarrow K$), extreme levels of intraspecific competition occur, and no growth is predicted. As $N \rightarrow K$, the term of Equation 6-9 in parentheses approaches zero, and will be zero when *N* reaches *K*. At this point, no further growth is predicted, so the population size in Figure 6-11 remains at the same value, *K*.

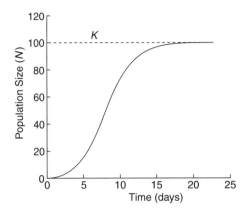

Figure 6-11 Pattern of population growth over time following a logistic model. The carrying capacity, *K*, provides an upper limit to population size.

Using methods described elsewhere (e.g., Hastings 1997), Equation 6-9 can be solved, giving the following solution over the time interval from time = 0 to time = t:

$$N_t = \frac{N_0 e^{rt}}{1 + N_0(e^{rt} - 1)/K} \qquad \text{(Equation 6-10)}$$

N_0 = Initial population size at time = 0
N_t = Population size at time = t
K = Carrying capacity
r = Intrinsic rate of increase
e = 2.71828 . . .

Figure 6-11 was produced from Equation 6-10 using values of N_0 = 1.0, K = 100, and r = 0.5. The logistic equation and graphs, such as that shown in Figure 6-11, are useful for predicting population growth and determining a stable population size *(K)* reached under a specific set of environmental conditions.

Assumptions for Logistic Growth

The logistic equation (Equations 6-9 and 6-10) is based on a number of assumptions:

1. Constant environment;
2. Constant *K*;
3. Instant response to increase in population density, no time lags;
4. Stable age distribution;
5. All members and stages of the population affected equally by crowding; and
6. Probability of mating does not depend on population size.

Application 6-4 *Violating the Constant Environment Assumption*

In reality, environmental conditions are rarely constant. So the assumption of a constant environment, critical to the logistic equation, is usually violated. Since K depends on a specific set of environmental conditions, this means that K will often change as well. For example, the number of weeds supported per m^2 may be very different in wet and dry seasons (Figure 6-12A). Most environmental changes may not be as drastic as this, and if they are relatively minor, effects on K may be minimal and a logistic model may still be useful for obtaining rough approximations of population growth.

Since management is a characteristic of agricultural systems, it can be useful to manage some aspects of the environment to deliberately affect K. One common example is the use of a resistant crop cultivar to manage a pest (Figure 6-12B). The resistant cultivar may be less desirable as a pest food source, may result in less reproduction of the pest, or may interfere with pest biology in some other way. The result is that lower numbers of pests are supported than on a more susceptible cultivar, providing a lower K when the resistant cultivar is used. This approach to managing pests is described in more detail in Chapter 11.

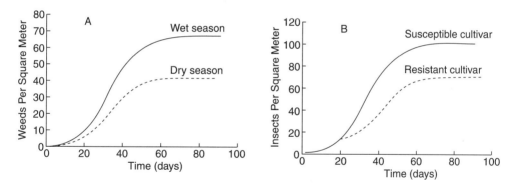

Figure 6-12 Changes in carrying capacity under different conditions. (A) Emergence of a weed population in a wet or a dry season. (B) Population density of an insect pest on susceptible and resistant crop plant cultivars.

In reality, some of these assumptions may be violated, causing important deviations in forecasts from logistic models. Violations of the first two assumptions are frequent and of particular consequence to agriculture (see Application 6-4). Under field conditions, two of the other assumptions (#3, #6) are often violated, producing characteristic responses.

The third assumption, dealing with instantaneous response to population growth, is unrealistic. Population increase is not an instantaneous process. Eggs or unborn offspring may develop in the body of the female for a period of time before they are laid or born, then

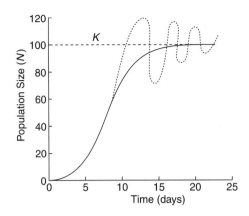

Figure 6-13 Population growth with time lags (dotted line) compared to instantaneous logistic growth, illustrating population overshoot of K and oscillations around K.

become counted as part of the population. The next generation is being developed as the previous one feeds. The previous generation may not detect that food will run short or that K is being approached because population size is still safely below K. Then large numbers of eggs are laid and hatched, and suddenly the population jumps beyond K (Figure 6-13). The number of individuals is then too great for the environment, so the population crashes, and a kind of roller-coaster effect develops over time as the population size oscillates around K.

The validity of assumption #6 is hard to generalize, since different kinds of organisms show a variety of types of reproduction. Two common types of reproduction are parthenogenesis and amphimixis. **Parthenogenesis** is reproduction without fertilization, so that parent and offspring are genetically identical females. Many bacteria use this method of reproduction to increase their population size quickly. **Amphimixis** is reproduction originating from cross fertilization, as is typical of most bisexual species. Most animals and plants use this method of reproduction. Assumption #6 is not of much concern if a population is reproducing by parthenogenesis. In amphimictic populations, reproductive success depends on the sex ratio and the probability of finding a mate. If population numbers are very low, potential mates may be rare, and so the growth of such populations may be unusually slow. This phenomenon is known as **underpopulation** and results in a slower growth rate at low densities than that predicted by the logistic equation. In extreme cases, it can result in extinction of a local population that is too small to maintain itself.

r versus *K* Strategies

Ecologists recognize two different types of life strategies among organisms, using terms from the logistic equation to describe the relative amount of energy a population devotes to growth or maintenance. An ***r* strategist,** or colonizer, devotes more energy to reproduction than to maintenance. The *r* strategists tend to maximize *r* and have high rates of increase, responding to disturbance and rapidly colonizing favorable environments. The ***K* strategist,** or persister, devotes more energy to maintenance of a stable population size. *K* strategists favor more stable habitats and their populations tend to be more persistent. Reproductive rates are usually lower and life cycles are longer than those for *r* strategists. Although many organisms show intermediate characteristics, many have been selected to excel at one strategy or the other. Bacteria, aphids, or annual weeds in a cleared field are good *r* strategists. Most mammals, bees, or shrub weeds in orchards show characteristics of *K* strategists.

Modeling and Forecasting Population Growth

Predictions of population growth or plant development are applicable in many areas of agriculture, from forecasts of potential yields and profits to anticipation of pest outbreaks and damage. The previous sections outlined some basic models for population growth. In reality, population growth is affected by many different factors, and growth patterns may appear erratic or unclear. Many alternative models have been developed for predicting population dynamics (the Bibliography at the end of this chapter can provide a starting point for further information). Some of these models are very complex, involving many equations and steps. **Simulation models** attempt to update population estimates over small time intervals, using current weather and environmental information. Today, there are models to describe thousands of situations in population growth or development or organisms. Some scientific journals, such as *Agricultural Systems,* contain many articles on this subject. A few examples, applicable to a variety of agricultural situations, are shown in Table 6-5.

Models of population growth or development have two main uses in science. The most obvious use is to make predictions that may be helpful in making future management decisions. Models may also serve as a useful framework for organizing our ideas about how a complex ecosystem works and for "testing" hypotheses in virtual experiments. The quality of the many available models varies greatly. Presumably, most models were tested extensively during development to verify that they work correctly and that their results make sense. However, the most critical test of model performance is *validation,* or the testing of model performance in the field to see how well model predictions fit a real data set.

Even the best models cannot be 100% accurate. Some error is always introduced into model results from sampling error in collecting any measured data on population size or plant size (see Chapter 11) or from variability in weather. A **deterministic** model does not include statistical variation, whereas a **stochastic** model considers sources of error and variation. For instance, a deterministic model may predict that an average population density of 15.2 weeds per m² emerges within a one-week time interval. This may be useful information, but it probably provides a false sense of accuracy and confidence. A stochastic model might be more realistic if it provided a forecast such as:

20% chance that <12 weeds/m² will emerge
60% chance that 12–16 weeds/m² will emerge
20% chance that >16 weeds/m² will emerge

After all, since weather forecasts are not 100% accurate, we certainly cannot expect events based on weather events (e.g., plant growth, pest population growth) to be 100% accurate either! Nevertheless, the use of reliable models and forecasts can take some of the guesswork out of agricultural planning and decision making, as well as improve our understanding of population growth and dynamics.

Table 6-5 Examples of some models used in agriculture to predict crop development, yield, or population growth.

Crop/Situation	Description	Location	Reference
Apple	General review on forecasting several apple diseases	Eastern United States	Sutton 1995
Bean	Predicting bean rust epidemics	Brazil	Amorim et al. 1995
Cassava	Spread of African cassava mosaic virus into cassava plantings	Ivory Coast, Tanzania	Fargette and Vie 1994
Corn	Effects of plant population and leaf area on yield	Nebraska	Piper and Weiss 1990
Corn	Predicting yield response to N and irrigation	Kansas	Llewelyn and Featherstone 1997
Dairy	Reproduction and revenues expected in dairy herds of various composition	Ontario	Plaizier et al. 1998
Fruit crops	Development of San Jose scale insect in apple and peach orchards	North Carolina	McClain, Rock, and Stinner 1990
Onion	Forecasting *Botrytis* leaf blight	New York	Lorbeer 1995
Potato	Crop growth and development	New York	Ewing et al. 1990
Potato	Virus infection of tubers due to virus transmission by aphids	Peru	Bertschinger, Keller, and Gessler 1995
Potato	Textbook, includes various situations	—	Haverkort and MacKerron 1995
Rangeland	Spring hatch of grasshoppers on rangeland grasses	Montana, Wyoming	Fisher, Kemp, and Pierson 1996
Rice	Development and crop production	Asia	Matthews et al. 1995
Soybean	Predicting flowering and maturity dates	United States	Piper et al. 1996
Tomato	Predicting various kinds of diseases	Eastern North America	Gleason et al. 1995
Various	Nematode populations on various crops	—	McSorley and Phillips 1993
Weeds	Forecasting emergence of weed cohort in spring	Quebec	Leblanc and Cloutier 1995
Wheat	Predicting crop growth response to N and irrigation	Victoria (Australia)	O'Leary and Connor 1996
Wheat	Postharvest development of beetle species in stored grain	Kansas	Hagstrum and Flinn 1990

Summary

During the growing season, the key events in the development of crop plants are biomass accumulation during vegetative growth and the change to the reproductive stage with flowering. Although the population densities of crop plants are controlled to some extent, weed biomass can increase as plant population density or size of individual plants increase. Models of varying degrees of complexity may be used to forecast increases in numbers of weeds and pest populations. Exponential growth models may be useful in describing short-term increases of some rapidly growing populations. Logistic growth models allow for a carrying capacity appropriate for a particular set of environmental conditions. Many other types of models have been developed to describe the complex behavior of populations in response to a range of physical, biological, and environmental factors.

Topics for Review and Discussion

1. The amount of seed planted per unit of land can be controlled easily, and provides some measure of control over plant density. What factors may affect the establishment of crop plants during the time from planting to emergence?
2. For most crop plants, an optimum plant spacing can be found that results in maximum yield. Yet this optimum plant spacing may not be used by all growers of a particular crop. What are some good reasons for *not* using the optimum plant spacing?
3. Discuss the relationship between Figure 6-5 and a common weed species in your area. Which features make this weed particularly troublesome for crop plants? During which parts of its life cycle might this weed be most susceptible to management?
4. What kinds of information would you need to make a model describing growth of a crop plant? What kinds of experiments could be conducted to obtain these data?
5. A dilemma in the development of a predictive model is determining when the model is ready for field use and prediction. A simple model might be brought to the field quickly, maybe at the risk of accuracy. Or a model may be perfected by years of refinement and testing, which might delay its use in the field for years. Realizing that no model would ever be 100% accurate, how would you decide when a model is ready for field use and "reliable" predictions?

Literature Cited

Aldrich, S. R., and E. R. Leng. 1972. *Modern Corn Production*. Urbana, Ill.: F and W Publishing.

Amorim, L., R. D. Berger, A. B. Filho, B. Hau, G. E. Weber, L. M. A. Bacchi, F. X. R. Vale, and M. B. Silva. 1995. A simulation model to describe epidemics of rust of *Phaseolus* beans. II. Validation. *Phytopathology* 85: 722–727.

Beasley, C. A., and C. J. Adams. 1996. Field-based, degree-day model for pink bollworm (Lepidoptera: Gelechiidae) development. *Journal of Economic Entomology* 89: 881–890.

Bertschinger, L., E. R. Keller, and C. Gessler. 1995. Development of EPIVIT, a simulation model for contact- and aphid-transmitted potato viruses. *Phytopathology* 85: 801–814.

Carey, J. R. 1993. *Applied Demography for Biologists*. New York: Oxford University Press.

Cuomo, G. J., D. D. Redfearn, and D. C. Blouin. 1998. Plant density effects on tropical corn forage mass, morphology, and nutritive value. *Agronomy Journal* 90: 93–96.

Evensen, C. I., R. C. Muchow, S. A. El-Swaify, and R. V. Osgood. 1997. Yield accumulation in irrigated sugarcane: I. Effect of crop age and cultivar. *Agronomy Journal* 89: 638–646.

Ewing, E. E., W. D. Heym, E. J. Batutis, R. G. Snyder, M. B. Khedher, K. P. Sandlan, and A. D. Turner. 1990. Modifications to the simulation model POTATO for use in New York. *Agricultural Systems* 33: 173–192.

Fargette, D., and K. Vie. 1994. Modeling the temporal primary spread of African cassava mosaic virus into plantings. *Phytopathology* 84: 378–382.

Fehr, W. R., and C. E. Caviness. 1977. Stages of soybean development. *Agriculture and Home Economics Special Report,* No. 80, Iowa State University, Ames, Ia.

Fisher, J. R., W. P. Kemp, and F. B. Pierson. 1996. *Aulocara elliotti* (Orthoptera: Acrididae): Diapause termination, postdiapause development, and prediction of hatch. *Environmental Entomology* 25: 1158–1166.

Gallaher, R. N., and R. McSorley. 1993. Population densities of *Meloidogyne incognita* and other nematodes following seven cultivars of cowpea. *Nematropica* 23: 21–26.

Gleason, M. L., A. A. MacNab, R. E. Pitblado, M. D. Ricker, D. A. East, and R. X. Latin. 1995. Disease-warning systems for processing tomatoes in eastern North America: Are we there yet? *Plant Disease* 79: 113–121.

Hagstrum, D. W., and P. W. Flinn. 1990. Simulations comparing insect species differences in response to wheat storage conditions and management practices. *Journal of Economic Entomology* 83: 2469–2475.

Hastings, A. 1997. *Population Biology.* New York: Springer-Verlag.

Haverkort, A. J., and D. K. L. MacKerron, eds. 1995. *Potato Ecology and Modelling of Crops Under Conditions Limiting Growth.* Dordrecht, The Netherlands: Kluwer Academic Publishers.

Holm, L. G., D. L. Plucknett, J. V. Pancho, and J. P. Herberger. 1977. *The World's Worst Weeds.* Honolulu, Haw.: University Press of Hawaii.

Leblanc, M., and D. Cloutier. 1995. Prédiction des levées des mauvaises herbes. In *International Symposium on Agricultural Pest Forecasting and Monitoring,* eds. G. Bourgeois and M. O. Guibord, 113–116. Quebec: Réseau d'avertissements phytosanitaires du Quebec.

List, R. J. 1945. Smithsonian meteorological tables, 6th ed. *Smithsonian Miscellaneous Collections,* Volume 114. Washington, D.C.: Smithsonian Institution Press.

Llewelyn, R. V., and A. M. Featherstone. 1997. A comparison of crop production functions using simulated data for irrigated corn in western Kansas. *Agricultural Systems* 54: 521–538.

Lorbeer, J. W. 1995. Disease monitoring and forecasting in vegetable crops emphasizing *Botrytis* leaf blight of onion. In *International Symposium on Pest Forecasting and Monitoring,* eds. G. Bourgeois and M. O. Guibord, 185–189. Quebec, Can.: Réseau d'avertissements phytosanitaires du Quebec.

Matthews, R. B., M. J. Kropft, D. Bachelet, and H. H. Van Laar, eds. 1995. *Modeling the Impact of Climate Change on Rice Production in Asia.* Wallingford, U.K.: CAB International.

McClain, D. C., G. C. Rock, and R. E. Stinner. 1990. San Jose scale (Homoptera: Diaspididae): Simulation of season phenology in North Carolina orchards. *Environmental Entomology* 19: 916–925.

McSorley, R., and M. S. Phillips. 1993. Modeling population dynamics and yield losses and their use in nematode management. In *Plant Parasitic Nematodes in Temperate Agriculture,* eds. K. Evans, D. L. Trudgill, and J. M. Webster, 61–85. Wallingford, U.K.: CAB International.

Nickel, S. E., R. K. Crookston, and M. P. Russelle. 1995. Root growth and distribution are affected by corn-soybean cropping sequence. *Agronomy Journal* 87: 895–902.

O'Leary, G., and D. J. Connor. 1996. A simulation model of the wheat crop in response to water and nitrogen supply: II. Model validation. *Agricultural Systems* 52: 31–55.

Pinkerton, J. N., G. S. Santo, and H. Mojtahedi. 1991. Population dynamics of *Meloidogyne chitwoodi* on Russet Burbank potatoes in relation to degree-day accumulation. *Journal of Nematology* 23: 283–290.

Piper, E. L., K. J. Boote, J. W. Jones, and S. S. Grimm. 1996. Comparison of two phenology models for predicting flowering and maturity date of soybean. *Crop Science* 36: 1606–1614.

Piper, E. L., and A. Weiss. 1990. Evaluating CERES-maize for reduction in plant population or leaf area during the growing season. *Agricultural Systems* 33: 199–213.

Plaizier, J. C. B., G. J. King, J. C. M. Dekkers, and K. Lissemore. 1998. Modeling the relationship between reproductive performance and net-revenue in dairy herds. *Agricultural Systems* 56: 305–322.

Rao, S. C., and T. H. Dao. 1996. Nitrogen placement and tillage effects on dry matter and nitrogen accumulation in redistribution in winter wheat. *Agronomy Journal* 88: 365–371.

Rowe, R. C., and G. A. Secor. 1993. Managing potato health from emergence to harvest. In *Potato Health Management,* ed. R. C. Rowe, 35–40. St. Paul, Minn.: The American Phytopathological Society.

Sutton, T. S. 1995. Disease forecasts: Tools for more efficient management of apple diseases. In *International Symposium on Pest Forecasting and Monitoring,* eds. G. Bourgeois and M. O. Guibord, 135–142. Quebec, Can.: Réseau d'avertissements phytosanitaires du Quebec.

Zhu, Y., C. C. Sheaffer, M. P. Russelle, and C. P. Vance. 1998. Dry matter accumulation and dinitrogen fixation of annual *Medicago* species. *Agronomy Journal* 90: 103–108.

Bibliography

Charles-Edwards, D. A., D. Doley, and G. M. Rimmington. 1986. *Modelling Plant Growth and Development.* Sydney, Aust.: Academic Press.

Cox, W. J. 1996. Whole-plant physiological and yield responses of maize to plant density. *Agronomy Journal* 88: 489–496.

Gizlice, Z., T. E. Carter, Jr., T. M. Gerig, and J. W. Burton. 1996. Genetic diversity patterns in North American public soybean cultivars based on coefficient of parentage. *Crop Science* 36: 753–765.

Gutierrez, A. P. 1996. *Applied Population Ecology.* New York: John Wiley and Sons.

Salisbury, F. B., and C. W. Ross. 1985. *Plant Physiology.* 3d ed. Belmont, Calif.: Wadsworth.

Steele, C. C., and L. J. Grabau. 1997. Planting dates for early-maturing soybean cultivars. *Agronomy Journal* 89: 449–453.

Weiner, J. 1900. Plant population ecology in agriculture. In *Agroecology,* eds. C. R. Carroll, J. H. Vandermeer, and P. Rosset, 235–262. New York: McGraw-Hill.

Chapter 7

Niches and Competition

Key Concepts

- Definition of an ecological niche
- Distinctions between interspecific competition and intraspecific competition
- Movement of species into a region
- Directional change in a community

Ecological Niches

An **ecological niche** is an abstract term that is difficult to define since it encompasses a wide range of factors and variables that all work together to describe the "role" of an organism in the ecosystem. It depicts where an organism is able to live (physical factors) and what it can do within that area (biological factors). It specifies physiological ranges for an organism that determine its distribution and migration patterns. It characterizes the interactions that an organism has with other organisms of the same species or of different species. It takes into account all of the resources that an organism needs both for survival and for reproduction.

For example, if we were to try to describe the physical niche of a corn plant, we would need to determine the ranges of temperature, salinity, moisture, nutrient levels, pH, and other physical factors within which the corn plant would survive and reproduce. If we were to look at biological components, we would also need to look at how the corn plant interacts with other organisms in the agricultural field. These variables are very difficult to describe, and there are almost an unlimited number of them, so the concept of a niche becomes a very theoretical one, rather than a practical one. Still, we can use this concept in agroecology in determining what plants should be grown where (within what physiological ranges) and with what other plants.

No two species can occupy exactly the same niche within an ecosystem. One species will always outcompete the other for the shared resources. For this reason, when we discuss niches, we must distinguish between a *fundamental niche* and a *realized niche*. A **fundamental niche** refers to the niche that an organism occupies in an ecosystem in the absence of competition or other such restraints to using resources. In the case of our corn plant, the fundamental niche would define the plant's distribution in terms of many of its physical limitations (it cannot grow in the Arctic Circle, on top of Long's Peak in the Rocky Mountains, or in Death Valley, for some extreme examples). Within the fundamental niche, all of the minimum requirements for the species must be available (recall the law of the minimum from Chapter 5). The **realized niche** is the niche that an organism actually occupies when modifying factors are present in a community. The realized niche for a corn plant depends on such determinants as the success of its competitors and the distribution and abundance of the herbivores that feed on it. Corn would have a difficult time growing in a forested area, even in a light gap, since most forest plants would quickly outgrow it and overshadow it.

How does the concept of a niche relate to agroecosystems? In the first place, it helps to determine what crops will and will not grow in a certain region. Temperature, day length, moisture, salinity, etc. influence what you should and should not plant in a particular region. An agroforestry system of oranges, bananas, and avocados will not work in Montana. Pineapple grows best in dry conditions, turnips grow best when it is chilly, and sweet potatoes need hot temperatures for a long period of time. Thinking about niches within an agroecosystem may also govern your choice of agricultural systems. You may choose to intercrop two crops that differ widely in their nutrient, light, and water requirements to reduce competition between plants and to increase yield in a field. You may choose to rotate one crop with another that modifies the environment by improving the soil conditions or nutrient composition of the soil, thus making it more favorable for the first crop to be planted there again. For example, if you have a field that is infested with the fungus *Cercospora sojina,* which causes frogeye leaf spot in soybean, you may choose to plant a different, resistant crop to reduce the fungal population (making the environment more favorable for soybean over time), while still getting a good yield (recognizing that while your field may provide a fundamental niche for soybean, the realized niche will be much reduced by the presence of *Cercospora*).

Plant Competition and Allelopathy

Competition between plants essentially refers to an overlap in resource requirements between plants of the same species (**intraspecific** competition) or plants of different species (**interspecific** competition). It occurs when the environment cannot supply the resources needed in quantities that are sufficient to provide for both competing plants (or plant species) at the same time. Plants compete primarily for water, light, nutrients, space, and CO_2. The normal effects of competition may include the reduction in growth rate and productivity of one or both competitors, and may also include a reduction in overall plant or harvest quality in the case of agricultural crops. Competition may involve either direct interaction between plants, or the indirect effects of one plant on another through modifications in the environment (**allelopathy**).

Intraspecific Competition

Intraspecific competition refers to any competition that occurs among organisms of the same species. Intraspecific competition tends to be quite intense since all members of a particular species are likely to require very similar biological, physical, and chemical resources.

The degree of competition among individuals depends strongly on the availability of these resources. As resources become more limited, competition becomes more intense. Similarly, as the number of competitors for a limited resource increases, the effect on any given individual also becomes greater. This is known as density-dependent population regulation, whereby the effects of competition on a population size become greater as population density increases.

In agroecosystems, as in any natural system, intraspecific competition occurs when plants of the same species compete for necessary resources. For this reason, it is important to carefully consider spacing when designing an agroecosystem to allow each plant the opportunity to obtain the resources that it needs for maximum growth and productivity.

We can look at this mathematically using the same logistic equation that we have already examined for population growth in Chapter 6. Population growth of an individual species was defined as:

$$\frac{dN}{dt} = rN\left(\frac{K - N}{K}\right) \quad \text{(Equation 7-1)}$$

where dN/dt is the change in population size over time, r is the intrinsic rate of natural increase of the population, N is the population size, and K is the carrying capacity of the environment for that species. If we increase the number of individuals in the population (N), we approach the carrying capacity, leading to competition among individuals for limited resources. We can observe this mathematically by simply multiplying Equation 7-1 out:

$$\frac{dN}{dt} = rN - \frac{rN^2}{K} \quad \text{(Equation 7-2)}$$

where rN is the rate of growth of the population if resources are unlimited, and rN^2/K is the decrease in that population growth that is caused by competition for limited resources.

An example from an agroecosystem illustrates this point:

- Suppose that you plant a field at a population of 200 plants/ha. The carrying capacity of the field is 300 plants/ha. Assume that $r = 1$. In this case:

$$\frac{dN}{dt} = (1 \times 200) - \frac{(1 \times 200)^2}{300} = 200 - 133 = 67$$

- Now suppose that you plant the field at a population of 400 plants/ha. The carrying capacity remains the same.

$$\frac{dN}{dt} = (1 \times 400) - \frac{(1 \times 400)^2}{300} = 400 - 533 = -133$$

Obviously, in agroecosystems, we are not interested in how a population grows (we plant a particular number of seeds and expect that a certain proportion of those will grow and provide us with our harvest), but these equations illustrate an important point. In the first case, there is still room for growth in the population. The fact that more individuals can be added indicates an abundance of natural resources, which would likely lead to good productivity in the crop. In the second case, the negative number indicates that we have exceeded the

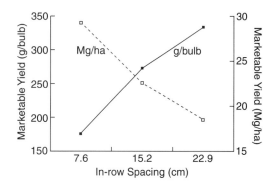

Figure 7-1 Relationship between plant spacing and marketable yield (Mg/ha, 1 megagram = 1 million grams) and onion bulb size (g/bulb).
Data from Stofella (1996)

carrying capacity of the environment, and instead of there being room for population growth, we would see a decrease in the number of individuals in the field over time. This indicates that there is intense competition for resources, and the plants are likely to be unable to get all they need, perhaps leading to decreased productivity and economic loss.

Decreasing spacing between plants in a field may actually increase quantity but decrease quality of a marketable product. As plant population size in a field increases, row spacing decreases, placing plants closer together and forcing them to compete with each other more intensely for resources. For example, as row space decreases and plant density increases, the total yield of onions in a field increases (increased quantity), but the size of each individual bulb decreases (decreased quality) (Figure 7-1). For this reason, farmers must understand that increased plant density has opposite effects on yield quality and quantity, and must weigh the market advantages and disadvantages accordingly.

Interspecific Competition

Interspecific competition is far more common in agricultural systems because it includes the competition between crops and weeds. Interspecific competition refers to competition between two or more different species. The intensity of this competition varies according to the resources being competed for, and the availability of those resources in the system.

Mathematically, we can introduce a second species to Equation 7-1, and look at the effect on species 1:

$$\frac{dN_1}{dt} = r_1 N_1 \left[\frac{(K_1 - N_1 - \alpha N_2)}{K_1} \right] \quad \text{(Equation 7-3)}$$

where α is the competitive effect of each individual of species 2 on the population of species 1. We can also look at the effect on species 2:

$$\frac{dN_2}{dt} = r_2 N_2 \left[\frac{(K_2 - N_2 - \beta N_1)}{K_2} \right] \quad \text{(Equation 7-4)}$$

where β is the competitive effect of each individual of species 1 on the population of species 2.

In agroecosystems, we are usually far more concerned with the effects of weeds on the growth of our crop plant than with the effects of our crop plants on the growth of weeds. Weeds in agroecosystems are not "spaced" an ideal distance from the crop plants, so they may provide serious competition for light, water, nutrients and other resources (see Chapter Endnote 7-1). For this reason, weed control in agricultural systems is a multi-million dollar business, with approximately 80% of all pesticides sold belonging to the herbicide family (see Chapter 11). Again, we are less concerned with population growth than with productivity.

An example from an agroecosystem illustrates this point:

- Suppose that you plant a field at a spacing of 200 plants/ha. The carrying capacity of the field is 300 plants/ha. Assume that $r_1 = 1$. However, in addition to the crop plant in your field, you also have 250 weeds/ha, which have a competitive effect of 0.25 on your crop plants (every four weeds in the field use the resources that one crop plant would use), In this case,

$$\frac{dN}{dt} = (1 \times 200) \times \left[\frac{300 - 200 - (0.25 \times 250)}{300}\right] = 200 \times 0.125 = 25$$

- Now suppose that this same field is planted to carrying capacity (300 plants/ha). Your weed problem remains the same (250 weeds/ha, with a competitive effect of 0.25 on your crop plants). In this case,

$$\frac{dN}{dt} = (1 \times 300) \times \left[\frac{300 - 300 - (0.25 \times 250)}{300}\right] = 300 \times -0.21 = -62.5$$

In this example, the weeds in the field are competing with the crop plants, reducing the amount of resources available for use by the target crop, and thus resulting in decreased yield and productivity.

While weeds in a field may reduce crop growth, the opposite may be true as well. Planting density can sometimes be manipulated to suppress weeds. When barley is planted at a seeding density twice that of a low-density treatment, weed biomass in the field is reduced by approximately 37% (Mohler and Liebman 1987). While intercrops may provide the same reduction in weed biomass in a field, the effect may be due to plant density, rather than to plant diversity (Liebman and Dyck 1993).

Weeds and crops clearly compete with each other for resources, which can seriously affect the productivity of crop plants, but this is not the only situation in agriculture that should be considered when thinking about competition. In some cases, particularly in intercropping situations, the effects of each plant on the other should be considered. Interspecific competition can result from poor agricultural planning. Intercropping species that need light with other species that tend to shade the ground below them results in a situation in which the plant that is able to grow fastest will outcompete the crop below it for light. Species that obtain their water from the same depth in the soil, or that have similar nutrient requirements, may also provide serious competition for each other, reducing overall productivity of both plants in the field.

Allelopathy

A third means of plant competition involves indirect effects of one plant on another through modification of the environment. **Allelopathy** is a common means of such interference. It is defined as the production of chemical substances, by a living plant or through the decomposition of plant residues, that interfere with the germination or the normal growth of a neighboring plant. Agroecologists must take allelopathy into consideration when planning cropping systems. Species should be chosen in combinations that are not allelopathic to each other. However, allelopathy can also be a useful tool in weed management, since some crops produce allelopathic chemicals that are toxic to some common weeds (see Chapter Endnote 7-1).

Migration, Dispersal, and Colonization

In agroecosystems, migration, dispersal, and colonization of organisms usually refers to the migration, dispersal, and colonization of plant pests. Weeds, pathogens, and insects are all commonly dispersed into a field, often leading to problems and economic loss for the grower. Knowing how these organisms arrive in an agricultural field and how they get established can aid us in forming strategies for their management.

Migration and Dispersal

When we think of migration, we usually envision flocks of geese migrating south for the winter, or salmon returning to the streams of their birth to spawn. At first glance, it seems to have few implications for agricultural systems. However, migration is also responsible for mass movements of plant-feeding insects such as locusts, which can devastate an agricultural field in a matter of hours. For this reason, we need to consider migration as well as dispersal when examining the movement of pest organisms into a region.

Migration is a mass movement of members of one species of organism from one location to another. It may be triggered by intrinsic factors, environmental factors, or a combination of both. *Dispersal* is a more individualistic process. Individual members of a species often disperse away from other members of their species, either actively or passively. **Active dispersal** indicates that the organisms themselves choose where they will go, often clustering around food supplies or in an environment suitable for their continued growth and development. **Passive dispersal,** on the other hand, implies that the organisms being moved passively (perhaps by wind or water) have no choice in where they end up. Their ability to survive once they land is determined by whether or not there is an unoccupied niche that they are able to exploit in their new location. Passive dispersal is the most common means of dissemination of most plant diseases and weed seeds.

Passive dispersal mechanisms can be divided into several categories. Bacteria, seeds, fungal spores, and even nematodes are commonly carried from one field to another by wind and rain, or by a combination of the two (windblown rain). Insects can also be blown from field to field, allowing the spread of any disease they may be carrying. Water dispersal is a common means of dissemination for bacteria, nematodes, and the spores or living fragments of fungi. In addition to carrying these organisms from one field to another, the moisture provided by the water provides an ideal environment for the growth and establishment of the pathogens once they have landed in a new location. Irrigation water can also serve to

Random distribution | Regular distribution | Aggregated distribution

Figure 7-2 Possible distribution patterns of organisms in a field.

disperse organisms, especially if water is allowed to run from a contaminated field to another site either by runoff or flood irrigation.

Humans are also responsible for dissemination of pathogens from one site to another. We can carry infective organisms or seeds from field to field on our boots and clothing, or on agricultural equipment (from small hoes and shovels to large tractors or plows). We can transmit diseases from sick plants to healthy plants by using infected pruning shears, knives, saws, or other hand tools. We may also introduce weeds and pathogens by using contaminated seeds, or by transplanting infected seedlings into our fields.

The distribution of colonizing organisms in a new environment may fall into one of three major categories (Figure 7-2). **Random distribution** occurs when an organism is no more likely to land at one location in a field than at another location. The presence of other individuals does not influence whether or not another individual will disperse to that location. **Regular distribution** occurs when individuals are equally spaced within a field. This could occur when individuals avoid close contact with other individuals. **Aggregated, or clumped, distribution** occurs when individuals are attracted to other individuals, and is thus more likely to occur in a field at locations very close to where other individuals are located.

In practice, we may see a combination of these distribution patterns, depending on the scale that we study. For example, plant parasitic nematodes are distributed in an agricultural field in an aggregated manner, since they are commonly found around plant roots where they feed. However, their distribution along the surface of the plant roots may be more regular to ensure that each is exploiting different cells along the same root (Figure 7-3).

The distribution patterns of organisms in agroecosystems may be determined by their mode of dispersal and by any intrinsic factors that govern their relationships with other individuals. Insects, for example, tend to move actively and have more control over their final destination. They may be more likely to be aggregated in a field, clustering around food sources.

The distribution patterns of weed seeds, bacteria, fungi, and viruses that are passively dispersed can change over time. In many cases, the colonization of a field may be random. However, large populations of these organisms can grow and spread out around each **focus,** or initial point of infection or colonization. As a result, the distributions of these organisms that we observe in practice are usually aggregated.

Establishment and Colonization

Dispersal of organisms into an area means little if the environment they land in is not suitable for their growth and reproduction. For example, while many fungal spores may be

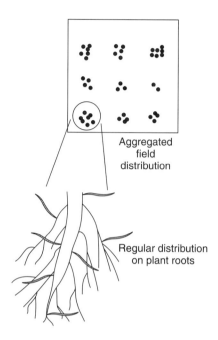

Figure 7-3 Aggregated nematode distribution in a field and more regular distribution on plant roots within the field.

blown from a contaminated field into a clean, neighboring field, most of those spores will not land on a susceptible host. Others will not survive the trip from field to field, while still others are too fragile to handle desiccation and may die before they are exposed to moisture.

Colonization, then, is dependent on a variety of factors, some of which are biological, and some of which are physical. All of them, however, are based on the ability of the organism to find a niche. Unfortunately, in many agroecosystems, a number of niches are available for exploitation because only one plant species is present when monocultures are planted. With so many other niches available for exploitation, the field then becomes vulnerable to the colonization of any number of organisms.

In plant pathology, the inoculum responsible for any plant disease (spores, eggs, infective juveniles, etc.) must go through a series of steps before it becomes established in a field or within a plant. It must first come in contact with the plant, which is generally accomplished by the dispersal and migration methods described previously. Once contact has been established, the pathogen must penetrate the plant, occasionally breaking through plant defenses, and finally set up a parasitic relationship with the host plant (infection). It is only after this relationship has been set up that the pathogen can actively grow and reproduce, so successful colonization is important in the spread of disease.

A similar situation occurs with the invasion of weeds. Weed seeds that land in fertile soil will germinate and grow, but how much of a competitive effect they will have depends on their own ability to obtain resources, and how well-established the crop plants in the field are (see Chapter Endnote 7-2). Weeds that are already present when a crop is put into the ground will have a better likelihood of successfully competing for resources against crop plants than those that germinate after the crop is well established. Established plants have the advantage of more extensive root systems and taller shoots, so they are better able to obtain water, nutrients, and light than the plants that are just beginning to grow. Weeds that are successful at competing for resources are able to put a great deal of energy into growth and reproduction, so successful competition for weeds is practically synonymous with successful colonization.

Extinction

Extinction, at first glance, may seem to be an odd topic for a textbook on agroecology. However, if we first distinguish between global and local extinction, we will see that one of the major goals in managing agricultural systems has always been the extinction of species.

Global extinction occurs when all members of a species have disappeared from the face of the earth, and no member of that species remains alive anywhere in the world. **Local extinction** occurs when a species no longer can be found in a region where it was once found, but can still be found elsewhere on the face of the earth. In ecology and conservation biology, local extinctions are usually discussed on a much larger scale than an agricultural field, but the principles of species eradication are the same.

Global extinction of species has wide-reaching effects on agricultural systems everywhere. For years, the pest resistance, improved performance, and high yields of many of our common crop plants have been genetically introduced into our plants by scientists who have crossed these crops with their wild relatives. The loss of these relatives to extinction may lead to the loss of potentially important genes that could improve productivity under future environmental conditions. We discuss this subject more fully in Chapter 8.

Local extinction, on the other hand, can be potentially good for an agroecosystem. For example, a field that has always been infested with soybean cyst nematodes can be useless for the production of soybeans without careful management, crop rotations, and occasional nematicide applications. If a combination of management tools could somehow completely eradicate soybean cyst nematodes from the field, it could again be useful for the production of soybean. Indeed, the absolute removal of any pest species from a field could be considered an improvement, since the threat of that organism to plant productivity has been eliminated. In practice, however, the local extinction of a pest from a field can be difficult to accomplish (see Chapter 11).

Local extinction can be detrimental to an agroecosystem if biological control methods are used to manage pest populations. Total eradication of a pest from a field may decrease the food supply available to the biological control agent, which can drive its extinction from the field as well. If this happens, and a resurgence of the pest occurs, there will no longer be a biological control agent present to prey on the pest. We address this topic in greater detail in Chapter 9.

Plant Succession

Succession refers to natural and directional changes in community structure over relatively long periods of time (Figure 7-4). When an area is disturbed, the ecological niches within that community are altered or destroyed, often causing the disappearance of the organisms that were present in that area. At the same time, the environment is modified to create new niches for other organisms to occupy.

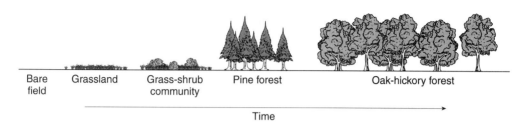

Figure 7-4 Directional changes (succession) in a community over time.

In natural systems, succession occurs when a cleared field is allowed to return to a previously forested state, when a new volcanic island appears in the Pacific Ocean and is slowly vegetated, or when a glacier recedes, leaving behind an area of rock and soil to be colonized by plants and animals. In agroecosystems, the natural processes are the same, but the length of time over which succession is allowed to occur is much shorter.

Successional processes are often used in agroecology to benefit both the grower and the land. Shifting cultivation, for example, takes advantage of the fact that succession will occur in a greatly disturbed area and, over time, organic matter and nutrients will slowly accumulate in the plants that invade the land. Thus, when the area is slashed and burned many years later, those nutrients will be available to be added back to the soil. In other cases, a grower might need to fight successional processes. For instance, letting a field lie fallow for a period of time can be a valuable way to increase soil moisture in some arid regions, but the growth of invasive weeds in the area might detract from the usefulness of the method.

We can describe succession as primary or secondary, depending on the initial state of the land on which colonization occurs. In primary succession, communities are established on land that has not previously supported life. Volcanic areas, terrain left open after glacial movement, and a new island that appears in the ocean are examples of areas where primary succession occurs. Secondary succession occurs when land has been previously vegetated, but that vegetation has been destroyed by some disturbance, such as fire, heat, water, or even tillage. Shifting cultivation takes advantage of secondary succession.

There is still considerable controversy over how succession actually occurs, but most ecologists now accept that the idea of a successional process leading up to a stable "climax community" in an area is greatly oversimplified. Three basic models of succession are currently being discussed in the scientific community (facilitation, tolerance, and inhibition; see Figure 7-5) (del Moral and Bliss 1993). It is not likely, however, that these models are mutually exclusive; they may in fact all be acting on a community either at the same time or in sequence.

Perhaps the most basic model of succession in a community describes the replacement of *r*-selected species by *K*-selected species over time. This is the facilitation view of succession, which occurs when those species best adapted to poor environmental conditions move into an area, and by virtue of their activities, make the area more suitable for other species to move in and colonize the region. In the initial stages of succession, the biological community tends to be simple, but unstable. As succession progresses, the community becomes more diverse and complex, leading to an inherent stability of the system.

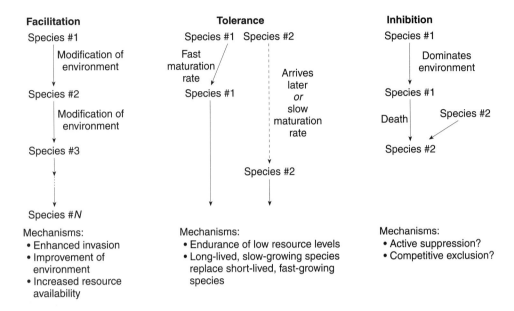

Figure 7-5 Three current models of succession. *Facilitation* occurs when initial arrivers to a system modify the environment to make it more habitable for other species to invade. *Tolerance* occurs when one species that can tolerate low resource levels arrives in an area, and is eventually replaced by longer-lived species that arrive later or arrive simultaneously but mature more slowly. *Inhibition* occurs when the death of one species opens a niche for a second species to move into a region, or when the second species competitively excludes the first species from the area.

Recall from Chapter 6 that *r*-selected species are organisms that are short-lived, have high reproductive rates, are generally controlled by density-independent factors, and often do well in unstable areas. *K*-selected species tend to be long-lived, have low reproductive rates, are controlled by density-dependent factors, and generally require stable environments in order to establish and survive. In the process of succession, it is generally accepted that *r*-selected species will colonize an area first, since they are thought to be most able to handle adverse and unstable conditions.

In terms of agriculture, *r* and *K* selection can have important implications for the practice of fallowing. Like any system, a fallow field is open to colonization by organisms from nearby fields or forests. Those plant species that invade the field and quickly establish are the *r*-selected species, which have very rapid reproductive rates and short life cycles. These plants will proliferate quickly, adding a good number of seeds to the seed bank of the field, and setting the stage for severe weed problems in the future. In natural systems, *r*-selected plants would, over time, be replaced and outcompeted by *K*-selected species, which would have longer life cycles and lower reproductive rates. These *K*-selected plants could also be serious agricultural pests since many of these long-lived plants are actually harder to eradicate from a field. However, successional processes are interrupted by agriculture. Farmers do not normally leave their fields fallow for more than a year or two at a time, which does not allow enough time for the establishment of a secondary community.

With slash-and-burn systems, the weed growth and much of the weed seed bank in the upper few centimeters of soil are destroyed by fire and high heat. In these systems, crop plants are usually seeded into the ground shortly after burning, providing them with a competitive advantage over any invasive species. The diverse polycultures that are normally planted in these systems are also less likely to face serious weed problems because most available niches for plants are already filled by crop plants.

Weed Growth as a Successional Process

Weed growth in agroecosystems is essentially succession in action. It involves the colonization of a disturbed area by r-selected species (high reproductive rates, relatively short life cycles). Unfortunately, since most agricultural plants are also r-selected, the competition between weeds and crop plants can be devastating to the system's productivity.

How often a system is disturbed will directly affect the kinds of weeds in the field. For example, suppose that you have a field that you plant to an annual crop every year. The field stays fallow in the winter, but you use a moldboard plow on it every spring to prepare the ground for planting. As you plow the field, you disturb it by essentially destroying the ecological niches that were present, and replacing them with new niches that are ready to be colonized. Those first organisms to move into the field will be the r-selected species. For the most part, then, you will be dealing with annual weeds with amazing reproductive capacities (for management options, see Chapter Endnote 7-1). If, however, you have an orchard where very little disturbance occurs to the system, apart from the yearly harvest of fruits, the early colonizing weeds in this area will likely have been replaced over time by longer-lived, slower reproducing weeds, or K-selected species. In this area, your management options would shift to cover perennial weeds, rather than strictly annuals.

The problem of weeds in an agroecosystem is incredibly complex because any change made to the system as a whole will also cause changes in the weed community over time. Since humans are constantly modifying the environment of annual agroecosystems through agricultural production practices such as tillage, fertilizer use, pesticide use, etc., we are essentially causing serious disturbance to the system on a regular basis. In ecological terms, we are essentially attempting to maintain the agroecosystem in an initial state of succession (very unstable), which can lead to serious weed problems that could become extremely costly in the long run.

One of the frustrating things about weed management is that the weed community is not static; it changes through time. The same management practices that we use to attempt to manage weeds can cause changes in the weed community that we might not have expected, such as the replacement of one weed by another. It is often the very practices that humans use to control weeds that can determine which weeds will be present in our fields. Therefore, in order to manage weeds responsibly, we must have a good understanding of how weeds fit into our agroecosystems, and how basic successional forces modify those weed communities over time (see Chapter Endnote 7-2).

Summary

The ecological niche refers to the requirements of an organism in an environment, including the resources it uses and the position it plays in the ecosystem. Organisms that occupy similar niches are likely to experience intense competition as each individual attempts to obtain sufficient resources for growth and reproduction. The more the niches overlap, the more intense competition tends to be. For this reason, intraspecific competition (competition within a species) tends to be much more intense that interspecific competition (competition between species), so growers should pay as much attention to crop spacing as they do to weed suppression. Over time, organisms in an ecosystem modify the environment (increased resources, changes in resource partitioning, etc.), opening up niches for new organisms to move into a system. A directional change in a community is known as succession, which commonly follows disturbance in both natural and agricultural systems. Weeds are good competitors for resources and are able to colonize a new area well, so they are a particular problem in areas that are disturbed often (annual systems) or early in the establishment of a more stable system.

Topics for Review and Discussion

1. What are the advantages and disadvantages of increasing crop density to manage weeds in a field?
2. Why do growers care about migration and dispersal? What problems are likely to be exacerbated by these mechanisms?
3. Why are *r*-selected species often the first species to move into a field following succession? What life history characteristics make them especially well-adapted to colonization?
4. A field is planted to 140 crop plants/ha, and the carrying capacity of the field is 200 plants/ha. Weed growth in the field averages 100 weeds/ha, and these weeds have a competitive effect of 0.36 on the crop plant. From these values, indicate whether the crop is in danger of lowered productivity due to weed growth and why. For your calculations, assume that $r_1 = 1$.
5. Based on the following data and your knowledge of plant competition, explain what might be happening in a field when mustard is grown alone or as a weed in association with barley, pea, or a barley/pea intercrop.

Treatment	Above-Ground Mustard Biomass (g/m²)	
	70 Days After Planting	212 Days After Planting
Control: mustard alone	132.2	1387.1
Mustard + barley	35.1	183.3
Mustard + pea	58.2	636.3
Mustard + barley + pea	28.2	185.9

Data from Liebman 1986, published in Liebman and Dyck 1993

Literature Cited

del Moral, R., and L. C. Bliss. 1993. Mechanisms of primary succession: Insights resulting from the eruption of Mount St. Helens. *Advances in Ecological Research* 24:1–66

Holm, L. G., D. L. Plucknett, J. V. Pancho, and J. P. Herberger. 1977. *The World's Worst Weeds.* Honolulu Hi.: The University Press of Hawaii.

Liebman, M., and E. Dyck. 1993. Crop rotation and intercropping strategies for weed management. *Ecological Applications* 3:92–122.

Mohler, C. L., and M. Liebman. 1987. Weed productivity and composition in sole crops and intercrops of barley and field pea. *Journal of Applied Ecology* 24:685–699.

Stofella, P. J. 1996. Planting arrangement and density of transplants influence sweet Spanish onion yields and bulb size. *HortScience* 31:1129–1130.

Bibliography

Altieri, M. A. 1987. *Agroecology: The Scientific Basis of Alternative Agriculture.* Boulder, Colo.: Westview Press.

Carroll, R. C., J. H. Vandermeer, and P. Rossett. 1990. *Agroecology.* New York: McGraw-Hill.

Odum, E. P. 1983. *Basic Ecology.* Philadelphia: Saunders College Publishing.

Radosevich, S., J. Holt, and C. Ghersa. 1997. *Weed Ecology.* 2d ed. New York: John Wiley and Sons.

Smith, R. L. 1996. *Ecology and Field Biology.* 5th ed. New York: HarperCollins.

Tivy, J. 1992. *Agricultural Ecology.* Essex, UK: Longman Scientific & Technical.

Chapter Endnote 7-1 *Weed Ecology and Management*

Weed management in agroecosystems is one of the biggest problems that many producers face, which explains why more than half of all agricultural chemicals sold are herbicides. But why are weeds such a problem? What is it about their ecology that makes them particularly disruptive for agricultural systems?

The main problem that weeds cause is their competition with cultivated plants for nutrients, light, and water. The more similar the needs of the weeds are to the needs of the cultivated plant, the more intense that competition will be. If the weeds are more efficient in harvesting water or nutrients, or if they grow quickly enough to overshadow crop plants and outcompete them for light, they may cause a serious and drastic decrease in the yield of the system.

However, weeds can cause other problems as well. Some weeds that are left in a field of forage can harm cattle, or reduce the quality of the animal forage. This can result in decreased quality or quantity of animal products like milk. Other weeds harbor diseases or pests that can spread to cultivated crops. Still others produce allelochemicals that reduce the productivity of other plants grown near them. But perhaps the most important reason that weeds are considered an agricultural nuisance is that they are so difficult to get rid of!

Perennials, Biennials, and Annuals

Perennial, biennial, and annual weeds differ in their ecology and in their management. More than 80% of weeds in most fields are **annuals,** which grow, reproduce, and die in a single season. **Perennials** live for three years or more. **Biennials** live for two years, growing vegetatively during the first year, then reproducing during the second.

- Annuals reproduce only by seed, but they produce such an incredible number of seeds (many thousands) that a good number end up in soil suitable for their growth.

(continued)

In addition, many annual seeds can remain dormant in the soil for years, until environmental conditions are suitable for their germination. Good management techniques for annual plants aim for their removal from the field before they set seed. If they are allowed to reproduce, many of their seeds will be ready to begin growing at the same time a cultivated crop is put into the ground, causing potentially serious competition for the target crop.

- Many perennial weed species live for much longer than three years. Almost all of these weeds reproduce by setting seed, but some kinds also self-propagate by reproducing vegetatively from underground nodes on roots and stems. Unfortunately, what this means is that if you simply destroy the aboveground part of the weed, the perennial will produce even more plants than before by sending shoots up from its underground parts. Good management of these weeds often means physically digging out the plants and roots, sometimes for more than one season, or using a chemical control. If these weeds are common in agroecosystems, they will have a competitive edge on the target crops because their root systems are well developed and they have stored nutrients to use for growth of shoots and leaves.
- Biennials will remain low to the ground during the first year as they grow only roots, stems, and leaves. The second year, they will grow upright and produce flowers and seeds. Good management strategies aim to treat biennials like a perennial weed the first year (remove it completely from the ground) and like an annual the second (destroy it before it sets seed).

Dissemination of Weed Seeds

Why do weeds show up in a field to begin with? You've just plowed a field and you're ready to go in and plant your crop, but you can already see small, green sprouts pushing up through the topsoil, ready to outcompete your target crop from the start. Where did they come from?

Under normal conditions, weed seeds are simply always present in agricultural fields. They are a part of the **seed bank,** which is essentially a seed storage mechanism in all ecosystems. The seeds are deposited into the soil, either directly from plants present in the field, or indirectly by wind, water, contaminated equipment, etc. Once they are present in the soil, they can remain there for several years before germinating and growing. When you plow a field, you may simply bring weed seeds up to the surface of the soil, where environmental conditions for germination are good. In addition, the newly sprouted weeds have a competitive advantage over your crop plants, since they have germinated first and have ample opportunity to lay down roots and grow shoots and leaves before your target crop gets established.

Weeds can also enter a field by wind or water dissemination. Runoff of water from other fields or natural areas surrounding your fields can bring large numbers of weed seeds onto your land. Wind dispersal is a common means of seed dissemination in the wild, and it works just as effectively in an agricultural field as in a pasture or a savanna. Movement of equipment from one field to another can also track weed seeds along with it.

Weed Management Strategies: Gaining the Competitive Edge

Now that we know that weeds will always be present, what can we do to suppress weed growth and protect our crop plants? How we manage our agroecosystems can strongly affect

(continued)

weed growth, either by promoting weeds or suppressing them. The major goal of management is to give cultivated crops a competitive advantage over weeds, and this can be accomplished by a combination of several different methods. These methods are discussed in more detail in later chapters (see Chapters 10, 13, and 14).

- *Crop density:* Increasing the seeding rates of the target crop can promote its dominance over weeds. This can backfire, though, because interspecific competition with weeds is often less intense than intraspecific competition with plants of the same species.
- *Crop diversity:* The role of diversity is still being investigated, but it apparently does play a part in reducing competition with weeds. It may be that overall diversity in an area leaves less available niches open for weed species to exploit. It may also be partially linked to plant density. Crop rotation (diversity over time) may also be important in weed control since it disrupts weed life cycles and prevents the buildup of adapted weeds in an area, which are even tougher to eradicate.
- *Crop spatial arrangement:* Controversial results have been obtained in studies looking at how the arrangement of crops in a field is related to weed growth. In some studies, scientists have found that crops arranged in clumps have fewer problems with weed competition than row arrangements, but the opposite is true in other studies.
- *Choice of crop species and cultivar:* Some crop plants have growing habits that make them better competitors against weeds. Early canopy formation and rapid shoot growth are two examples of such growing habits. Growth characteristics that potentially contribute to cultivar competitiveness are still being studied.
- *Herbicides:* Herbicides have several advantages over other weed control methods. Not only are they quite effective and reliable, but they can also save time and labor in the field. They also allow for reduced tillage, which has its own advantages (see Chapter 14). However, herbicide use is being closely examined because of the off-site effects of these chemicals. Studies have shown that some herbicides can cause groundwater and surface water contamination in areas quite remote from where they were applied. In addition, over-reliance on chemical control methods may be risky since abrupt removal of a commonly used product from the market may leave growers unprepared. Still, herbicides can give target crops a competitive edge by allowing them time and space to establish and grow.
- *Mechanical weed control:* This method is probably the oldest method used in the world today, except for hand-weeding. Changes in mechanical weed control methods have come about as mechanization has changed, but hand-weeding is still the most widely used method in developing nations. Simply removing the weeds from the field will reduce competition with your target crops.
- *Timing of planting:* Timing can be crucial to weed control and to maintaining a competitive edge over weeds. Plowing a field and then waiting two weeks before seeding the land will likely result in some serious competition problems between target crops and the weeds, since the weeds will already be well-established. Planting a crop when it is not well-adapted to temperature or moisture regimes will decrease its ability to grow and be productive. In this situation, weeds that are growing along with crop plants are likely to be very well adapted to the environment, and will easily outcompete your target crop.
- *Biological control:* Classical biological control is used for control of perennial weeds in some specialized situations. It involves the introduction of a biological agent to an

(continued)

area in small quantities. The agent builds up over a period of years to keep weed growth in check. If biological control agents feed primarily on the weeds and not on the crops, this will give a strong competitive edge to the crops because they are not dealing with pest infestation in addition to competition.
- *Use of cover cropping systems* (living mulches): Low-growing cover crop species can be seeded between rows to provide competition for weeds, or to negatively affect weeds through allelopathy (see following section). These systems must be carefully managed so that they outcompete weeds without outcompeting crops. While the biggest potential drawback to this system is simply that the cover crop can provide serious competition to the target crop, this risk can be partly avoided by either manipulating planting times, or by selecting competitive crop varieties.

Weeds and Conservation Tillage

The idea behind reduced tillage is to (logically) reduce tillage operations in the field. Normal tillage can lead to serious soil compaction over time, as well as reduced aeration and poor drainage. However, tillage has been used conventionally as a means of mechanical weed control. So while conservation tillage (see Chapter 14) is being adopted rapidly in the United States and in many other countries, weed control methods that are compatible with these reduced tillage systems are now needed. Most farmers who use conservation tillage as a part of their soil conservation program are relying solely on herbicide use to manage weeds, which can lead to problems just listed.

Farmers who switch from conventional tillage to reduced tillage often find a change in the weeds that predominate their fields (see the preceding discussion on weed growth as a successional process). This change is usually seen as a shift toward more perennial weeds and fewer annuals (since the system is disturbed less frequently, it is inherently more stable).

Weeds and Allelopathy

Because the effects of allelopathic chemicals can be residual (still present in decaying plant tissue or soil), mulches made from plants that have allelopathic effects on weeds can be incorporated into a weed management system. In other cases, weed growth can be suppressed by establishing a cover crop that is allelopathic to competitive weeds prior to actually planting the target crop. Care must be taken, however, because some cover crops may also exhibit allelopathic inhibition of the cash crops themselves! For example, various legumes contain an allelopathic compound that is released into the soil from their roots. This exudate not only causes reduced growth in many grass species, but can reduce yield of some economically important crop plants as well.

When Do Weeds Play a Useful Role in Agroecosystems?

In some cases, weed management involves letting a number of weeds grow in the field. Complete eradication of weeds is not only costly, but can also cause some consequences that you may not have planned. Besides that, eradication is next to impossible! Weeds are a nor-

(continued)

mal part of any ecosystem. Weeds can be beneficial in some situations, and here are some ways in which weeds can actually improve the overall health of your system:

- Weeds can help to control erosion. Soil runoff from bare land is much higher than from land with plants present. Roots and shoots help to capture soil particles that are eroding from a system, and can also help to slow water movement. This is particularly important during times when crops are not being grown on a particular site.
- Weeds can increase the amount of organic matter present in the soil and help to recycle nutrients. Weeds are made up of the same biological components as crop plants, so when they die and decompose, the nutrients and organic matter that had been tied up in the weeds are available for future crop use.
- Weeds can increase the diversity of the system, providing some systems with higher stability. In some cases, they can provide an alternative food source and habitat for many biological control agents, thus giving the overall system a better natural control of diseases and insects.
- Weeds can increase the amount of genetic material available for breeding. Some of the tomato varieties that are on the market today have been dramatically improved by using related weeds as a source of genes!

Chapter Endnote 7-2 *The Ten Worst Weeds of the World*

Sources such as Holm et al. (1977) have published lists and descriptions of the world's worst weeds for years. Following is a list of probably the most troublesome weeds in agriculture on a worldwide basis. Of course, any list will vary locally. For example, several amaranths (*Amaranthus* spp., Figure 7-6) are near the top ten on most lists.

1. Purple nutsedge and yellow nutsedge (*Cyperus rotundus* and *Cyperus esculentus*)
   Most authorities consider purple nutsedge the world's worst weed. Purple nutsedge is a perennial weed with a very extensive underground system of rhizomes and tubers. These tubers can remain dormant when environmental conditions are unfavorable, then produce new shoots when conditions improve. The weed propagates mainly by vegetative methods because its seed has a very low germination rate. Since purple nutsedge is sensitive to shade, it can be outcompeted by early growing crops. However, shade does not prevent further tuber growth, so this method should not be relied on as a control method. Additional practices should be aimed at the prevention of new tubers, and may include physically digging the weeds from the soil, solarization to kill shallow tubers, improving soil drainage (since nutsedge likes wetter soils), or shallow and frequent tillage (deep tillage can bring buried tubers to the surface, and cultivation should be frequent to continually kill the growing shoots so that more nutrients are not available to produce more tubers). The competitive effects of purple nutsedge can often be reduced dramatically by controlling the plant early, thus allowing your crop to become well established. Any emerging nutsedge, then, will be easily outcompeted by the target crop. Unfortunately, once purple nutsedge is present in your field, it is almost impossible to eradicate, so you will be battling this

(continued)

Figure 7-6 Pigweed (*Amaranthus* spp.), one of the ten worst weeds in some parts of the world.

(continued)

Figure 7-7 Whether a plant is a weed is sometimes a matter of opinion. Bermudagrass, one of the world's worst weeds, is used for pastures and for urban lawns and golf courses in many areas of the United States.

   weed indefinitely. The related yellow nutsedge is abundant in many areas, and often may be found growing along with purple nutsedge.
2. Bermudagrass (*Cynodon dactylon*)
   A perennial weed, bermudagrass produces a large number of underground rhizomes and stolons, which spread fairly rapidly below ground. It is a very aggressive weed that is invasive and difficult to control. Bermudagrass grows most rapidly during warm weather, and dies back to its dormant underground stems when environmental conditions are unfavorable for growth. Mechanical attempts to remove the weed can simply result in formation of new shoots. Certain cultivars of bermudagrass are used as a turf grass or a forage in many parts of the world (Figure 7-7)
3. Barnyard grass and jungle rice (*Echinochloa crusgalli* and *Echinochloa colonum*)
   Barnyard grass and jungle rice are annual grasses that commonly compete with rice as well as with many other temperate and tropical crops. They are well adapted to the aquatic environment, and can grow rapidly and multiply swiftly, which makes them strong competitors with rice. The similarity of these weeds to rice at early stages of

(continued)

growth makes it very difficult for farmers to tell them apart while hand-weeding. Leaving these grasses in the field until they are recognizable can actually cause great damage before they are removed. Barnyard grass and jungle rice can germinate at any time in the growing season, so control usually involves allowing the plants to emerge before rice is planted, then destroying the weeds by cultivation. These are large grasses that can cause problems in many other crops in addition to rice.

4. Goosegrass (*Eleusine indica*)
Goosegrass is a summer annual weed that presents serious problems in tropical and subtropical areas of the world. It reproduces by seed, and is often confused with crabgrass (see # 9). Goosegrass can compete very well in compacted soils, but once the compaction is alleviated, it can be outcompeted by a well-maintained and healthy crop.

5. Johnsongrass (*Sorghum halepense*)
Johnsongrass is an extremely competitive perennial weed that is spread both by seed and by rhizomes. It can reach heights of 2.5–3 meters, and produces hundreds of seeds per seedhead. It reproduces both by seed and by rhizomes, most of which are found in the upper 20 cm of soil, so it can be very difficult to eradicate once it moves into a field. Conventional tillage may be one of the best management techniques since it allows for the breakup of the underground rhizomes and, if done at the proper time, can also prevent seed from setting. Johnsongrass can also cause problems after it has been killed—if it is allowed to remain in a field, it emits an allelochemical during decomposition that can inhibit the growth of other plants.

6. Cogongrass (*Imperata cylindrica*)
A perennial weed, cogongrass reproduces vegetatively by underground rhizomes, as well as by producing large numbers of seed sporadically. This weed is so aggressive that it has caused farmers to abandon shifting cultivation sites that have been invaded by it. Cogongrass does well in all soil types from sand to clay. Since the rhizomes of this weed penetrate deeply into the soil, it does quite well on poor soils with low soil fertility and low moisture. Most ecotypes of cogongrass do not tolerate shade well at all, and for that reason, can be outcompeted if your target crop is able to establish well before cogongrass emerges. In addition to competition problems caused by cogongrass, the weed also exudes an allelopathic chemical that can inhibit the growth of other plants. Eradication of the weed requires the destruction of the rhizomes, which is often done by frequent cultivation. Burning of the aboveground biomass can help to prevent some of the potential allelopathic effects.

7. Common purslane (*Portulaca oleracea*)
An annual herb, purslane is a succulent weed that has been used as pig feed in many areas of the world. It grows very well in hot, dry weather, reproducing by seed and by stem fragments if the soil is humid. Purslane prefers loose, nutrient-rich sandy soil, but can grow in a wide variety of soil types and conditions. It is so succulent that it can continue living and can even produce seeds after it has been cut, so it can be very difficult to kill.

8. Lambsquarter (*Chenopodium album*)
A summer annual herb, lambsquarter is one of the most widely distributed weed species in the world. It does well on all soil types. It has a temperature-dependent germination cycle, requiring soil temperatures between about 10–30° C to germinate.

(continued)

9. Large crabgrass (*Digitaria sanguinalis*)
   A summer annual grass, large crabgrass is a problem both in temperate and tropical regions. It can root at its nodes, so it does show some perennial growth, but reproduction is primarily by a prolific production of seed. Crabgrass matures during the summer and then produces seeds in the late summer or fall.
10. Field bindweed (*Convolvulus arvensis*)
    Field bindweed is a creeping perennial plant, often referred to as morning glory. It has vine-like stems that allow it to remain prostrate on the ground, or to climb. The roots of this plant can penetrate to over 5 meters in soil, enabling it to survive and compete well in dry soils that are low in nutrients. Field bindweed can reproduce by both rhizomes and seeds. It is very difficult to eradicate because its rhizomes are so deep and because its seeds can remain viable in soil for up to fifty years.

Chapter 8

Adaptation of Plant Cultivars

Key Concepts

- Biodiversity and genetic diversity
- Genetic variation in agriculture and its importance in breeding efforts
- Genetic drift within populations
- Herbivory and breeding for pest resistance

Biodiversity and Ecosystem Health

Biodiversity is defined by the Worldwide Fund for Nature (1989) as "The wealth of life on earth, the millions of plants, animals, and microorganisms, the genes they contain, and the intricate ecosystems they help build into the living environment." Biodiversity, then, not only addresses the variety of species of living organisms on our planet, but also the genetic diversity within those organisms and the diversity of ecosystems in which they live. It originates in natural systems, where each organism is exposed to a variety of conditions and events that shapes its evolution and drives its distribution and success.

Biodiversity is often quantified by simply enumerating the species present in an area, but that is only a small component of the health of an ecosystem. Preservation of biodiversity involves more than merely placing a few token members of a species into conservation. This approach is insufficient in both natural and agricultural systems; simply growing a few token plants of a certain variety in a botanical garden will in no way preserve the range of unique genetic diversity that allows the plants to adjust and flourish as conditions change. Plant populations vary in their ability to resist pests and disease. For example, in order to find one variety of rice with resistance to grassy stunt virus, which is a serious pest on rice in all parts of the world, many thousands of varieties of rice from many different locations all over the world were carefully screened (Myers 1983). Simply saving a few token "rice plants" could in no way recover the genetic diversity of rice that would be lost if its wild relatives were unprotected and eventually killed. Similarly, one of the great dangers of relying on a small population of plants to maintain genetic diversity over the long term is the danger of inbreeding. A small population is more likely to exhibit deleterious genes, and the good genes that are present with low frequency are more likely to be lost over time. This cannot be undone by producing larger populations in succeeding generations. Once genes are lost, they cannot be restored.

Genetic diversity is an important part of the definition of biodiversity. It is this genetic diversity that provides the basis for all agricultural productivity across the world. Without the inherent ability of plants to adapt to changing environmental conditions, we would be limited in our pursuits of higher-yielding, pest-resistant plants. The ability of scientists to transfer genes from resistant wild relatives of our staple crops to new cultivars is estimated to account for an annual increase in crop productivity of about 1% (National Research Council 1992). In addition to providing a gene pool for the improvement of existing crops, the biodiversity of plants on earth could include different crops that may be utilized in the future to provide food for a growing world population (see Application 8-1).

Genetic Variation and Agriculture

In the natural world, plants grow in those areas to which they are well adapted, but may grow poorly or not at all in other regions. In agricultural systems, plants face similar limitations. What we grow in our agroecosystem in any given region is based first and foremost on the plants that are adapted to that region. The adaptations of plants to different climatic regions or to various environmental pressures such as disease or insect pests occur as a result of *natural selection.* Natural selection has an important influence on biodiversity in both natural and agricultural systems. Perhaps one of the major differences between an agricultural system and a natural system, however, is the agricultural system's added component of purposeful breeding to produce crop plants that are better adapted to a region, higher yielding, disease-resistant, or perhaps simply more visually appealing. These genetic changes in a plant can do wonders for agricultural production. The danger, however, is in the universal acceptance of these new plants at the cost of losing all of the old varieties, and thus losing the genetic diversity that may be important for breeding programs in the future. Before we can discuss this potential problem, as well as other problems related to exclusive use of one variety of a crop plant, we must first address genetic diversity and basic population genetics.

Genetic variation in agricultural plants is nowhere more visible than in a seed catalog. There are dozens of different varieties, or cultivars, of almost any crop plant imaginable, based on size, color, flavor, disease resistance, days to maturation, and numerous other characteristics. For example, just one seed catalog from a supplier in Virginia touts the following cultivars of tomatoes: Beefsteak, Golden Jubilee, Homestead 24, Long Keeper, Marglobe, Oxheart, Ponderosa Pink, Ponderosa Red, Red Cherry, Roma VFN, Rutgers, Sunray, Yellow Pear, Beefmaster, Better Boy, Big Beef, Big Boy, Big Girl, Celebrity VFN, Early Girl, Empire, Floramerica VFN, Golden Boy, Lemon Boy VFN, Mountain Delight, Olympic, Patio Hybrid, Pilgrim, Pink Girl VF, Small Fry VFN, Supersteak, Super Bush, and Super Sweet 100. All of these varieties are tomatoes, but each is adapted to different environmental conditions, exhibits resistance to a particular pest, or has been bred for a particular market. There is no longer any such thing as a simple "tomato."

The source of this variation is found in the genes of the plants. Each cell in every plant contains a characteristic number of **chromosomes,** which occur in pairs. One set of chro-

Application 8-1 Is Genetic Engineering the Solution to World Hunger?

Optimists say "Yes." Most of the agricultural advancements in the past have relied on technological changes in production methods, increases in fertilizer or pesticide use, or traditional plant breeding techniques. We have barely begun to tap into the kinds of advancements possible with molecular technology and genetic engineering. The potential for yield increases in all staple crops is enormous. With the kinds of methods currently available, we should be able to produce crops that have higher yield, faster growth rates, increased resistance to diseases and pests, and tolerance to a wide variety of environmental conditions, and even higher levels of certain nutrients. We are standing on the threshold of great changes in agriculture.

But before we breathe a great sigh of relief and stop concerning ourselves with world population increases . . . pessimists say "No." While it is true that genetic engineering may be able to improve some crops for certain regions of the world, those crops will still require soil, light, nutrients, and moisture. In many parts of the world, water is scarce, soil is degraded, and soil nutrient content is alarmingly low. Genetic engineering provides a tool for the development of new plant varieties. However, these varieties would be subject to many of the same problems that affect varieties produced by more traditional methods. In time, pests and pathogens should be able to overcome resistance mechanisms (discussed later in this chapter and in Chapter 11). In addition, there is the possibility that new genes introduced into crop plants may eventually enter related weed species, resulting in genetically enhanced weeds. With all of these problems, changes in the genetic makeup of crops may not be sufficient to feed a growing world population for long.

What is the truth? It probably lies somewhere in between the two extremes. Genetic engineering is indeed promising for producing new strains of crops that are well adapted to different regions of the world and may be able to produce good yields even under stressful conditions. These methods can save time by increasing the rate at which new varieties can be produced. But without good stewardship of the agricultural land that now exists in the world, and proper management of **germplasm** (genes available for crop improvement) these improvements will only serve to hold us at a steady state, rather than allow us the increases in production necessary to feed a growing population. The Green Revolution experience (see later in this chapter) has shown that unexpected risks and ramifications, as well as benefits, can arise from well-intended programs and technologies. While the Green Revolution and genetic advances are important for increases in agricultural productivity, it is only with concurrent advancements in the areas of human health and population stability, and improved global resource management, that we will be able to meet the needs of a hungry population.

---

mosomes comes from the maternal parent, and one set comes from the paternal parent. Each chromosome in a given pair contains instructions for the same traits. These are the units of heredity, or **genes.** Each gene for any given trait is found at a certain position **(locus)** along the chromosome, and those genes that occupy the same locus on different chromosomes are called **alleles** (Figure 8-1A). If a pair of alleles (one on each paired chromosome) affects any given trait in the same way, the alleles are homozygous. If the alleles affect the trait in different ways, the alleles are heterozygous (Figure 8-1B).

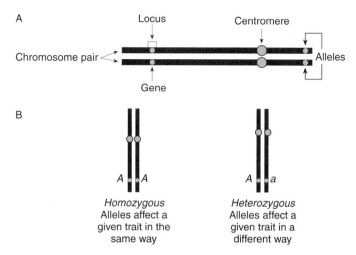

Figure 8-1 Chromosomes occur in pairs, and the alleles on each member of a given pair determine how a trait is expressed. (A) Each gene is located at a particular locus on the chromosome. Genes occupying the same locus and coding for the same trait on different chromosomes are called alleles. (B) If the alleles affect a given trait in the same way (A and A), they are homozygous. If they affect the trait in different ways (A and a), they are heterozygous. The centromere is the region on the chromosome that spindle fibers attach to during cell division, and is also known as the area of primary constriction.

During meiosis, or the production of sexual gametes, the pairs of chromosomes within the parent cells split to produce cells that contain only one chromosome from each pair, effectively separating the alleles for each trait. During this process, however, the alleles can sort randomly onto different chromosomes, providing great potential for variation in the offspring. When pollination of a crop plant occurs, the sexual gametes of two different plants combine. When this happens, the offspring then have different pairs of alleles for each trait, which can lead to differences in appearance, disease resistance, flavor, or other variables that affect the quality of the plant. Such an observable trait of a plant is referred to as the **phenotype**, which is the outward expression of the **genotype**, the internal genetic components of the plant.

The Hardy-Weinberg Law and Genetic Drift

Because every gene has two alleles coding for a specific trait (one from each parent), any individual may be homozygous for the trait (*AA* or *aa*), or heterozygous (*Aa*). If an individual is homozygous for a given trait, all of the gametes it produces will be identical (all *A* or all *a*). If an individual is heterozygous for a given trait, half of the gametes it produces will contain the dominant allele (*A*) and half will contain the recessive allele (*a*). If the gametes from two heterozygous individuals randomly recombine, then, the offspring may be homozygous dominant (*AA*), homozygous recessive (*aa*), or heterozygous (*Aa*) (Figure 8-2).

When a homozygous dominant male (*AA*) is crossed with a homozygous recessive female (*aa*), all members of the first generation of offspring (F1) will be heterozygous (*Aa*) for the

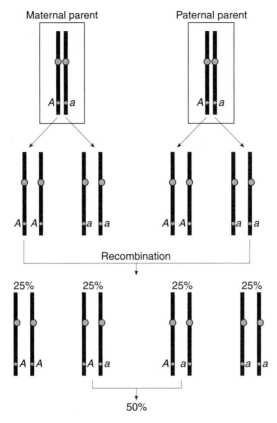

Figure 8-2 Fifty percent of the gametes produced by a parent that is heterozygous for a given trait will contain the dominant allele (A), while 50% will contain the recessive allele (a). If gametes from two heterozygous individuals recombine, the offspring will receive either a dominant or a recessive allele from either parent. Random recombination will produce offspring, 25% of which is homozygous recessive (aa), 25% is homozygous dominant (AA), and the remaining 50% is heterozygous for the trait (Aa).

trait. When the F1 generation produces gametes, 50% of them will contain allele *A* and 50% will contain allele *a*. When these gametes recombine, 25% of their offspring will be homozygous recessive *(aa)*, 25% will be homozygous dominant *(AA)*, and 50% will be heterozygous *(Aa)*. When this F2 generation produces gametes, however, the frequency of alleles will remain the same: 50% of the alleles present in the *population* will be allele *a*, and 50% will be allele *A* (Figure 8-3).

In most situations, the allele frequencies are not 50%–50%. However, since there are only two alleles coding for each trait, the frequencies with which those alleles appear in the population must add up to 1 (100% of the population will contain one or the other of the two alleles). Theoretically, these gene frequencies would remain constant from generation to

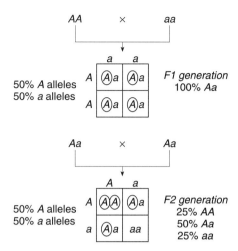

Figure 8-3 A crossing of a homozygous dominant (*AA*) individual with a homozygous recessive (*aa*) individual will produce an F1 generation of heterozygous individuals. Fifty percent of the alleles in the F1 generation will be dominant (*A*) and 50% will be recessive (*a*). If the F1 generation is then crossed, the resultant F2 generation will contain 25% homozygous recessive (*aa*) individuals, 25% homozygous dominant (*AA*) individuals, and 50% heterozygous (*Aa*) individuals. The allele frequency, however, will still be 50% dominant (*A*), and 50% recessive (*a*).

generation, assuming that mating is random, the population is closed (gene flow does not occur outside of this population), the population size is not limited, no natural selection occurs, and mutations do not occur (or if they do, the frequency of mutation from *A* to *a* is the same as the frequency of mutation from *a* to *A*). In this case, the *Hardy-Weinberg Rule* states that in a population at genetic equilibrium (when all of the preceding conditions are met), the proportions of genotypes at one locus will be:

$$p^2 \, AA + pq \, Aa + q^2 \, aa = 1$$

where *p* is the frequency of allele *A* and *q* is the frequency of allele *a*.

Suppose that you wish to examine whether the allele frequencies of a given population will remain constant from one generation to another. You choose to follow a wild population of mustard plants, some of which have alleles that code for wide leaves (*A*), and some that have alleles that code for narrow leaves (*a*). The heterozygous (*Aa*) condition results in a leaf of intermediate width. You start by assuming that the population is at genetic equilibrium. In the population as a whole, the frequencies of *A* and *a* must add up to 1, since only two alleles are coding for this trait. For example, if *A* occurs at 65% of all the loci for the gene in this population ($p = 0.65$), *a* must occur at the remaining 35% ($q = 0.35$) of all the loci ($0.65 + 0.35 = 1$). The proportions of the alleles may vary, but $p + q$ must equal 1.

During meiosis, the alleles in one parent plant separate from each other and segregate into two gametes. Therefore, the allele frequency is also the frequency of gametes that will contain a particular allele (p = frequency of allele *A and* proportion of gametes with allele *A*).

The frequencies of alleles in the next generation will add up to 1 as well (p^2 AA + pq Aa + q^2 aa = 1):

	pA	qa
pA	AA (p^2)	Aa (pq)
qa	Aa (pq)	aa (q^2)

Suppose you now want to see whether the allele frequencies will remain the same from generation to generation in a field of these mustard plants. You examine 2000 of the mustard plants, each of which produces gametes with the same frequency at which the alleles occur at a given locus (assume two gametes/plant). You count:

- 845 AA individuals with wide leaves, which produce 1690 A gametes
- 910 Aa individuals with intermediate leaves, which produce 910 A gametes and 910 a gametes
- 245 aa individuals with narrow leaves, which produce 490 a gametes

The frequency (p) of A among the 4000 gametes = (1690 + 910)/4000 = 0.65, and the frequency (q) of a = (490 + 910)/4000 = 0.35.

The gametes will combine randomly during fertilization. Assuming that the population size of the next generation remains the same,

p^2 AA = (0.65 × 0.65) = 0.42   845 AA individuals
2 pq Aa = (2 × 0.65 × 0.35) = 0.46   910 Aa individuals
q^2 aa = (0.35 × 0.35) = 0.12   245 aa individuals

Using the Hardy-Weinberg Rule, the allele frequencies and genotype frequencies in the succeeding generation hold constant:

$$p^2 AA + pq\ Aa + q^2\ aa = 1$$
$$(0.42 + 0.46 + 0.12) = 1$$

You could continue calculating gene frequencies from generation to generation, but as long as the initial assumptions just given are met, there will be no deviation in the allele frequencies of the population. However, these conditions are *never* met in any natural system. Natural selection occurs, mutation occurs, and population size is limited by resources and competition. In this respect, agroecosystems are no different from natural systems. This becomes obvious when the genotypes and phenotypes that you calculate based on the preceding equation do not show up in your study population in the proportions predicted. At that point, you can examine which of the assumptions are being violated, and begin searching for the evolutionary force that is driving the change in your population.

How does this relate to an agroecosystem? It illustrates one of the major reasons that farmers buy seed from year to year, rather than producing their own seed and using it from generation to generation. If allele frequencies were held constant from generation to generation, we would be able to predict with certainty the genetic composition of each succeeding crop. However, we know that mutation occurs, and outbreeding occurs, etc. Because of this, predicting the outcome of crop production from year to year can be unreliable. While natural selection may work over time to produce a plant that is highly adapted to a particular region, it may not produce a plant with all of the characteristics that make a good *crop* plant. If it is not high yielding, good tasting, and generally of good quality, it will not sell well on the market, no matter how well it does in Kansas during a mid-summer drought.

Natural Selection and Plant Breeding

In natural systems, crosses of plants occur to produce offspring with many different phenotypes and widely variable genotypes. Some of these offspring will have the traits needed to succeed in their environments, while others may lack a needed adaptation. At any rate, natural selection determines which plants survive to pass their genes on to the next generation. For example, suppose that in a field of wild barley, the offspring of two plants have a particular combination of genes that code for resistance to verticillium wilt, a fungal disease. If verticillium wilt becomes a problem in the area, the majority of plants that survive to pass their genes to the next generation will be those plants that have resistance to the disease. In this way, nature has "selected" this gene combination as a trait that will be passed on to other barley plants in the area. Of course, unless verticillium wilt simultaneously becomes a problem in all areas where these plants are grown, other plants without this gene combination will survive in other regions. This is a very important, because some of these plants that lack verticillium wilt resistance may have other favorable traits that could be important for their survival in the future. Genetic biodiversity is critical for the long-term survival of a species.

In agricultural systems, natural selection works hand in hand with artificial selection, or the input of the breeder or farmer. Selection of favorable traits in crop plants is not a new development in agriculture. For centuries, farmers and peasants have been selecting plants with desirable traits and growing those preferentially. Within the past century, plant breeders have been crossing these selected plants with other plants to create new varieties that grow well under certain conditions (see Application 8-2). In recent years, the process of crop improvement has become more complicated as our understanding of genetics increases. Through genetic engineering, scientists are now able to select specific genes and insert them into the chromosomes of certain plants, rather than relying solely on breeding efforts and tedious tests of offspring.

Unfortunately, because of the great success of breeding programs and genetic engineering, farmers in many areas of the world now rely on relatively few varieties of crop plants, many of which have been bred for high yield (see Application 8-3). As the genetic diversity in our agricultural fields decreases, our food supply becomes very vulnerable to diseases and other problems. For example, the famous Irish Potato Famine of the 1840s was the result of a late-blight fungus *(Phytophthora infestans)* that affected potatoes growing in Ireland in 1845–1846. Potatoes are clonally propagated, produced from tuber seed pieces rather than from seeds produced by crosses of two plants, so they have very low genetic diversity. At the time of the potato famine, potatoes in Ireland consisted of only two genotypes, which were introduced to the country in the 1500s. Unfortunately, both clones were highly susceptible to late blight, leading to the mass destruction of the majority of potatoes growing in Ireland, and causing the death of more than 1 million people. More recently, when a new strain of a potato fungus entered the United States from Mexico in the early 1990s, it resulted in serious crop losses from potato blight that exceeded $100 million dollars. Such serious crop losses could be at least somewhat alleviated by planting more than one variety of potato, some of which may have resistance to the disease. Such problems are uncommon in the

Application 8-2 *Hybrid Varieties*

In many cases, desired traits are guaranteed to be present in the offspring of a particular cross when a homozygous dominant parent is crossed with a homozygous recessive parent. In this way, all of the offspring will receive one gene for the dominant trait, and the trait will be expressed in 100% of the resulting population: However, this is only guaranteed for the F1 generation. If the F1 generation is allowed to mate randomly and produce seed, there will be a significant number of F2 plants that do not express the desired trait:

Initial Cross to Produce the F1 Generation

All members of the F1 generation will contain one dominant allele, so all offspring will express the desired trait:

	A	A
a	Aa	Aa
a	Aa	Aa

Hybrid seed is produced from the F1 generation, so that all offspring will show the desired trait.

If the F1 Generation Is Allowed to Reproduce

In the F2 generation, 75% of the offspring will contain the dominant allele and express the desired trait, but 25% will be homozygous recessive *(aa)* and will not express the trait:

	A	a
A	Aa	Aa
a	Aa	aa

If the trait being expressed is resistant to a particular disease, approximately 25% of the F2 population will be susceptible to that disease, which could result in fairly dramatic yield and economic losses for the grower. Buying new hybrid seed each year can prevent that loss and ensure that all plants are resistant.

---

potato-producing regions of South America because farmers there plant a great diversity of potato clones, thus preventing the establishment of epidemics.

Conservation of Genetic Resources. Conservation biologists are quite concerned about the lack of genetic diversity in our food supply. *Gene banks* exist in some areas of the world, including international research centers and research stations, etc. Private organizations are forming in an attempt to preserve some of the genetic diversity of our crops and crop relatives. However, while these efforts are commendable, they may need the backing of national and international governments to coordinate worldwide efforts and movement of germplasm, and to provide the finances necessary to keep gene banks running. Operating

Application 8-3 *The Green Revolution*

Worldwide food production can be increased by one of two ways: We can plant more land to food crops, or we can increase yield on land already planted to food crops. Between 1950 and the late 1970s, total world grain production increased dramatically, even as world population continued to climb. Between 1950 and 1984, food grain production increased at a rate of 2.1% (Brown 1998), while population growth averaged between 1.8 and 2.1% per year (Brown et al. 1994). This astonishing increase in food production, resulting from increasing yields per acre of cropland, was due to what is now known as the Green Revolution.

What Was the Green Revolution?

Beginning in the 1950s and continuing into the 1970s, scientists began to apply technology and management to world agricultural systems. They began by developing new strains of wheat, corn, rice, and other staple crops. These new strains were genetically bred both to produce high yield and to reach harvestable stages faster, but they had very high genetic uniformity. Growers all over the world were encouraged to plant these new strains as monocultures, promising huge economic returns and great increases in regional productivity.

The Green Revolution involved more than crop breeding, however. In order for these potentially high-yielding crops to produce their maximum yields, they needed intensive management. This meant that systems of irrigation, fertilizer management, and pesticide application were necessary, and this generally meant a move away from animal and human labor in fields to increased mechanization. Many small farms began to be swallowed up as larger organizations and wealthier individuals found value in planting large fields to increase farm efficiency. Finally, because higher yields were common, farmers needed better storage and marketing facilities for their crops.

What Were the Advantages of the Green Revolution?

The most significant advantage of the Green Revolution was the dramatic increase in food production during the 1960s and 1970s. Almost 90% of the increase in grain production worldwide in the 1960s was due to Green Revolution principles. Production efficiency was greatly increased. Crop surpluses became common. In developing nations, electricity became more widely available, housing and transportation improved, and people were able to afford more modern conveniences and products.

What Were the Problems of the Green Revolution?

Unfortunately, the Green Revolution did not produce the same benefits in all nations and regions. Prior to the introduction of these supercrops to developing nations, agriculture tended to be very diversified. Farmers planted a wide variety of crops in one small field, and each farm was distinctly different. They were often planted according to different schedules and needs, and growers tended to use different genetic strains for each crop. As a result,

(continued)

while overall regional yields of particular crops were low, farmers had insurance of a sort. If one crop or strain failed for any reason, there were usually plenty of other food crops available for their use. In addition, the farms were usually managed entirely with animal or human power, and farmers did not use inorganic fertilizers or pesticides. This meant that the small farms were completely dissociated from world markets, petroleum prices, and issues of chemical availability. They were, in a sense, economically removed from the rest of the world.

When the Green Revolution began, much of this changed. One genetically uniform strain of a crop was planted on the same schedule over a large region. This extremely narrow genetic base meant that the crop was much more vulnerable to natural disasters such as disease epidemics or insect outbreaks. Rather than having a highly diverse farm where some crops were likely to survive any natural disaster, farmers now risked losing an entire monoculture when a disaster occurred.

In addition to increased vulnerability to disease, farmers were now more vulnerable to global economic changes. Increased capital expenses were common, since farmers needed to add tractors and irrigation equipment to their farms. Increased mechanization also meant increased reliance on petroleum, as did the increased use of fertilizers and pesticides. Because large yields of single crops were produced, they needed to be marketed over a large region instead of being used locally, which necessitated storage and transportation facilities. As inflation and increasing production costs drove prices up, farming became prohibitively expensive for many small growers, who were often forced to abandon their fields to wealthy landowners who could afford the costs of such intensive farming.

An additional problem with the Green Revolution was that it lacked ecological sustainability in some areas of the world due to resulting pollution, soil salinization (from irrigation), or health problems associated with pesticide use.

Assessment of the Green Revolution

It is impossible to unequivocally state that the Green Revolution was good or that it was bad, largely because of its variability of success across the world. In many regions of the world, application of the principles of the Green Revolution produced incredible yield increases that led to better food supply and increased standards of living. In other regions, it must be termed a failure.

Wheat, corn, and rice production have all increased, worldwide, both from increased farmland planted to those crops and from increased yield per land unit. Many countries that were once grain importers, such as India, China, and Vietnam, became self-sufficient, and then exporters following the Green Revolution (although population growth in China is again leading to lower per capita production of grain [Brown 1995]). The worst failures of the Green Revolution occurred in Africa, which has shown a consistent decrease in per capita grain production since the late 1960s. Some of this was due to ecological and natural disasters, and some was due to the unstable political climate in some African nations.

Perhaps one of the most significant legacies of the Green Revolution is the increased dependence on fossil fuels as a result of the industrialization of agriculture. Fossil fuels are needed to provide energy for farming machinery, and to produce the high volume of inorganic fertilizers and pesticides used around the world. Because of this increased reliance on Green Revolution technology, the amount of fossil fuel used to produce each ton of grain has more than doubled since the 1960s.

costs for such facilities are high. Seeds lose their viability over time and need to be occasionally germinated and harvested to keep viable seeds in a gene bank. If this occasional germination does not occur, seeds can lose viability rapidly, and precious genetic material can be lost. Genetic resources are also conserved in less formal settings. In many areas of the world, habitat- and community-based conservation is occurring as growers maintain wild crop relatives in nearby natural habitats, borders of fields, and other locations. Sometimes these efforts are intentional and sometimes not; natural habitats can be an important repository for germplasm of some crops.

Nevertheless, this issue of genetic conservation has potential for serious political controversy. Who owns and controls genetic resources? Can one country or company maintain control over wild relatives of a certain crop if it is the only one with the foresight to conserve them? How much control might one nation have over world hunger in the event of an environmental catastrophe? These questions are worth considering now, before it is too late to take action. Because the development of resistant plant cultivars is a continuous process, scientists are already in the process of developing new cultivars of standard crop plants with different resistance mechanisms to pests and adaptations to different environmental conditions. Therefore, if pests are able to overcome resistance in an existing cultivar, or if an environmental event causes an existing cultivar to experience lower productivity, a new cultivar may be available to meet the needs of growers. For this reason, scientists and growers are continually looking for new sources for resistance and for cultivar development, and it is the responsibility of all nations and all people to ensure that these genetic sources will be available in the future.

Genetic Resistance to Pests and Diseases

The conscientious efforts of plant breeders and farmers over the centuries has resulted in a wide variety of plants that are well-adapted for human use. Most commonly, plants are bred for productivity (increased yield), performance (resistance to diseases, insects, etc.), or some trait that makes the product more marketable to the general public (improved flavor, color, size, or shape). All of our crop plants that we rely on for food and sustenance have been subject to some degree of breeding effort in the past, and will likely be the subject of further breeding effort in the future. The development of new crop cultivars has been accelerated by genetic engineering, which offers the opportunity to introduce new genes (from unrelated species or organisms) into crop plants.

The main objective when breeding for a new cultivar is to accumulate as many favorable alleles as possible in one genotype. Resistance to a pest will not mean much if the final product is unacceptable to the consumer. Similarly, wonderful flavor and appearance of a crop will mean little if it is destroyed by a pest before it can produce its final product. Growers and breeders are therefore well aware that it takes more than one good trait to make a final cultivar that can be widely used and accepted.

Given that public concern over pesticide use has increased in the past decades, and organic fruits and vegetables are starting to appeal to a wider consumer market, plant breeding for pest resistance is becoming more and more important over time. Plant diseases and infestations by insects or nematodes can reduce biomass (and therefore yield) by killing plants, stunting growth, killing branches, or by destroying leaf tissue (Simmonds 1979). In addition, some insects and diseases can lower crop value by reducing overall quality of a product, such as by causing blemishes or rots. In general, all crops are exposed to some dis-

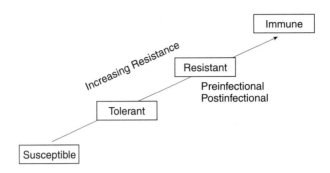

Figure 8-4 The range of host plant response to pests or diseases. Susceptible plants have no defense against a pest or pathogen, while immune plants are not affected by a particular pest or pathogen. Plant response can fall into many categories between these two extremes.

ease organisms or insects over their lifetimes, and some yield reduction will likely occur. Plant resistance becomes essential, however, when an epidemic occurs. An **epidemic** is an outbreak of a disease or insect population that severely reduces yield in an area. If an epidemic goes unchecked and no resistance is available, the disease organism can essentially destroy a crop.

Resistance is usually expressed on an arbitrary scale ranging from susceptible to immune (Figure 8-4). **Susceptible** plants have no defense against a particular disease or pest organism. If an epidemic hits, these plants will decline in productivity significantly. Resistance can take two major forms: resistance related to the complete inhibition of infection, or resistance related to inhibition of the growth of the pathogen once infection has occurred. Within these categories are several other subcategories of resistance:

Tolerant: Plants do nothing to defend themselves against a disease or pest organism, but remain productive despite an infection. In this case, the disease organism can reproduce without restraint, but productivity may not be significantly altered. Many insects and diseases are minor pests because crop plants can tolerate them to some extent. Of course, the negative side to tolerance is that it allows a larger population of the pest to build up in the field, possibly limiting productivity of the next crop used in rotations.

Resistant: Plants may or may not "respond" to an infection, but they have some mechanism in place that actually decreases infection rates. These mechanisms vary in both effectiveness and response time.
- *Preinfectional:* Some mechanism is in place prior to infection that prevents serious damage to the plants. It may be a physical barrier of some sort (thick outer cuticle, waxy coating of leaves and stems, thorns, etc.), or it may be a chemical that the plant naturally produces that is distasteful or poisonous to pests.
- *Postinfectional:* These mechanisms occur in response to infection. The plant may respond by producing a toxic chemical, or may quickly kill off tissue surrounding the introduced organism to prevent further spread.

Unfortunately, since these mechanisms tend to affect reproductive rates of pests and disease organisms, natural selection will often favor those organisms that can reproduce despite the defenses of the plant. Over the long term, this can lead to the ability of the pests to "break" the resistance of the plant, making the resistance mechanism relatively obsolete.

The Genetics of Resistance

Resistance to pests or pathogens may be due to either one or more major genes, or to polygenes. Resistance conferred by *major genes* is usually dominant, often gives immunity to the plant, and is usually very specific to certain pathogens. **Polygenic resistance** involves many genes, each having small individual effects, but collectively providing resistance to the plant. Polygenic resistance is often less specific to pathogens.

Major-gene resistance is often referred to as **vertical resistance.** Since it is highly pathogen-specific, its presence in a plant determines whether the plant will be completely immune or completely susceptible to an epidemic. This resistance is very effective at preventing damage or drops in productivity, even in epidemics. However, if the pathogen is slightly different from that which the plant is resistant to, disaster can ensue. On the other hand, polygenic resistance provides a plant with **horizontal resistance.** Because many genes are responsible for the overall resistance effect, plants with horizontal resistance have some level of resistance to many different pests. Plants can have some level of both horizontal and vertical resistance. In fact, if a plant with vertical resistance to Pathogen A is hit by Pathogen B, the effects of vertical resistance may be irrelevant, and the crop will be saved or lost on the basis of its horizontal resistance. If, however, the plant is hit by Pathogen A, vertical resistance will provide immunity to the crop.

One cannot breed for both horizontal and vertical resistance in the same plant, since one cannot select for horizontal resistance in the presence of vertical resistance (Simmonds 1979). Most breeders are therefore choosing to move away from vertical resistance, and are concentrating on horizontal resistance. Indeed, it is the horizontal resistance inherent in plants that allows most crops to survive and adapt to pest and disease problems worldwide, particularly where monocultures of the same genetic makeup are avoided.

It must also be remembered that resistance is only one characteristic that breeders work with. The challenge in breeding resistant plants is to produce plants that are not only hardy and able to withstand a particular pest in the field, but also plants with high yield, quality, and other advantageous characteristics. In general, it is probably better for plants to have a little resistance to several major diseases, rather than to one particular pathogen.

When Resistance Can Be a Problem

For most common crop plants, resistant cultivars are available against many different kinds of plant pests. Growers can also rely on a variety of chemicals to protect their crops from destruction if the cultivar they are growing is susceptible to a pest that occurs in the area. However, the pests themselves are also genetically diverse, and are well able to adapt to their environment, which may mean overcoming the resistance of the cultivars themselves, or developing resistance to pesticides. Once that occurs, the method used to prevent pest outbreaks is often rendered ineffective. For this reason, it is important to minimize selection pressure in order to slow development of resistant pests. This can be done by a variety of ways, including integrated pest management (see Chapter 11), reduced dosage of pesticides, or rotation of pest management methods. Examples of resistant pests and their management are discussed in Chapter 11.

Genetic Diversity and Crop Breeding

The development of resistant or high-yielding cultivars is a costly one, in terms of time, resources, and genetic diversity. All crop species available today have descended from wild relatives, each having its origin of diversity in some particular region. Over time, these wild relatives have been selected for high performance or for resistance to a particular pathogen or environmental problem, making them well adapted to other areas of the world. As these plants were being grown as crops, growers began to select for those traits that were most important to them. This eventually led to more intensive plant breeding efforts, which have given us most of our popular crop cultivars.

As this process progressed, agronomic performance of the cultivars increased. The plants were selected for those traits most important for high yields and high quality, leading to cultivars that are widely used in the United States and abroad. However, along with this increase in agronomic performance came a decrease in the genetic diversity of the crop. The wide genetic base of the wild relatives was no longer an asset since it often meant that yields were not as high as they could be. Therefore, while the new cultivar has many desirable attributes (e.g., high yield, resistance to a particular pest), it may become susceptible to other problems because of this loss of genetic diversity. Fortunately, most breeding programs that exist today now recognize this as a potential problem, and include a broader genetic base in any new cultivars produced. The unfortunate exception to this is found in those plants that are commonly vegetatively or clonally propagated (e.g., banana). Because recombination never occurs, the offspring are genetically identical to the parents, resulting in very little genetic diversity, and therefore a very narrow genetic base.

Because genetic diversity within a plant cultivar does not necessarily go hand in hand with high productivity, scientists now advocate the maintenance of genetic diversity in a region, rather than in a specific cultivar (Figure 8-5). This can be done on a single farm, if mixed crop genotypes are planted. In this case, genetic diversity will be high *within* the plant population in a field. This is often done in tropical countries, with a single farmer planting several varieties of one crop in a field. Alternatively, regional diversity can be maintained if different farms plant different cultivars, leading to high genetic diversity *among* plant populations. This is common in some U.S. crops such as corn. When farmers in an area plant different cultivars of corn, high diversity can be maintained in the region, even though each farm or field is homogeneous.

This kind of regional diversity has marked advantages: Mixed genotypes of a crop can provide a wealth of different genes that can serve as a future source for developing new resistant cultivars. In addition, these mixed crop genotypes can slow epidemics, reduce yield loss due to invasion of new pests, and can often withstand serious biological and environmental events. Their use provides a kind of insurance against overall crop failure. However, yields of mixed genotypes are often lower than yields of crops selected for high yield in a particular region, which can result in decreased revenue for the grower. These mixed genotypes can also be less effective than resistant cultivars against single pests that may be common in a particular region.

Herbivory and Plant Defense Mechanisms

Herbivory in an agricultural system involves the removal of plant tissue such as leaves, stems, bark, or roots. This activity may not kill the plant, but it will certainly affect its fitness and productivity, and possibly its ability to survive. Loss of foliage may limit the ability of the plant to harvest photosynthetically important sunlight, thus decreasing its carbon

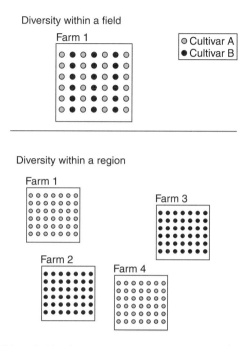

Figure 8-5 Diversity within a field refers to the planting of several different cultivars or species in a field on a single farm. Diversity within a region refers to the planting of several different cultivars or species on many different farms within a region.

stores and ultimately lowering its ability to produce the harvestable portion of the plant. When plants are defoliated by herbivores, they respond with a flush of new growth that ultimately drains nutrients and carbon from reserves that would have gone into growth and reproduction. Herbivores that concentrate primarily on young leaves remove considerable amounts of nutrients from the plant. Some insects such as aphids penetrate the phloem of a plant and literally suck the sap of the plant, significantly reducing growth as well.

Herbivores can be divided into generalists and specialists. *Generalists* feed on many different kinds of plants and even different parts of plants, although they may have an order of preference for what they will consume. *Specialists* tend to feed on specific plants or parts of plants, and may have various methods to overcome specific resistance mechanisms. Of these, specialists are perhaps more common and more devastating to agroecosystems, but generalists can cause serious damage as well, and can often survive on neighboring plants until a field is replanted to a new crop.

Herbivory remains an important problem in agroecosystems because many of the production systems in place today are monocultures. If an adult European corn borer lays its eggs in a field that is subsequently planted to a susceptible variety of corn, there will be ample food available for all of its offspring, as well as a few generations of descendants. Insects traveling in swarms, such as locusts, can rapidly defoliate huge areas of susceptible plants in very little time. Herbivore damage is one of the most common means of yield reduction in agricultural systems today.

When grazing by an herbivore first occurs, the biomass of the vegetation in a field will decrease in proportion to the amount removed by the herbivore, which is a function of the number of herbivores and their rate of consumption. This is why a few insects in a field will not likely cause economic damage to a crop, but the arrival of thousands of insects can cause rapid mass destruction. In addition, as the herbivores continue to graze, their numbers will likely increase as reproduction occurs, which will often lead to an overgrazing of their environment. In natural systems, this could result in the extinction of the plant, or the plant may eventually recover if the herbivores move on or if they themselves are controlled by predators (see Chapter 9). In agricultural systems, however, this can be a disaster. Severe defoliation and grazing of agricultural plants can lead to enormous reductions in yield and productivity, and can also result in low-quality products that are either unacceptable to markets, or can only be sold at very low prices for feed or factory use.

Because plants are immobile and are therefore at a disadvantage in an attack by herbivores, some plants have derived different modes of defense to protect themselves. Primary defense mechanisms are quite common, and generally require very little energy output. These mechanisms are used to make penetration or attack by herbivores difficult. They include tough leaves, spines or thorns, hairs on leaves, or hard seed covers. Secondary defense mechanisms tend to be more energetically costly, and often involve chemical defense, such as the accumulation of a wide range of products that may be toxic or untasty to herbivores. In some cases, these chemicals are produced upon attack, but may be present to inhibit attack in other plants.

If this is the case, could we simply breed these defense mechanisms into all of our crops and let the plants themselves fight it out with their attackers? Why does it seem that instead of being less susceptible to insects, many of our crops are more susceptible than their natural relatives? The answer is that we, as humans, are also herbivores in a sense. Many of the defense compounds that would be distasteful to herbivores feeding on plants would also be distasteful or even toxic to us. Therefore, agriculturally important crop plants are selected for production of low levels of these compounds, rather than high levels. This, of course, makes them even more susceptible to herbivory.

Summary

Conservation of genetic biodiversity is critical to maintain high agricultural productivity levels in the future. A variety of plants with different resistance mechanisms, abilities to withstand various environmental conditions, and many different desirable traits are needed to ensure that breeders will have great amounts of genetic diversity to incorporate into crop plants in the future. Genes that provide plants with resistance to herbivores are frequently incorporated into crop plants in an attempt to increase productivity by decreasing loss to herbivory.

Topics for Review and Discussion

1. What is the difference between horizontal and vertical resistance? Which would you rather have incorporated into a plant if you have a field that is always infested with a particular plant pest? Which do you think would be more effective against generalist herbivores and why?
2. Why is the conservation of genetic diversity so important? What could happen in the future if we allow different cultivars of crops and their wild relative to go extinct?
3. What is inbreeding depression? How does it pertain to agroecosystems?

4. From the point of view of a single herbivorous insect, what resistance mechanisms by plants would be most difficult to overcome? Easiest? Does your answer change when you consider the entire population? Why or why not?
5. *Meloidogyne incognita,* a plant-parasitic nematode, invades plant roots and feeds on plant cells from within the roots. The number of nematodes within a root can be estimated by counting the number of galls that appear on root surfaces (see Chapter 10). You plant several varieties of tomato, and observe the following results:

Tomato Variety	Number of Galls/Plant	Yield (kg/ha)
A	20	240
B	55	60
C	450	20
D	500	238

For each of the varieties given, state whether they appear to be susceptible, tolerant, or resistant to *Meloidogyne incognita* and why. If they are resistant, do you think that they are using primary defense mechanisms or secondary defense mechanisms? Why?

Literature Cited

Brown, L. R. 1995. *Who Will Feed China?* New York: Worldwatch Institute.
Brown, L. R. 1998. Struggling to raise cropland productivity. *State of the World 1998.* New York: Worldwatch Institute.
Brown, L. R., H. Kane, and D. M. Roodman. 1994. *Vital Signs 1994.* New York: W. W. Norton.
Myers, N. 1983. *A Wealth of Wild Species.* Boulder, Colo.: Westview Press.
National Research Council (NRC). 1992. *Managing Global Genetic Resources: The U.S. National Plant Germplasm System.* Washington, D.C.: National Academy Press.
Simmonds, N. W. 1979. *Principles of Crop Improvement.* Essex, England: Longman Scientific & Technical.
Worldwide Fund for Nature. 1989. *The Importance of Biological Diversity.* Gland, Switzerland: World Wildlife Fund.

Bibliography

Collins, W. W., and G. C. Houtin. 1999. Conserving and using crop plant biodiversity in agroecosystems. In *Biodiversity in Agroecosystems,* eds. W. W. Collins and C. O. Qualset, 267–282. Boca Raton, Fla.: CRC Press.
Daily, G. C., S. Alexander, P. R. Ehrlich, L. Goulder, J. Jubchenco, P. A. Matson, H. A. Mooney, S. Postel, S. H. Schneider, D. Tilman, and G. M. Woodwell. 1997. *Ecosystem Services: Benefits Supplied to Human Societies by Natural Ecosystems.* Issues in Ecology Series, no. 2. Washington, D.C.: Ecological Society of America.
Gollin, D., and M. Smale. 1999. Valuing genetic diversity: Crop plants and agroecosystems. In *Biodiversity in Agroecosystems,* eds. W. W. Collins and C. O. Qualset, 237–265. Boca Raton, Fla.: CRC Press.
Klug, W. S., and M. R. Cummings. 1999. *Essentials of Genetics.* Upper Saddle River, N.J.: Prentice-Hall.
Primack, R. B. 1995. *A Primer in Conservation Biology.* Sunderland, Mass.: Sinauer Associates.
Wilson, E. O. 1992. *The Diversity of Life.* Cambridge, Mass.: Harvard University Press.

Chapter 9

Predation and Parasitism

Key Concepts

- Predator-prey dynamics
- Foraging efficiency by predators
- Parasitism and biological control

Interactions

Recall from Chapter 3 that a food web exists in all ecosystems, and energy is thus passed from member to member according to a hierarchy of producers and consumers. We have seen the effects that herbivores can have on plants (the primary producers), and now turn our attention to the interactions that occur farther along the food web when higher-level consumers begin to impact populations of herbivores.

Both predation and parasitism can play important roles in the development of agroecosystems. As we shall see in this chapter, the interactions that occur between herbivores and the predators and parasites that consume them can directly affect crop growth and productivity. In addition, these effects can be altered radically by management methods that are chosen and employed by farmers and growers.

Predation and parasitism differ from each other in one fundamental way. *Predators* feed on other living organisms (prey), but never take up residence in or on them. Predation generally results in the death (consumption) of the prey. On the other hand, **parasites** take up residence in or on their prey, feeding on living tissue of the host animal, and may or may not kill their host animal. The **host** is the organism infected by a parasite. Some parasites live on animal hosts, while others live only on plant hosts. The **parasitoids** are intermediate between the extremes of predation and parasitism. Parasitoids are insects that lay their eggs within the bodies of developing insects. As the parasitoid eggs develop, they feed on the tissue of the developing larvae. This always results in the death of the host. All of these interactions occur within agroecosystems, and their importance at any given time varies according to what insect species are present in the field at any given time.

Why are these interactions important in agroecology? After all, the predators and parasites feed on other animals, and we are primarily interested in plant productivity. So why would the presence of communities of predators or parasites make a difference? The answer is found in **biological control,** a pest management method used widely throughout the developing and developed worlds.

Biological control has been practiced for centuries by growers who have recognized that when natural enemies of plant pests are present in agricultural fields, the damage due to those pests is significantly less than if the natural enemies are not present. Within the last few decades, biological control has received considerable attention from scientists and growers in North America. Recently, there has been a great deal of research conducted on the uses of biological control in agriculture. While many kinds of pests have been studied, the majority of research attention has focused on insect control.

Despite renewed interest in biological control as a pest management system, it is important to keep in mind that it is based on the ecological principles of parasitism and predation. The introduction of a beneficial insect to a field to prey on an insect that is harmful to crops will work only under certain conditions. By developing a good understanding of the ecology of predation and parasitism, you will be better prepared to develop a working biological control system in most situations.

Predation

Predation in any ecosystem involves more than the effects of predators on prey abundance. The prey themselves can influence predator abundance and activity, and competition among predators can lead to complex interactions within the system, depending on the specificity of the competing predators. Simply attempting to eradicate all prey species from a field by introducing all of their natural enemies may actually lead to decreased productivity in the system if the most competitive natural enemy also preys on insects that are keeping other plant pests in check.

In agroecosystems, as in any natural system, predation occurs when an organism of one species feeds on an organism of another species, assimilating its energy and nutrients for its own use. For simplicity of discussion, we limit our attention to generalities concerning insect populations, but we do look at specific examples of predation as a biological control mechanism in Application 9-1. Predation involves two groups of organisms: predators and prey. We cannot discuss the dynamics of one without also considering the dynamics of the other.

We can look at the dynamics of predator-prey interactions mathematically using the Lotka-Volterra model, in which the populations of predators and prey species are assumed to be directly related. This model is admittedly simple, but serves as a good initial point of reference. With this model, two paired equations are used to examine predator and prey population densities. Recall from Chapter 6 that a population will increase exponentially, provided that there are sufficient resources available and no predators to limit its growth:

$$\frac{dN}{dt} = rN$$

However, when predators are present, prey individuals will be consumed by the predators at a rate proportional to both the population size of the predators and the population size of the prey. It will also depend on the efficiency of the predators at finding and attacking their prey (consumption rate). The equations used to describe predator and prey population dynamics, then, must take into account the rate of population increase of the one species

Application 9-1 *Examples of Some Predators Important in Biological Control*

Ladybird Beetles

Ladybird beetles (Figure 9-1) belong to the insect order Coleoptera, and are general predators that are quite widely distributed throughout the world. As adults, they are approximately 5 mm long, oval-shaped, and have orange or red wings with brown or black spots. Eggs of ladybird beetles are typically laid in clusters on leaves, are yellow to orange in color, and are elongated. Larvae tend to be grey in color, brightly spotted, and rather flat-looking.

Both adult and young ladybird beetles feed on mites and a wide range of insects including aphids, whiteflies, scales, thrips, and mealybugs. Many scientists consider ladybird beetles to be one of the most beneficial of all natural enemies in the world. Since ladybirds feed on pollen and flower nectar during times when few pests are present, it is not difficult to keep a good population of these natural enemies in a field.

Figure 9-1　Ladybird beetles are one of the most beneficial groups of insects in the world.

Green Lacewings

Lacewings belong to the insect order Neuroptera. Adult lacewings are approximately 2.5 cm long, with a green body, lacy wings, long antennae, and large, golden compound eyes. Eggs are usually deposited at the ends of tiny stalks on vegetation, are pale-white, and measure about 1.5 mm in diameter. Larvae range in size from about 1.5–3 mm long, have long jaws (earning them the common name of "aphid lions"), and have bodies that taper down to the tail. They live on vegetation and feed on aphids, mites, whiteflies, and other soft-bodied insects. While green lacewings are very good predators, they are unfortunately also very good prey. They are a common food of other predators, and of each other. If released in a field, they should be released far from other predators and spread around the field, rather than placed all in one spot.

in the absence of the other, and the rates of population change that occur as a result of interaction between the two species. For prey population growth, then:

$$\frac{dN}{dt} = aN - bNP \qquad \text{(Equation 9-1)}$$

For predator population growth:

$$\frac{dP}{dt} = cNP - dP \qquad \text{(Equation 9-2)}$$

where N and P are population densities of prey and predators, respectively, a and d are the rates of population change in the absence of the other species ($a = r$ for the prey; d = death rate for the predator), and b and c are the rates of change in population growth that occur as a result of interaction between the two species.

These changes in predator and prey populations can be depicted graphically (Figure 9-2). If the number of predators and prey population are plotted against time, the Lotka-Volterra model predicts that prey populations increase as predator populations decline. When the prey populations build up, they can support a larger predator population, so the predator population begins to increase again, thus driving down the prey population. As the prey population decreases, less food is available for the predators, which therefore decline in number, again leading to an increase in prey populations.

This model, however, involves a number of assumptions that are seldom met in any natural or agricultural system. The model assumes that: 1) the prey population will increase exponentially in the absence of predators; 2) the predator population will decrease exponentially when no prey are present; 3) predators are not limited by their own ability to search for and capture prey, regardless of prey population density; and 4) there is no lag time for handling, consuming, and digesting prey. In reality, factors such as age structure of predators, interaction between prey and their food supply (in this case, the plants in the agricultural field), ability of predators to find and consume prey, the mortality of predators due

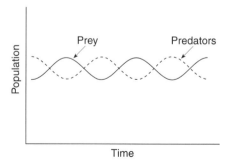

Figure 9-2 Changes in predator and prey populations over time. As predator populations increase, more prey will be consumed and their populations will decrease. As prey populations decrease, fewer resources will be available to support a large number of predators, leading to a decrease in the number of predators, which will again allow an increase in prey species.

to crowding and lack of resources, genetic changes, predator specificity, and immigration/emigration can all affect population growth of predators and prey. Nevertheless, the Lotka-Volterra equations are somewhat useful for illustrating the tendency for prey and predator populations to interact directly with each other.

The Lotka-Volterra equations suggest that as prey density increases, predators may react in one of two ways, depending on their foraging efficiency. If predators are primarily limited in their consumption of prey by their ability to find the prey, an increased prey density may simply increase their ability to find prey. In this case, their consumption rate will increase in proportion to prey density until satiation is reached. For example, if ladybird beetles are released into a field containing aphids, and the only thing limiting the beetles' ability to consume the aphids is their ability to locate their prey, an increased number of aphids in the field will increase the likelihood of an encounter, so the beetles will be able to find and consume more of their prey. This is known as a **functional response.** In agroecosystems, this means that pests could potentially reach a level that is far too great to be controlled by the number of predators present in a field. Effective control would be possible only if more predators were introduced to the system. Functional response has been classified into three different types by Holling (1959):

Type I response: This response is most similar to the response expected from the Lotka-Volterra model. In an agricultural field, Type I response would be indicated if, as the number of herbivorous insects (prey density) in the field increased, the number of those insects eaten by each individual predator per unit time increased linearly to a maximum satiation level (Figure 9-3). This does not take searching and handling time into account, but simply assumes that the predatory insect will immediately eat the herbivorous insect and continue to its next prey item.

Type II response: This response would be indicated if, as the number of herbivorous insects in a field increased, the number of those insects eaten by each individual predator per

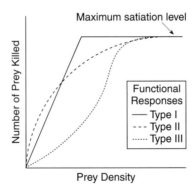

Figure 9-3 Classification of functional response of predators in response to prey density. Type I response: As prey density increases, the number of prey eaten by each predator will increase proportionately until the predator is satiated. Type II response: As prey density increases, the number of prey eaten by each predator will increase, but that rate of increase will slow over time, allowing for handling time. Type III response: As the prey density increases, the number of prey eaten by each predator will increase in a sigmoidal fashion, allowing for both search and handling time. (After Holling 1959)

unit time increased, but at a *decreasing rate,* until a maximum satiation level is reached (Figure 9-3). This recognizes that the predators need to spend time pursuing, capturing, and consuming each of the insects they encounter, leading to an increased handling time as more prey items are found.

Type III response: This response would be indicated if, as the number of herbivorous insects in a field increased, the number of those insects consumed by predators was low at first, then increased in a sigmoidal fashion until a maximum satiation level was reached (Figure 9-3). This response is similar to a Type II response, but recognizes that search time for the herbivorous insect is greater at lower prey densities (they will be fewer and harder to find) than at higher prey densities (when handling time may become the limiting factor).

However, functional response assumes that predators are searching at random for prey that are evenly distributed throughout an area. In reality, prey are quite unevenly distributed in a field. Large pockets of prey tend to produce patchy distribution of their predators as well, leading to an **aggregative response.** In an agricultural field, this may mean that predators will tend to congregate in areas of high prey density. This would be beneficial to prevent massive crop destruction in an area that is heavily populated with a crop pest, but may also lead to more widespread (but far less intense) damage across the remainder of the field, since small populations of the pest would be unlikely to attract predators.

Nevertheless, if a predator is not limited by its ability to find prey, an increased number of prey in the field will simply support a larger population of predators over time. In some cases, more predators will occur as a result of immigration into the region. In other cases, the life cycle of the predators may speed up, resulting in higher reproduction that leads to a larger population of predators. This is known as a **numerical response.** Higher predator populations can result in better control of pest populations, although once the pest population has decreased significantly, there will no longer be enough food to support such a high population, unless the predator is able to *switch* to a different prey species.

Switching occurs when a predator changes its diet in response to prey abundance. When the abundance of its preferred prey decreases to the point that the predator is no longer able to hunt for it profitably (in other words, it spends more energy seeking and handling the prey than it obtains from the prey itself), it may change its diet to concentrate on a more abundant food source. Those predators that feed preferentially on only one food source are known as *specialists;* those that consume many different kinds of prey, largely based on abundance, are known as *generalists*. Most predators have relatively wide diets, which is fortunate for farmers who rely on the presence of natural enemies to keep pest populations at a manageable level. If all predators were specialists, they would die out when their prey populations decreased to very low levels; if the prey then resurged, there would be no natural enemies present to control them. Since many predators are generalists, they are often able to move on to an alternative food source, and are therefore present in fields (though often in low numbers) when a pest population resurges.

Diet Width and Optimal Foraging Theory

Suppose that your agricultural field has several insects that are potential food sources for a predator, some of which are serious pests to your crop and others that are fairly harmless—which insect species will the predator choose to eat? The answer depends to a large extent on the predator's foraging strategy.

In general, animals are able to consume a wider range of foods than they actually choose to eat. The actual diet of the animals is determined by many factors, and may be predicted by **optimal foraging theory,** which was first introduced by MacArthur and Pianka in 1966. They suggested that a predator must expend energy both to *search* for its prey, and to *handle* it after finding it (the predator must still capture its prey, subdue it, and consume it). While searching for one particular prey species, a predator may come across individuals of other species. The response of the predator to these alternative food sources influences the diet width of the predator, and, in turn, determines how much damage a pest may do to a crop when other insects are present in the field.

Generalists tend to handle most of the prey that they find in their search. This cuts down on search time, but increases handling time since both profitable and unprofitable prey species are handled (Figure 9-4A). Specialists tend to ignore prey species if they are not a preferred species. In this case, search time is much greater, but they do not waste time handling prey species that are unprofitable (Figure 9-4B). The optimal diet of any predator, then, would include less profitable food items only when the energy obtained from the food source is more than the energy required for handling that food item (in other words, it must increase the overall rate of energy intake of the predator).

If the predator in your agricultural field feeds preferentially on the pests that could seriously damage your crops, it will likely continue to feed on those pests, as long as it is energetically efficient for it to do so. Once those pests decrease in number, the predator may choose to move on to a less profitable species, but one it may find in more abundance. In this case, the predator will still eat the serious pest species as they are encountered, but may

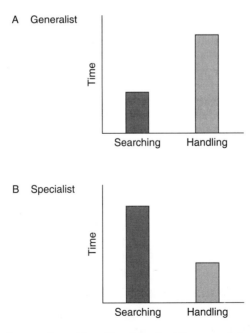

Figure 9-4 Differences in search and handling time for (A) generalist and (B) specialist predators.

spend less time hunting for them specifically. Therefore, management of these pests is likely to be quite good, since they are the preferred choice. If, however, the predator in your field feeds preferentially on some of the less innocuous species of insects and only feeds on your serious pests when it can no longer encounter its preferred prey, management may be less than optimal, since serious damage to the plants can occur before the predator runs out of its preferred food source and turns to other serious pests. For this reason, it is important that releases of natural enemies into a field for the purpose of biological control are managed carefully. Accurate identification of your pest species is essential in choosing the best natural enemy for the job, and the food choices of these potential predators should be considered before release (see Application 9-1)

Foraging Efficiency

Predatory insects in an agricultural field are not confined to that field. They can easily migrate from field to field in search of prey. How long they choose to stay in a field depends on how much food is available to them at that location. The **marginal value theorem** (Charnov 1976; Parker and Stewart 1976) suggests that the length of time that a predator remains in a food patch depends on how much food is available to it, time spent in travel, and time spent to obtain prey. In other words, the length of time that beneficial insects remain in the field where they are released depends on how many prey insects are present to keep them fed, how far the next field with a good prey population is from the target field, and how much time the insects need to spend searching for and handling prey. Energetically speaking, if a predatory insect stays too long in a field, it will deplete the prey to the point at which it is no longer energetically profitable to search and capture prey. If the predator leaves too soon, it has not used the resources in that field efficiently.

Because insects migrate, and will often leave a field when prey density reaches a certain minimum threshold, there are several factors to be considered when releasing beneficial insects into an agricultural field. First, it may be necessary to release insects into a field at more than one time during the growing season. The first release may decrease the population of plant pests to a fairly low level, but it is unlikely that the pests will be eradicated totally from the field. Once the beneficial insects move on, the pest population can resurge, creating the need for a second release of beneficial insects to prevent serious crop damage. Alternatively, many growers choose to create borders around their fields that contain a variety of different plants from many different families. In this way, they hope to attract many different kinds of insects, some of which may serve as an alternate food source to the beneficial insects when pest populations in the field have been depleted. Because some of the beneficial insects may then be able to live in the area, they will be available immediately to prey on pest insects once pest populations begin to build up again.

Parasitism

Parasites differ from predators in that they may or may not ultimately kill their hosts. Parasitoids are parasitic in the immature stage, but free-living as adults. Unlike parasites, parasitoids always kill their hosts, although the host may be able to complete much of its life cycle before dying. In agricultural systems, some parasites may be beneficial agents of biological control, but others may themselves be pests, causing many of the most serious plant diseases in the world. Plant diseases such as bacterial galls, fungal wilts, and viruses are caused by "parasitic" organisms that invade and destroy plant tissue, decreasing productivity, and often causing serious problems for growers. Plant-parasitic nematodes are responsible for huge losses in yield in some parts of the world. Even some flowering plants such as dodders are parasitic on other higher plants.

Parasites and Parasitoids as Beneficial Control Agents

In agricultural systems, parasites such as insects and nematodes frequently feed on plant pests. Parasitoids are often more beneficial, since many species kill insect larvae before they can mature or reproduce and cause serious damage to a crop species. For many species of parasitoids, eggs are laid within the body of a host where they develop and feed on internal tissues, eventually killing the host. Of course, this means that the life cycle of the parasitoid must be closely synchronized with the life cycle of the host insect for the parasitoid to successfully complete its reproductive cycle. Some specific examples of parasites frequently used as biocontrol agents are discussed in Application 9-2.

Application 9-2 *Examples of Some Parasites Important in Biological Control*

Insect-Parasitic Nematodes

Nematodes in the families Steinernematidae and Heterorhabditidae are parasites of insects. They enter insect hosts through natural openings, including the mouth, anus, or spiracles, and penetrate the body cavity. Upon entering the body cavity, the nematodes release symbiotic bacteria into the insect, which develop on the insect tissues and cause an infection known as **septicemia,** which rapidly kills the insect. The nematodes reproduce within the insect, and the offspring continue to feed on this bacteria until resources begin to decline. At that point, the nematodes enter a resistant stage known as a **dauer** stage, which provides the worms with an extra protective cuticle, and exit the body of their host, looking for new hosts. Under humid conditions, a nematode in the dauer stage is able to survive for long periods of time outside of a host. Once it finds and enters a new host, it sheds its extra cuticle and the cycle repeats.

Nematode parasites of insects kill their host so rapidly that they effectively prevent or minimize damage from the parasitized insect. Such a rapid infection also can prevent the reproduction of the host, which also serves to decrease insect damage to plants over the long term.

Braconid Wasps

Braconid wasps belong to the insect order Hymenoptera, and the family Braconidae. The wasps are brownish to black in color, and are quite small (1.5–10 mm in length). The young larval stages of these wasps are parasitic on aphids, or on larval stages of many butterflies and moths. These parasitoids deposit eggs directly on the host; the larvae develop, cut through the exoskeleton of the insect, feed, and complete pupation within the host. A fully developed adult will eventually emerge. Like parasitic nematodes, braconid wasps effectively halt the life cycle of their hosts, preventing both further plant damage and reproduction.

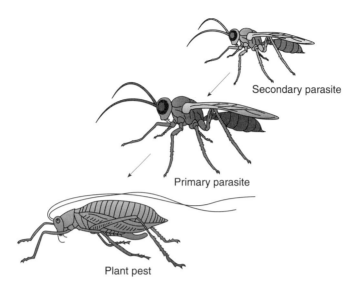

Figure 9-5 Hyperparasites (secondary parasites) are parasitic on primary parasites, which are themselves parasitic on plant pests.

Hyperparasitism: Parasites of Parasites

In some cases, insects may actually be parasites of parasites of plant-feeding insects! This is known as **hyperparasitism** (Figure 9-5). In this case, the natural enemy of the plant pest is called the *primary parasite,* and the hyperparasite is referred to as the **secondary parasite.** Because secondary parasites can weaken or kill natural enemies of plant pests, they are not usually beneficial to an agroecosystem.

Parasites as Crop Pests

Parasites are defined as organisms that derive their nutrients from one or more hosts, normally harming them but not necessarily killing them. We can further classify such organisms as microparasites or macroparasites. **Microparasites** are organisms that multiply directly within their host, usually within host cells. **Macroparasites** grow in their host, but multiply by producing infective stages that are then released from the host to infect new hosts. They commonly live between host cells of plants, rather than within them (May and Anderson 1979).

Common microparasites include bacteria, viruses, some fungi, and protozoa. They may be transmitted by wind, water, contaminated farm equipment, or by vectors. Some viruses are spread from plant to plant by insects (e.g., aphids) or nematodes. Macroparasites include many higher fungi, nematodes, and some plant species. Transmission may be by wind, water, contaminated farm equipment, or direct movement of the parasite from plant to plant. Some of the more common plant parasites and their characteristics are as follows:

Bacteria: Bacteria are single-celled organisms that can be parasitic on a wide range of
organisms from plants to insects to livestock. Bacterial diseases can be particularly

devastating because one, single bacterial cell can produce an enormous number of cells in a very short time. As they multiply, the cells induce chemical changes in their environment that lead to the common disease symptoms associated with bacterial infections. In agricultural systems, bacterial diseases tend to occur under fairly moist and warm conditions, and affect all kinds of plants.

Fungi: Parasitic fungi can cause a variety of disease symptoms on plants, including powdery mildew, blights, and wilts. Almost all fungi that are pathogenic to plants spend part of their lives on their host, and the other part in soil or plant debris. Many fungi cause serious damage to the plants they infect, often blocking the vascular system or secreting toxins that interfere with plant function while simultaneously feeding on host plant cells. As with bacteria, fungi tend to thrive in moist conditions, and have a wide host range. In addition to parasitizing plants, some fungi are parasitic on nematodes and insects, and thus can act as biological control agents against these plant pests.

Viruses: Viruses are unique in the world of parasites because they do not divide or produce reproductive structures to spread from host to host. Instead, they induce host cells to form more of the virus by altering the genetic code of the plant, and rely on vectors such as aphids or nematodes to transmit them from host to host. Viruses, therefore, multiply only in living hosts, and cause disease by upsetting the metabolism of their host cells, which leads to the development of metabolic products that are abnormal for the plant.

Nematodes: Plant-parasitic nematodes are microscopic roundworms that feed on roots, stems, or bulbs of plants. They have long, needle-like stylets that they insert directly into host cells to remove nutrients and water. Nematodes can also be parasitic on insects, and are occasionally promoted as biological control agents for soil-dwelling insects or larvae.

Parasitic flowering plants: These plants can be further classified as holoparasites or hemiparasites. Holoparasites have no chlorophyll, and thus rely completely on their host for nutrients, water, and carbon. Hemiparasites do have photosynthetic cells, but may rely completely on their hosts for water and nutrients. These plants tend to have a poorly developed root system or none at all. Fortunately, these parasites are not very common in agricultural systems.

Parasites of Animals

Many different kinds of organisms (e.g., tapeworms, nematodes, flukes, etc.) can be parasitic on livestock, and can lead to significant losses in production, particularly in warm climates. Despite the availability of many drugs to combat these internal parasites, the parasites continue to thrive in many regions of the world. Resistance to these drugs is also growing, but producers are beginning to rely on other methods of management including sanitation and hygiene. Examples of some animal parasites are discussed in Chapter 10.

Biological Control

Biological control includes the manipulation of predators, parasites, and pathogens to maintain pest populations in a field at levels below those that will cause economic injury to the crop. A precise definition of biological control is a subject of some controversy (see Chapter 11), but its application often involves human intervention to manipulate populations of organisms in a field. However, naturally occurring biological control has been an important means of pest control for centuries. Farmers often see its importance when they use a pesticide that kills both beneficial and pest insects in a field, since pest populations often rebound quickly, followed much more slowly by their predators. In such cases, damage can be much worse than before the pesticide was applied.

Biological control involves the use of such methods as rearing and mass release of natural enemies, increasing numbers of natural enemies already present in a field, and keeping alternative food sources and habitats available for natural enemies. It is probably one of the most successful approaches to pest management, especially when combined with other methods of management, including cultural control and use of resistant plants (see Chapter 11). Once established, populations of biological control agents in a field can provide good control of pests that will last far beyond the control provided by the application of an insecticide.

Nevertheless, biological control does not get the kind of widespread attention given to chemical control, partly because it is not a "quick-fix" management mechanism. Mass release of a huge number of ladybird beetles into a field will not result in an immediate reduction in the number of aphids. Such control takes time. Establishment of a good population of natural enemies in a field, whether it be mobile beneficial insects or fairly sedentary nematode-parasitic fungi in the soil, may take years to reach a level that will provide satisfactory management of plant pests.

Returning to the admittedly oversimplistic Lotka-Volterra equations given earlier in this chapter, the goal of biological control is to minimize dN/dt for the pest population (see Equation 9-1). In order for this to occur, the predatory removal of these pests from the environment (bNP) should be greater than the rate of natural increase of the pests (aN). When this occurs, the rate of population growth for the pests will fall. This can be accomplished by mass release of natural enemies, or by simply providing good alternative habitats for natural enemies so that they are present in favorable numbers before the pests ever reach great numbers in a field.

Despite its success rate, biological control does not always work, at least not to the extent that growers wish it to. The natural ecology of the system indicates that predators often move on before all prey are eradicated from a field, which can still allow some damage to occur. Competition between predators and between pest species can upset the balance of the system. Switching can allow some of the more destructive pest species to survive while some of the less important pests are satiating predators. In short, biological control is not inherently predictable, which can be unsettling for many growers who are suddenly faced with massive numbers of pests in a field.

Advantages and Disadvantages of Biological Control

Because biological control encompasses a range of techniques from introduction of nonindigenous species to augmentation or conservation of natural enemies, the advantages of this overall method of management are many. In some cases, biological control may simply involve modifying an existing crop culture to provide a suitable environment for natural enemies. In other cases, the introduction of a natural enemy to an area for eradication of an introduced pest may show promise. A great deal has already been accomplished in the area of biological control, and many greenhouses, small farms, and organic producers have used biological control mechanisms effectively.

Still, some scientists caution against the acceptance of some biological control techniques as standard ways to rid ecosystems of unwanted pests. Simberloff and Stiling (1996) argue that the introduction of nonindigenous species to an area has potential to do great harm to an ecosystem by causing the extinction of nontarget organisms. They urge careful consideration and monitoring when introducing new species.

Summary

Predators and parasites can play very important roles in agroecosystems, feeding on plant pests that could ultimately lower productivity and yield. Predators have different foraging strategies, which not only determine their success in finding prey, but also control the amount of time that they remain in one field before moving to another. Parasites are highly effective for the management of some plant pests, since they often affect the reproductive capacity of their host, even though they may not directly cause its death. Biological control involves the use of predators and parasites in an agricultural field to reduce the number of plant pests, and, ultimately, to increase productivity.

Topics for Review and Discussion

1. Explain the differences among predators, parasites, and parasitoids.
2. If you are using biological control to manage pest populations in your field, would you prefer to have a population of generalist predators or specialist predators? Explain your answer.
3. We have discussed the interacting population trends of predators and their prey. Do you think that the same population fluctuations are likely to occur with parasites or parasitoids and their prey? Why or why not? Include a graph of the likely population trends in your response.
4. If a predaceous insect comes across a large population of its prey in a field, which functional response curve would you expect to see exhibited when it first encounters the pocket of prey? Why? Suppose that another predator enters the area after the prey population has declined significantly in that area. Would you expect to see the same functional response curve? Why or why not?
5. Most biological control efforts focus on the use of insects to manage insects. Why has most of the research concentrated on insect management, rather than management of bacteria, fungi, viruses, or nematodes? Do you think that the use of biological control agents in the United States will increase or decrease in the next ten years? Explain your answer.

Literature Cited

Charnov, E. L. 1976. Optimal foraging: The marginal value theorem. *Theoretical Population Biology* 9: 129–136.

Holling, C. S. 1959. The components of predation as revealed by a study of small mammal predation of the European pine sawfly. *Canadian Entomology* 91: 293–320.

MacArthur, R. H., and E. R. Pianka. 1966. On optimal use in a patchy environment. *American Naturalist* 100: 603–609.

May, R. M., and R. M. Anderson. 1979. Population biology of infectious diseases: Part II. *Nature* 208:455–461.

Parker, G. A., and R. A. Stewart. 1976. Animal behavior as a strategy optimizer: Evolution of resource assessment strata and optimal emigration thresholds. *American Naturalist* 110: 1055–1076.

Simberloff, D., and P. Stiling. 1996. How risky is biological control? *Ecology* 77: 1965–1974.

Bibliography

Agrios, G. N. 1997. *Plant Pathology.* 4th ed. San Diego, Calif.: Academic Press.

Pedigo, L. 1996. *Entomology and Pest Management.* Upper Saddle River, N.J.: Prentice-Hall.

Price, P. W. 1975. *Insect Ecology.* New York: John Wiley and Sons.

Romoser, W. S., and J. G. Stoffolano, Jr. 1998. *The Science of Entomology.* Boston, Mass.: McGraw-Hill.

Smith, R. L. 1996. *Ecology and Field Biology.* New York: HarperCollins.

Chapter 10

Agricultural Pests

Key Concepts

- Major types of agricultural pests
- Ecological factors affecting pests
- Outline of management methods for each pest group

The term "pest" is used subjectively to include almost any organism that is perceived as being annoying in some way. In this book, however, we consider it to mean any organism that causes economic damage or disease to crop plants or domestic livestock. Many organisms may not eat enough or be present in sufficient numbers to actually cause measurable damage or yield loss. Others may become pests only under certain conditions. Some fungi and soil invertebrates, such as isopods (see Chapter 3), which normally feed on dead plant material, may become facultative pests, feeding on and damaging living plant material in some cases. Such organisms may not distinguish whether suitable plant material is living or dead.

Many different kinds of organisms can be pests in agroecosystems. Several scientific disciplines focus on the study and control of agricultural pests, including weed science (weeds), plant pathology (fungi, bacteria, viruses, and other disease-causing agents), nematology (nematodes), and entomology (insects, mites, and other invertebrates). Animal science, parasitology, and, in particular, veterinary medicine, examine the rather specific pest and disease problems of domestic animals.

Weeds

Some of the characteristics that make weeds good competitors with crop plants were examined in Chapter 7, along with an introduction to some of the world's most problematic weeds. Many of these weeds are good *r* strategists, which are particularly well adapted for colonizing fields of annual or seasonal crops. The harvest and cleanup of these crops constantly present clean sites for colonization. Knowledge of the dynamics of weed species (Chapter 6) is critical in understanding which features of a weed's life cycle are most important in its establishment and competition with crop plants. Rapid colonization of bare areas beneath trees is also of concern in young orchards. But over time, the main weed problems in these more permanent agroecosystems often result from the establishment of persistent *K* strategists, such as perennial grasses, vines, and woody or deep-rooted weed species that are capable of regrowth despite efforts to remove them.

Although most weed species compete directly with crop plants for water, light, and nutrients, a few types of weeds are actually parasites on plants (see Chapter 9). Some parasites, such as dodder (*Cuscuta* spp.), are parasitic on the stems of other plants, while others, like broomrape (*Orobanche* spp.) are root parasites. Witchweeds (*Striga* spp.) are important root parasites in many tropical locations. The introduction of witchweed into the United States has been limited to some counties in North Carolina and South Carolina by quarantine.

Factors Affecting Competition of Weeds with Crops

Many ecological factors influence the outcome of the competition between crops and weeds. The relative importance of these factors varies depending on the situation, but it is possible to divide factors that affect competition into several main areas: weed density, timing, plant biology, and environment. However, these factors are interrelated and interact in most field situations.

Weed Population Density. Crop yield response to weed interference is related in part to the population density of weed species (Figure 10-1). If weed numbers are very low, there may be no measurable effect on yield. But as weed populations increase, yield declines, often in a predictable pattern. If a relationship such as that shown in Figure 10-1 is known for a particular crop, then it may be possible to make some prediction of crop loss based on weed densities present early in the season, seed bank data, or information on field history. Unfortunately, such relationships are not known for most weed-crop combinations. More importantly, weed infestations are rarely present as one species uniformly distributed in an area. A mixture of weed species of varying numbers and size is usually the case.

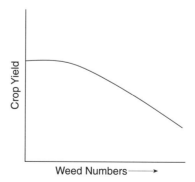

Figure 10-1 General relationship between crop yield and population density of a weed species.

Crop yield-weed density relationships are complicated by a number of issues. Weed size or biomass can be more important than the number of weeds since competition increases as weeds grow rapidly. Large weeds such as johnsongrass (*Sorghum halepense*) affect a crop much more than the same number of smaller species such as purslane (*Portulaca oleracea*).

Timing. The timing of the period of crop growth relative to weed growth is another critical aspect of competition. Early-season weeds that compete with crop establishment are particularly serious. Weeds that germinate when the crop canopy is well developed or when the crop is near harvest may be of little consequence. Therefore, the shape of the curve in Figure 10-1 will vary depending on crop age. A **critical period** is recognized for many crops, a time during which weeds will reduce yield and should be controlled. For rapidly growing field crops such as corn or soybean, a critical period of only a few weeks may be sufficient for the crop to establish and compete with weeds. At the other extreme, a crop such as onion, which does not grow rapidly or form much canopy, can have a critical period of several months.

Plant Biology. Population dynamics, seed dispersal, seed survival, seed bank ecology, and other characteristics examined in previous chapters (see Chapters 6 and 7) affect the competitive ability of weed species. Many serious weed problems arise due to differences in physiology of weed and crop species. Weeds that are C4 grasses are very difficult to manage in C3 crops during summer months or other growing seasons with high light intensity and temperature. On the other hand, corn, a C4 plant, competes well with many weed species during the summer in the midwestern United States, but may be affected during the early spring by C3 weeds like cruciferes, which are better adapted to cooler growing conditions and lower light intensity.

Environment. Many different environmental factors affect the outcome of the competition between crops and weeds. Local weed populations are often better adapted than crop plants to soil type, temperature, rainfall, and soil nutrient levels prevalent in a particular region (Figure 10-2). In some environments, crops that are adapted to the region may be more effective than recently introduced crops in competing with native weeds.

Methods for Managing Weeds

Many methods for managing weeds in agricultural crops are listed in Table 10-1. Some of these methods are very important and are used often, while others may be applicable only in limited or specialized situations. General information on managing weeds may be found in a variety of sources (Gill, Arshad, and Moyer 1997; Minotti 1991; Radosevich, Holt, and Ghersa 1997; Regnier and Janke 1990; Slife 1991).

Biological Methods. Biological control methods involve the suppression of a pest species by a living organism. Although some sources consider suppression of weeds by mulches or smother crops as biological control, we have grouped these methods with cultural practices, as consistent with most sources, and limit biological control to effects from herbivores and diseases. One of the oldest biological methods for weed management is selective grazing by goats, sheep, geese, or other domestic animals. The release of insects for control of weeds has produced spectacular results in some cases, such as the control of prickly pear cactus in

Figure 10-2 Although not generally considered one of the ten worst weeds on a worldwide basis, ragweed (*Ambrosia artemisiifolia*) is well adapted to the environmental conditions of the midwestern United States, where it can be a serious problem.

Table 10-1   Outline of methods for managing weeds in agriculture.

Method	Remarks
Biological	
Insects	Often requires regional effort
Mycoherbicides	Fungi causing diseases of weeds
Selective grazing	Goats, sheep, geese, etc.
Chemical	
Herbicides	Synthetic chemicals and natural products
Physical and Mechanical	
Hand weeding	Hoeing, pulling still widely used
Tillage	Mechanical method widely used, but demands energy
Mowing	Mechanical method widely used; but demands energy
Burning	Includes crop residues, prescribed fire
Flaming	Old method, heats weeds to lethal temperature
Soil sterilization	Greenhouse use, soil for transplants
Soil solarization	Heating soil under clear plastic
Mulching	Physically blocks weed emergence
Flooding	Limited to specific situations
Fallow	Undesirable effects on soil structure, fertility, erosion
Preventive	
Quarantines	Federal, state, regional levels
Seed inspection	Restricts weed seed among crop seed
Restrict seed dispersal	On equipment, animals, soil, manure, transplants
Cultural	
Cropping practices that improve crop competition	
Narrow row spacing	Increases crop density and minimizes open space
Increased seed density	Increases crop density and minimizes open space
Transplants versus direct seed	Head start for crop plant
Optimum planting date	Useful for some crops and weeds
Competitive crop cultivar	Allelopathic, herbicide resistant, etc.
Fertilizer management	Depends on N needs of crop versus weeds
Cropping practices that reduce weed numbers	
Cover crops	Breaks cycle of weed with competitive crop
Rotation crops	Breaks cycle of weed with competitive crop
Intercropping	Desirable plants fill open spaces
Living mulch	Includes groundcovers in orchards
Mulching	Decomposition of residues may affect weeds

Australia or St. Johnswort in the western United States. Many of these successes have been against exotic weed species, and were accomplished by introducing insect enemies from regions in which the weeds were native. As such, this method usually requires a regional approach conducted by government or research institutions rather than by individual growers. The insect feeding may not kill the weed outright, but may weaken the competitive ability of the weed so that it is ineffective or dies from other causes. Some fungi that are pathogenic to weeds have been formulated as **mycoherbicides,** which can be applied to control selected weed species in some crops.

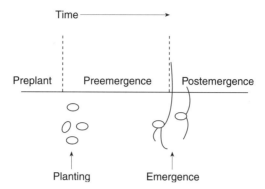

Figure 10-3 Timing of herbicide application relative to crop growth.

Herbicides. Herbicides are the most heavily used pesticides in U.S. agriculture. They include many different types of chemicals, such as natural products, inorganic compounds, and a wide range of synthetic organic products applied to soil or foliage. Some are plant growth regulators, while others are toxins that selectively exert various effects on plant species (see DiTomaso and Linscott 1991). Herbicides improve crop yield by reducing weed density and competition. Not only are they highly effective and fast-acting, but many herbicides are highly selective as well. Their main disadvantages are costs (in dollars and energy) and possible pollution and nontarget effects (see Chapter 12).

Many herbicides are sufficiently selective that they can be used against specific weeds or types of weeds in a particular crop. Information from product labels, the Cooperative Extension Service, or other advisory services is needed to determine which herbicides, among many possible candidates, are most effective in a particular situation and location. Because some herbicides can be applied only at certain times relative to crop growth, it is important to differentiate several stages of crop growth during which herbicides are frequently applied (Figure 10–3). For example, some herbicides that can be used successfully in preplant situations might damage crops if they are applied postemergence. Some frequent considerations about herbicide choice and specificity include whether the product is more effective against grasses (in a broadleaf crop) or against broadleaf weeds (in a graminaceous crop), and whether the material should be applied preplant, preemergence, or postemergence.

With any herbicide, knowledge of dosage and rates is critical to avoid injury to crop plants **(phytotoxicity).** Particular concerns with the use of *selective herbicides* include their effects on nontarget crops (from spray drift, etc.) and the roots of woody plants beneath treated areas, and carryover. **Carryover** is the persistence of a herbicide residue in soil where it may affect a later crop. The herbicide atrazine has been used widely for managing broadleaf weeds in corn. However, if soybean or other broadleaf crops are planted immediately after corn treated with atrazine, there is a possibility that enough atrazine residue may remain in the soil to damage the broadleaf crop. Whether such residues persist into the next crop depends on a variety of factors such as the rate applied, herbicide chemistry, soil type, rainfall, temperature, and soil chemistry (see Chapter 12).

Nonselective herbicides include a number of general-purpose synthetic herbicides, such as glyphosate and paraquat, as well as soil fumigants (see Application 10-1). Nonselective herbicides kill most types of vegetation and are used widely for broad-spectrum weed control

Application 10-1 ◼ Soil Fumigants

Soil fumigants are general-purpose biocides that kill a wide range of pest organisms, including germinating weeds and weed seeds, bacteria, fungi, nematodes, and insects, but also beneficial organisms. Fumigants are injected into the soil as gases or volatile liquids that diffuse through the soil. The soil may be covered with a plastic tarp to seal the volatile material into the soil and reduce evaporation. Because fumigants are toxic to living plants, a waiting period of several days to several weeks or more (depending on fumigant and soil conditions) is needed between the time of fumigation and planting. The performance of soil fumigants and ease of application vary with soil type. Soil fumigation can be a very expensive process; it may be economical only for high-value crops.

---

in agricultural fields and along roadways and fences, and for killing crop residues and emerged weeds in reduced tillage systems (see Chapter 14). A **desiccant** is a chemical that is applied to a crop before harvest to kill the crop foliage and weeds. Use of desiccants improves efficiency during mechanical harvest. Desiccants are used widely on some crops, such as cotton or potato.

Physical and Mechanical Methods. Hand-weeding, tillage, and mowing are among the most important methods of weed management today. Hand-weeding and hoeing, although very laborious, are still the most important weed management methods in many agroecosystems. Mechanical tillage substitutes energy from fossil fuels for energy from human labor. **Primary tillage,** which is conducted before planting or in the fall before spring planting, is a deep operation for loosening soil and other initial preparations for a crop. **Secondary tillage,** which is typically a shallow operation, is used for seedbed preparation and most routine cultivation. Secondary tillage operations can be timed to correspond with maximum emergence of weed populations, although effects differ with various weed species. Cultivation may stimulate some weed seeds to germinate and emerge. Tillage is a complex process that is heavily dependent on soil type and conditions. Equipment and environmental factors that are important in optimizing tillage practices are discussed in more detail elsewhere (Radosevich, Holt, and Ghersa 1997; Tivy 1992). The consequences of reduced tillage are discussed in Chapter 14.

Mowing can prevent seeding or can gradually deplete the energy reserves of weed species. Some crops, such as alfalfa, respond much better to mowing than do many of the weed species present in them. Mowing is often used to manage many dicot weeds while allowing grasses to survive. Soil solarization, which is used in managing plant pathogens in soil (discussed later in this chapter), is also useful in managing some weeds. Flooding may be useful against some weeds in rice or other crops adapted to extremely wet conditions, although several grasses that cause serious problems in rice are also adapted to flooding. Mulches can be made from plastic films and other synthetic materials, or from natural products such as wood chips, crop residues, leaves, composts, etc. Strong synthetic mulch or deep layers of organic mulch may reduce weed establishment significantly. Clean fallow (an extreme tillage practice) or burning can control weeds but may have undesirable environmental consequences. Serious fires often result from attempts to burn weeds and crop residues during dry seasons in the tropics. On the other hand, prescribed fire or controlled burning is an important component of some forest, grassland, and rangeland ecosystems, and is used in the successful management of these systems. In some forest ecosystems, scheduled controlled burning can reduce understory vegetation, brush, and fuel loads, reducing the chance for wildfires.

Preventive Methods. Prevention of movement of weeds and weed seed is critical in limiting the spread of weeds to new locations. Regulatory and quarantine programs from international to local levels conduct inspections to restrict certain weed species. Inspection of crop seeds is particularly helpful and critical, since crop seeds may be contaminated with small percentages of weed seeds. At a local level, seed dispersal can be restricted by cleanup of weed patches and seed sources, as well as by limiting unintentional spread of seeds by farm equipment, animals, animal manures, or other means.

Cultural Practices. Cultural practices include a variety of cropping practices that benefit weed management, and some sources include preventive, physical, and mechanical methods here as well. Many of these methods are aimed at improving the ability of a crop to compete with weeds (see Chapter 7). Useful characteristics in crop cultivars for competing with weeds include tall or rapid growth habits, formation of dense canopy, consistent uniform stand, allelopathy, or herbicide resistance. Crops that grow in dense stands, such as sorghum, clovers, and various small grains like rye, barley, and wheat, can be especially effective in smothering weed populations. Some crops, such as sunflower, oat, and sorghum, can be allelopathic. Of course, the timing of planting or transplanting crops influences not only the crop-weed competition, but also tillage operations.

A number of practices, such as crop rotation, cover cropping, intercropping, and mulching, which are useful in reducing the number of weeds present, are discussed in more detail later (see Chapters 13 and 14). Some of these methods are directed toward suppression of weed growth in spaces between crop rows. Others are used in efforts to break the cycle of crop weeds by replacing a crop that is prone to weed problems with a more competitive crop or a crop in which weeds are more easily managed (by tillage, herbicides, etc.). In addition to physically blocking weed emergence, mulches made from organic materials like crop residues may release various amounts of allelopathic or toxic substances during decomposition. Some crops may exert multiple actions against weeds. Small grains such as oat, rye, and wheat compete well with weeds due to their high plant densities and various degrees of allelopathic properties. When grown as cover crops, decomposition of their green residues may affect weeds, or mulch from their straw may provide some physical barrier to weed emergence.

Plant Diseases

A **plant disease** is a disturbance from a pathogen or environmental factor that interferes with plant physiology, producing changes in plant appearance (= symptoms) or yield loss. These changes may result from direct damage to cells, from toxins or by-products that affect plant metabolism, or from a pathogen's interference with nutrient and water uptake by the affected plant. Diseases can be caused by noninfectious agents or by **pathogens,** infectious agents that can reproduce or replicate. Infectious agents causing plant disease include:

Fungi
Bacteria

Viruses
Nematodes
Mycoplasmas
Parasitic plants
Protozoa

Insects and other herbivores are not generally considered to cause plant disease because much of their feeding simply involves removal of plant parts and cells. However, the feeding of some insects, such as aphids or whiteflies, may affect plant physiology. The field of **plant pathology** examines the causal agents of plant disease, although plant-parasitic nematodes may be covered by nematology and parasitic plants by weed science. **Mycology** covers both pathogenic and nonpathogenic fungi, while **bacteriology** examines both beneficial and harmful bacteria.

Noninfectious Agents of Plant Disease

Many symptoms of plant disease result from nonpathogenic causes. Deficiencies of important macro- and micronutrients are a common cause of characteristic symptoms (see diagnostic information in Bennett 1993). If deficiency alone is the cause, symptoms may be relieved by application of appropriate fertilizer elements. However, nutrient deficiencies may also arise as complications from other causes (root damage, pathogenic root diseases, nematodes, etc.), and in these cases, any relief from nutrient application will be only temporary. A number of other factors commonly cause noninfectious disease symptoms on plants. These include injury from frost, heat or water stress, air pollution, chemical burns (pesticides or surfactants), lightning, etc. Herbicide injury may occur from direct application or from residues in soil. Diagnosis can be complicated because plant tissue killed or injured by these causes may be colonized by secondary infections of bacteria or fungi.

Infectious Agents of Plant Disease

Nematodes and parasitic plants are discussed elsewhere in this chapter, and protozoa cause relatively few important plant diseases (see Agrios 1997). With the exception of parasitic weeds and some fungi, most plant pathogens are microscopic in size. Therefore, although disease diagnosis can be made from symptoms, often a diagnosis cannot be confirmed unless the pathogen is isolated and examined in a laboratory.

Fungi. Although most fungi are involved in decomposition of organic matter (see Chapter 3), several thousand species are plant parasites. Some may feed on living or dead plant tissue and may become facultative parasites on damaged or unhealthy host plants. The life cycles of many fungal parasites are complex and highly specialized. Fungal mycelia can be difficult to identify, and often a fungus must be cultured to produce reproductive structures before a positive identification can be made. Many fungi have alternate hosts on which they can persist in the absence of a susceptible crop plant. Moisture is important in the life cycles of many species (discussed later in this chapter). Fungi affect plants by causing leaf spots, rusts, root rots, vascular wilts, or seedling damp-off. Fungi cause many serious diseases on agricultural crops, such as cereal rusts, mildews, late blight of potato, southern corn leaf blight, Sigatoka disease of banana, coffee rust, etc. (Figure 10-4).

Figure 10-4 Powdery mildew on a cowpea leaf. Mildews are common foliar diseases on many different kinds of plants.

Bacteria. Most bacteria are beneficial as decomposers, but others cause diseases in many different kinds of organisms. Those that cause diseases in plants do not infect humans and animals, and vice versa. Due to their extremely small size, identification is difficult and usually requires chemical tests and growth on specific culture media. Their rapid reproductive rates and short life cycles make bacteria the ultimate r strategists, capable of rapid colonization of a susceptible host. Some bacterial plant pathogens may infect only one plant species while others have wider host ranges. Depending on the bacterial species and the host, plant-pathogenic bacteria cause blights, leaf spots, cankers, crown galls, vascular wilts, or soft rots. Citrus canker and fire blight of pear are examples of important bacterial diseases. Other bacterial species infect vegetable crops, causing a number of bacterial blights on foliage and soft rots on fruit.

Mycoplasmas. Mycoplasmas are closely related to bacteria, and are also classified in the kingdom Monera. Like bacteria, they are susceptible to certain kinds of antibiotics. Like viruses, they require a vector (see next section) or other mechanical means of transmission to infect a host plant. They cause a few important plant diseases, including lethal yellowing of coconut, which has killed many palms in the Caribbean.

Figure 10-5 Symptoms of bean mosaic virus on lima bean leaf.

Viruses. A number of different viruses attack plants, causing various diseases. They often cause mottling or other characteristic symptoms on infected plants (Figure 10-5), but biochemical methods are required to confirm identification of plant viruses. Viruses occur within host plant cells and are carried by the sap and vascular system within a plant, but cannot move to a new plant host unless transmitted in sap, plant material, or other debris. Wounds caused by human contact are a frequent method of virus transmission. Viruses may persist in sap on tools used in grafting or pruning operations. A **vector** is an organism that acquires a plant pathogen (such as a virus or mycoplasma) from an infected host and transmits it to another. Certain types of insects (e.g., aphids, leafhoppers) are the most important vectors of plant viruses. Feeding on infected plants causes them to acquire various viruses that they may then transmit to uninfected plants. Some species of mites and nematodes may also serve as vectors of plant viruses. Viruses can also be transmitted in seed or other propagating material. In some cases, they are even transmitted by pollen, by a few types of fungi, or by dodder, a parasitic plant. Incidental contact with plant wounds by living organisms or inanimate objects is of concern for possible virus transmission. The host ranges of particular plant viruses vary. However, some viruses have many weed hosts from which vectors can acquire the virus and move it into crop plants. Some of the more important virus diseases include tomato spotted wilt, tobacco mosaic, cucumber mosaic, sugarcane mosaic, sugarbeet yellows, citrus tristeza, and several bean and squash mosaics.

Ecological Factors Affecting Fungal and Bacterial Diseases

A number of key ecological factors characterize the development and spread of plant diseases.

Disease Triangle. Three elements—host, pathogen, and a specific set of environmental conditions—must be present and suitable for development of disease to occur. Microscopic fungal spores and bacteria with the potential to cause disease are often present in great numbers in air, on plant surfaces, and in soil. Yet disease does *not* develop in most situations! All three requirements of the disease triangle must be met (Figure 10-6). The *host* must be susceptible to the disease agent. Many plant species and cultivars have resistance to certain plant pathogens and will not be attacked, while other plants may be attacked only when in a weakened state. A specific *pathogen* must be present in sufficient numbers and, in some cases, must gain entry to the plant through wounds following pruning or feeding by other pests. Many plant pathogens exist as various *isolates,* or populations that have differing degrees of pathogenicity to a plant host. Finally, *environmental conditions* must be favorable for development of the pathogen.

Disease Progress and Dynamics. Under favorable environmental conditions, life cycles of many pathogenic fungi and bacteria are short; thus, there is potential for rapid population buildup. Exponential growth (see Chapter 6) may be a concern over the short term, but over time, the logistic model can be useful for describing disease buildup (Figure 10-7). It may be very difficult or impossible to count numbers of fungi or bacteria present in a crop, so disease symptoms are usually recorded instead. Disease levels may be expressed as **incidence,** the number or percentage of plants that are infected by the pathogen, or **severity,** the proportion of leaf or fruit surface that is covered by the disease symptoms. A curve such as that shown in Figure 10-7, in which disease incidence or severity is plotted over time, is called a **disease progress curve.** Disease incidence may level off below 100% infection, but if $K = 100\%$ infection, the entire crop may be destroyed, and the epidemic ends because there are no more hosts to infect. Obviously, prediction of such events is of considerable importance, and various types of models have been developed for predicting disease progress (see Mundt 1990). Because disease development is closely linked to environmental conditions, weather forecasting is important in predicting disease progress and epidemics, and is included in many predictive models.

Environmental Factors Affecting Disease Development. Although the growth rates of many fungal and bacterial plant pathogens are increased by temperature and heat unit accumulation, some are better adapted to cooler conditions. It is necessary to know the specific temperature requirements of a particular pathogen before making predictions about it. Wind is important in the spread of many plant pathogens, but in some cases may have a

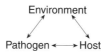

Figure 10-6 Disease triangle showing the three factors essential for development of plant disease.

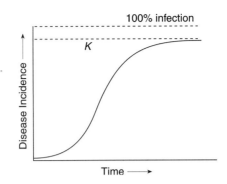

Figure 10-7 Progress of disease symptom development over time may resemble the logistic growth curve.

positive effect in drying out leaf surfaces. Some pathogens have requirements for particular levels of soil pH or light intensity.

One of the most important environmental factors affecting disease progress is moisture. Water may stimulate the ending of resting stages, germination of spores, or growth and penetration into a host. Many pathogen structures, such as fungal mycelia or germinating spores, are very fragile and dry out easily if not protected. Splashing water helps to distribute immobile fungal spores or bacteria. Therefore, rainy conditions are important for the spread and development of many diseases. Actually, diseases are favored by any factor that maintains increased wetness and moisture, such as sprinkler irrigation practices, high relative humidity, dew, and leaf wetness. Leaves or roots that remain moist for long periods of time are particularly susceptible to colonization by fungi and bacteria.

Methods for Managing Plant Disease

Recognizing the importance of the disease triangle in the expression of plant disease, methods for management of plant diseases are directed toward three areas:

- Improving the quality and ability of the plant host to withstand disease;
- Reducing or eliminating populations of the pathogen; and
- Establishing environmental conditions unfavorable for pathogen development.

As a first course of action, preventive strategies are used to delay the introduction of a pathogen into a new site, or eliminate or reduce any initial inoculum present. Once a pathogen is established in a site, emphasis shifts toward slowing its rate of increase and shortening exposure to favorable conditions for disease development.

Methods for managing plant diseases are summarized in Table 10-2. A number of these methods are very specific, and to use them effectively, it is important to accurately identify the targeted disease and its causal agent (fungi, bacteria, virus). For instance, application of fungicides or use of a plant cultivar that is resistant to a particular fungus would be ineffective against a virus disease. It is also critical to recognize whether a disease is **foliar** (aboveground plant parts) or **soilborne** (roots, below ground). Methods for

Table 10-2 Outline of methods for managing plant diseases in agriculture.

Method	Remarks
Biological	
Parasites	Versus bacteria or fungi
Cross protection	Versus viruses
Chemical	
Fungicides	Target fungi
Bactericides	Target bacteria
Insecticides	Target insect vectors of plant viruses
Physical	
Soil sterilization	Raises soil to temperatures that are lethal to plant pathogens
Soil solarization	Raises soil to temperatures that are lethal to plant pathogens
Hot water	Treatment of live plant material
Burning	Kills plant pathogens in crop residues
Flooding	Limited to specific situations
Radiation	Limited postharvest application
Refrigeration	Slows disease progress postharvest
Tillage	Varies depending on situation
Preventive	
Quarantines	Federal, state, regional levels
Certification programs	Disease-free seed and plant material
Sanitation	Many ways to limit disease introduction:
Site selection	Plant in sites free of disease
Tissue culture	Clean propagation material
Clean equipment	Avoid disease spread on tools, soil, tractors, etc.
Remove crop debris	Reduce inoculum source and buildup
Remove alternate hosts	Weed hosts of viruses, other pathogens
Roguing	Removal of infected plants
Pruning	Removal of diseased wood
Cultural	
Host—Cropping practices that improve plant health	
Resistant cultivars	Important method for disease management
Crop nutrition	Favors crop needs rather than pathogen
Transplants versus direct seed	Reduced exposure to soil diseases during seedling stage
Shallow planting of seed	Reduced exposure to soil diseases during seedling stage
Environment—Cropping practices that reduce factors favoring disease	
Irrigation type	Minimizes leaf wetness
Wide row spacing	Increases air circulation and drying of foliage
Low plant density	Increases air circulation and drying of foliage
Soil drainage	Wet soils favor many diseases
Timing of planting	Useful for some situations
Pathogen—Cropping practices that may reduce plant pathogens	
Liming	Affects pathogens that need low pH
Crop rotation	Useful versus soil pathogens
Cover crops	Varies with situation
Organic amendments	Varies with situation

reducing leaf wetness may be useful in the first case, but are irrelevant for soilborne problems. Crop rotation may be used against pathogens confined to soil of a specific site, but would be ineffective against wind-dispersed foliar pathogens or viruses vectored by insects.

Several methods, such as controlled environmental conditions or fungicide application, are particularly important in reducing postharvest losses to plant pathogens. Many fungi and bacteria can continue to develop in or on harvested fruit and other plant products. Refrigeration provides an environment that is too cold for the development of some pathogens and slows the development of others.

Some of the methods listed as preventive may overlap with and be considered as cultural methods by some authors. Additional information on disease management can be found in a variety of sources (Agrios 1997; Gould and Hillman 1997; Marshall 1997; Mink 1991; Rush, Piccinni, and Harveson 1997).

Biological Methods. Many biological agents are examined experimentally, but relatively few are available for disease management on a field scale (see examples in Cook 1993; Mehrotra, Aneja, and Aggarwal 1997). Various agents may affect plant pathogens in different ways. Some bacteria are hyperparasites (parasites of parasites) on bacteria or fungi that are parasitic on crop plants. **Mycoparasites** are fungi that are parasitic on other fungi. Some organisms may release toxins or by-products that affect pathogens. **Antibiosis** is the production of compounds that inhibit growth of microorganisms. Other organisms may compete with pathogens for access to plant roots or nutrients. Mycorrhizae may interfere with the ability of pathogenic fungi to colonize roots. It is possible that the numerous rhizobacteria surrounding root systems may also interfere with root colonization by pathogens. Some plants may respond to pathogens by developing **induced resistance,** which occurs when various biochemical changes confer some resistance in response to a pathogen. **Cross protection** is the inoculation of a plant with a mild strain of a virus to provide some protection against more pathogenic strains of the virus. Finally, many predators and parasites of plant pathogens are always present in soil where they may provide some level of management. Soils in which some natural control of plant pathogens is recognized or measured are referred to as "suppressive soils."

Chemical Methods. Chemicals for management of plant diseases include preplant soil fumigants, seed treatments, foliar sprays, postharvest applications, warehouse fumigation, and treatments for tree wounds and pruning operations. Most are specific for managing a particular type of plant pathogen. Insecticides may be used to manage the insect vectors that carry plant viruses. Bactericides include copper compounds and antibiotics such as streptomycin. An introduction to common fungicides and their mode of action is provided by Ragsdale and Sisler (1991). Many fungicides are used on a preventive basis to keep disease from developing under favorable environmental conditions, rather than to cure or reduce active epidemics.

Physical Methods. Several physical methods are used to reach temperatures that are lethal to pathogens, including use of steam, hot water, solar heat (see Application 10-2), fire, or flame. Live plant material may be treated in hot water, but it is critical to maintain a temperature that is high enough to kill the disease organism but low enough not to result in plant damage. Such temperature requirements may not be possible for all crop-disease combinations.

Application 10-2 *Soil Solarization*

Soil solarization is the heating of soil under clear plastic to kill soilborne pests. A variety of plant diseases, nematodes, and weeds can be managed to some extent by this method. A strong, but thin, clear plastic sheet or tarp is stretched over the surface of the soil and sealed (buried) around the edges. Solar energy penetrating the clear plastic warms the soil, achieving temperatures of 50°C (122°F) in the top 10–15 cm (4–6 in.) of soil. Prolonged exposure to these temperatures is lethal to many pests and pathogens. Best results are achieved when a thin tarp is used, when soil is moistened before tarping (water conducts heat), and when soil is free of sticks, debris, and large air pockets. The sealed tarp must remain unbroken for four to six weeks. The length of time depends on the ambient temperature and cloud cover. The method has been used most effectively in warm locations with minimal cloud cover, such as Israel and California. However, solarization can also be used in the southeastern United States and in tropical locations that have frequent thunderstorms, because warm, sunny conditions may still be available through much of the solarization period. It is possible to do solarization in the midwestern United States, for example, but most people do not want to sacrifice up to six weeks of the growing season. In Florida and California, yields of winter vegetable crops have improved following solarization during the summer (when cash crops are not usually grown).

Although soil solarization is an alternative to chemical control, plastic residues and disposal may be a problem. Another disadvantage is that beneficial organisms may be killed along with pests. Since high temperatures are reached only in the upper layers of soil, organisms (pest and beneficial) will recolonize from deeper soil layers. Therefore, the beneficial effects of soil solarization typically last three to four months, or about the length of the life cycle of most vegetable crops.

Preventive Methods. Sanitation is the critical first step in limiting the spread of disease. Plant disease can be introduced on any item used in establishing a crop—seed, planting material, tools, tractors, soil, potting media, containers, etc. Some operations such as cultivation or other routine work in wet fields may help to spread certain pathogens within the field.

Cultural Methods. Several cultural practices are aimed at maintaining healthy plants that can better resist or withstand disease. Use of resistant cultivars is a very important and effective method for managing plant disease (removes favorable host from the disease triangle). Disease-resistant cultivars are available for many crop-disease combinations (see Shaner 1991). Nutrition and fertilizer management remain the principal means of treating noninfectious disease symptoms arising from nutrient deficiencies.

Because many plant diseases are favored by excessive moisture and leaf wetness, several methods can be used to create an environment that is more favorable to crop than to pathogen. Many seedling diseases thrive in cool, wet soils. Planting at a warmer time of year, using well-drained soil, or using transplants are methods for reducing the time that seedlings are exposed to these conditions. Limiting moist, dense plant canopies through wide row spacing and reduced plant density can be helpful in disease management, although these are not the best practices for weed management. Overhead sprinkler irrigation may keep foliage wet more than drip or furrow irrigation, providing increased opportunity for foliar disease.

The use of cover crops and organic amendments for disease management is somewhat variable and unpredictable. While some cover crops may break the disease cycle or even be suppressive to some plant diseases, others may act as alternate hosts or provide residues favorable for disease development. The effect of crop residue management on plant disease is discussed in more detail in Chapter 14.

Plant-Parasitic Nematodes

Agroecosystems contain many different kinds of nematodes. Beneficial nematodes include predators, insect parasites, and a variety of types involved in decomposition (see Chapter 3). Nematode pests include animal parasites, which are often host-specific and may have complicated life cycles, as well as plant parasites. The plant-parasitic species are not harmful to animals or people, and animal or insect parasites do not affect plants, unless the parasite infects a plant pest, indirectly benefiting the plant.

Nematodes as Plant Parasites

Most plant-parasitic nematodes live in soil around plant roots and feed on root tissue. The mouthpart is a needle-like stylet used to puncture plant cells and withdraw fluid and nutrients. **Ectoparasites** remain in the soil while they feed, and do not enter plant roots. **Endoparasites** enter plant roots and may tunnel and feed within the root system. *Sedentary endoparasites* have the most complex life cycles among plant-parasitic nematodes (Figure 10-8). Migratory and sedentary endoparasites are not restricted to roots, but may

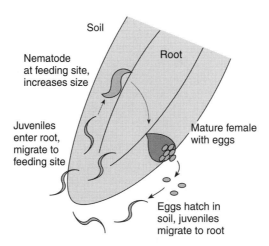

Figure 10-8 General life cycle of a sedentary endoparasitic nematode, such as the root-knot nematode.

inhabit bulbs, tubers, and other below-ground plant parts. *Foliar nematodes* include a few species that can be transported to foliage or may move up stems during times of high humidity. They may infect stems, leaves, growing tips, or seeds, causing diseases in rice and a few other crops.

The life cycles of most plant-parasitic nematodes are rapid (about one month at summer temperatures). Reproductive rates may be high, with some species producing several hundred eggs per female. As a result, they are good *r* strategists that can quickly build up to high numbers. *Winter* is important in interrupting the life cycle and population growth of nematodes. In tropical and other warm climates, nematode multiplication proceeds uninterrupted, resulting in especially severe problems.

Symptoms. Because plant-parasitic nematodes are microscopic and hidden in soil or plant roots, many people may be unaware of their presence. A few kinds, such as root-knot nematodes (*Meloidogyne* spp.), may produce direct damage such as knots or galls on roots (Figure 10-9), or misshapen or deformed potatoes, peanuts, or carrots. In most cases,

Figure 10-9 Root galls caused by root-knot nematodes on the root system of an infected plant.

however, damage is indirect, and the most common symptoms are reduced yield and quality, stunting, yellowing, erratic stands, and nutrient or water deficiency. Many of these symptoms can be attributed to root damage from a variety of causes. Soil or root samples are required for laboratory diagnosis of nematode problems in these cases.

Factors Affecting Nematode Damage

The severity of nematode damage depends on:

- Kinds of nematodes present;
- Numbers of nematodes present;
- Environment; and
- Other organisms present.

The type of nematode present is critical (see later in this chapter), but sufficient numbers to cause damage must also be present. Environmental conditions also affect damage, since the speed of nematode population growth depends on temperature, heat unit accumulation, and favorable soil moisture. Soil type may influence which nematodes are present as well as plant symptoms. Also, nematode damage to roots may provide entry points for bacteria and fungi, so that some plant diseases may be much more severe if nematodes are present.

Kinds of Plant-Parasitic Nematodes. Most agricultural soils contain several different kinds of plant-parasitic nematodes, but fortunately, some kinds are not particularly damaging, and others may damage only certain crops. Because of the impact of their life cycle on the root system, sedentary endoparasites are among the most damaging nematodes. Two kinds are especially problematic. The root-knot nematodes (*Meloidogyne* spp.) cause swellings or galls in the root system (Figure 10-9) and damage a wide variety of crops. While a few species are important in cooler climates, they tend to be especially severe in the southern United States and in tropical regions. Cyst nematodes (*Heterodera* spp. and *Globodera* spp.) are also sedentary endoparasites, but the body of the mature female hardens to form a protective case, or *cyst*, around the eggs, enabling them to survive harsh conditions. The potato cyst nematodes (*Globodera* spp.) and the beet cyst nematode (*H. schachtii*) are the most important nematode pests in much of Europe. The soybean cyst nematode (*H. glycines*) is of concern in soybean-production regions, and the cereal cyst nematode (*H. avenae*) affects small grain crops in many parts of the world. Other nematodes can be extremely damaging to certain crops (see Application 10-3).

Methods for Managing Plant-Parasitic Nematodes

Many of the principles and methods for managing plant-parasitic nematodes are similar to those for managing soilborne diseases (Table 10-3). Although nematodes can move (up to several meters a year), they are so small that they cannot get far without help. So the principles of sanitation and exclusion emphasized for plant disease are also critical in prevent-

Application 10-3 *Managing the Burrowing Nematode on Bananas*

Due to the lack of a cold winter to interrupt life cycles, nematode problems in the tropics can be especially severe. One of the most serious nematode problems worldwide is the burrowing nematode (*Radopholus similis*) on bananas and plantains. The burrowing nematode is a migratory endoparasite that feeds within banana roots, causing such severe damage that plants may fall over when roots are unable to support the weight of fruit.

The burrowing nematode problem illustrates several important aspects in the management of nematodes and plant diseases. First, in selecting a site for a new banana planting, the nematode may already be present in the soil, and therefore some methods may be needed for managing nematode numbers in the site (see methods in Table 10-3). Bananas are propagated by removing a small shoot or sucker (with roots) from an older clump of bananas. Nematodes can be carried to a new site in roots or soil, unless planting material is inspected and the nematodes are removed by paring or treatment in hot water. Production of sterile banana plantlets through tissue culture has helped to avoid this problem. If nematode problems develop later in an established commercial plantation, they can often be managed by application of nonfumigant nematicides. However, in some cases, root damage may be so severe that attempts to control nematodes are abandoned, and damaged plants are propped up by wires or poles so that plants with fruit will not fall over.

---

ing the movement of nematodes to clean sites. Once established in a site, nematodes may be impossible to eradicate, and the rise or fall of nematode population densities in that site depends on the susceptibility of any plants (even weeds!) grown in the field. For instance, population densities of the soybean cyst nematode in a field will increase on a susceptible soybean cultivar, but will decline if corn, a nonhost, is planted.

Host plant resistance is probably the most important method for managing plant-parasitic nematodes. Nematologists distinguish between resistance and tolerance, either of which may result in improved crop performance. A resistant plant allows only low nematode reproduction, so numbers do not build up. A tolerant plant is one that withstands nematode damage, even if nematode numbers are relatively high (see Chapters 8 and 11). Resistance is not an all-or-nothing phenomenon. Some cultivars of the same crop may show high levels of resistance, some may be susceptible, and many may show intermediate levels of resistance. Of course, to use resistant cultivars successfully, you must know which nematodes are present in a site and have information on which crop cultivars are resistant or tolerant to those nematodes. The use of crop rotation or cover crops provides an additional means of using crops that are highly resistant or are nonhosts to a particular nematode population.

Several methods for nematode management are controversial. Although many predators and parasites in the soil environment are known to attack nematodes in the laboratory, few commercial successes with biological control of nematodes have been achieved. The use of organic amendments (crop residues, manures, almost any organic residue) for nematode management has received much attention, but results remain inconclusive. It is hypothesized that addition of organic amendments may provide decomposition products that affect

Table 10-3 Outline of methods for managing plant-parasitic nematodes in agriculture.

Method	Remarks
Biological	
Parasites	Bacteria, fungi
Predators	Mites, nematodes, etc.
Chemical	
Nematicides	Fumigants and nonfumigants
Physical	
Soil sterilization	By steam, hot water, microwave, etc.
Soil solarization	See Application 10-2
Hot water	Kills nematodes in planting material
Flooding	Limited to specific situations
Tillage	Not much effect on nematodes
Fallow	Deprives nematodes of a host
Paring	Cut infected parts from planting material
Propping	Supports bananas with damaged root systems
Burning	Affects foliar nematodes
Preventive	
Quarantines	Government programs
Certification	Nematode-free planting material
Sanitation	Critical in many areas:
Site selection	Avoid nematode-infested sites
Tissue culture	Clean planting material (e.g., bananas)
Clean equipment	Avoid moving nematodes in soil
Remove crop debris	Reduces future buildup of nematodes
Remove weed hosts	Avoid points of nematode infestation
Cultural	
Cropping practices that improve plant health	
Resistant cultivars	Important method for nematode management
Crop nutrition	May improve crop tolerance to nematodes
Transplants versus direct seed	Gives head start to root system
Timing of planting	Use for some situations
Cropping practices that may reduce nematodes	
Crop rotation	Breaks nematode cycle by providing an unfavorable host
Cover crops	Breaks nematode cycle by providing an unfavorable host
Organic amendments	Inconsistent; results vary
Green manures	Suppressive effects possible from some residues
Intercropping	Inconsistent; often not much effect
Trap crops	Risky

nematodes directly or create an improved environment for biological control by naturally occurring organisms. If so, these responses should be verified by decreased numbers of nematodes in soil and root samples following treatment with the amendment. Just because an added amendment improves plant growth does not mean that nematodes are controlled. Organic amendments yield nitrogen and other nutrients that may benefit plant growth regardless of any effects on nematodes.

Because various nematode species respond differently to management methods and crop cultivars, detailed information may be required to use these methods successfully. Additional details about nematode management can be found elsewhere (Barker, Pederson, and Windham 1998; Evans, Trudgill, and Webster 1993; Luc, Sikora, and Bridge 1990; McSorley and Duncan 1995; Stirling 1991).

Insects

Insects are the largest and most diverse group of organisms, and certainly many different kinds are represented in most agricultural fields. Distinctions between insects and other arthropods were made in Chapter 3. However, noninsect arthropods, such as mites, spiders, millipedes, and terrestrial isopods, are often included in the science of entomology. Snails and slugs, which are molluscs rather than arthropods, are of similar size and may cause damage similar to that of insects. Some of the methods used for managing insect pests may also be applicable against snails, slugs, and noninsect arthropods.

Insect Life Cycles

After a young insect hatches from the egg, it passes through several stages before it becomes an adult. Insects and other arthropods have a skeleton that is external (an exoskeleton). This outer skeleton restricts the size to which an insect can grow. If a young insect is to grow further, it must shed this exoskeleton in a process called **molting.** Then the insect can grow for a time until the new exoskeleton confines it. These growth stages between molts are called **instars,** and may be distinguished by their differences in size. An insect molts several times during the course of its development, and the number of instars depends on the species.

Although immature stages of some insects, such as silverfish, resemble adults, most insects go through **metamorphosis,** or change in form, during their development. Although entomologists recognize several different types of metamorphosis, two are particularly important in agroecosystems:

Simple metamorphosis: Egg → nymph → adult

In simple metamorphosis, the immature stages, called **nymphs,** resemble the adults except that the wings are not fully developed or expanded. The number of nymphal instars varies with species, but four to six are probably most typical. Adults and nymphs generally prefer the same food source and live in the same habitat.

Complete metamorphosis: Egg → larva → pupa → adult

In complete metamorphosis, the appearances of young and adult stages are different. The immature stage is the **larva,** which may be a caterpillar, maggot, grub, or wormlike individual. Several larval instars occur before the insect becomes a pupa. The **pupa** is a resting stage that does not feed and may occur on the host plant or in soil or debris. The adult emerges with an appearance typical of a butterfly, moth, fly, beetle, etc. From an agricultural perspective, it is important to recognize that the food sources of the larvae and adults of the same species may be very different, which ecologically prevents competition between parents and offspring. For instance, some

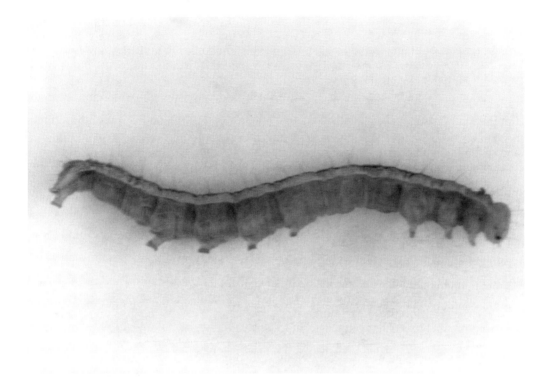

Figure 10-10 This caterpillar has three pairs of true legs near its head (at right). The remaining "legs" are really prolegs, outgrowths of the abdominal body wall that will disappear after the larva undergoes metamorphosis.

insects are damaging as caterpillars (Figure 10-10) but may be beneficial plant pollinators as adult butterflies or moths. Also, it is important to recognize that the feeding and damage of a large 5th-instar larva may be many more than five times as severe as that of a 1st-instar larva.

Pest and Beneficial Insects

So many different kinds of insects occur in a typical field that it may be difficult to know which are pests and which are beneficial. Many insects are beneficial to agriculture as predators and parasites of pest species or as pollinators of crop plants (Figure 10-11). Table 10-4 provides a general indication of which insects commonly found in agroecosystems are likely to be pests or beneficial.

More detailed identification of insects can be found in other sources (e.g., Borror, Triplehorn, and Johnson 1989; Dindal 1990; Romoser and Stoffolano 1998). Some kinds of insects are familiar to most people, but others may not be so well known (Figure 10-12). Near-

Figure 10-11 Effective management of bees and other pollinators is important in reaching full yield potential of some crops.

microscopic thrips are common in flowers and on foliage in many agroecosystems, but are often overlooked. The piercing-sucking insects include small members, such as whiteflies and aphids, which may colonize plants in great numbers. In addition to the direct damage these pests may cause to plants by their feeding, many are also important vectors of virus diseases to plants. Although larger ants, bees, and wasps are familiar to most people, many members of this group are microscopic wasps. Many of these wasps are parasitoids that lay eggs in or on a host insect, which the developing larvae consume (see Chapter 9). Larvae are the most common hosts of parasitoids, but some species may even deposit eggs into the eggs of other insects.

From Table 10-4, we can see that some insect groups are almost always beneficial (e.g., lacewings, dragonflies), while others are potential pests (e.g., aphids, fleas, grasshoppers, thrips). For other groups (e.g., beetles, true bugs), there are so many possibilities that we would need to identify the insect in more detail to know if it was a potential pest or predator. Some groups are extremely diverse. Flies are frequent pests in livestock operations, but in a cultivated crop, some species of flies may be pollinators, some may be parasitoids, and

Table 10-4 Major groups (orders) of agricultural pests or beneficial insects and insect relatives.

Insect Orders	Pests	Beneficials
Dragonflies (Odonata)[a]	—	Predators of flying insects
Grasshoppers, crickets, mantids, walking sticks (Orthoptera)	Grasshoppers, crickets can be serious pests	Some (e.g, praying mantids)
Thrips (Thysanoptera)	Most are plant pests, some carry viruses	Some pollination in flowers
True bugs (Hemiptera)	Some economic pests (stink bugs)	Some predators
Piercing-sucking insects (Homoptera)	Aphids, cicadas, whiteflies, scale insects, leafhoppers; some carry viruses	—
Lice (Anoplura, Mallophaga)	Livestock pests	—
Beetles (Coleoptera)	Many (weevils, grubs, wireworms, leaf-feeding beetles)	Many (ground beetles, ladybeetles, etc.)
Nerve-winged insects (Neuroptera)	—	Usually predators (e.g., lacewings)
Butterflies, moths (Lepidoptera)	Many kinds of caterpillars	Pollinators
Bees, wasps, ants (Hymenoptera)	Very few	Many parasitoids, predators, pollinators
Fleas (Siphonoptera)	Livestock pests	—
Flies (Diptera)	Livestock pests, some leafminers	Some parasitoids, some predators, some pollinators
Mites (Acarina)	Some plant, some livestock pests	Some predators
Spiders (Araneae)	—	Predators

[a] Scientific name of order in parentheses

others may be predators. Some are leafminers, tunneling within leaves of crop plants (Figure 10-13).

Methods for Managing Insect Pests

Many methods are available for managing insects (Table 10-5), but they vary in their effectiveness. Some methods are highly specialized, limited to specific situations that depend on the biology of a particular pest. For example, delaying the planting of wheat beyond the "fly-free" date is a classic method for avoiding damage from the Hessian fly in the midwestern United States. The screwworm, a serious pest of domestic livestock and wildlife, was eradicated from the United States and much of Mexico by release of large numbers of sterile male screwworm flies. Overwhelming numbers of sterile males meant that most matings

Figure 10-12 Typical appearance of one of the true bugs, a large group of insects that contains both plant pests and beneficial predators.

resulted in no offspring, and screwworm numbers declined quickly. Sterile male release is obviously a highly specialized control method, but it has been attempted against several insect pests (see Bartlett 1991).

Chemical and biological methods are used extensively in managing insects, and are discussed in more detail in the next chapter. The narrow host range of some parasitoids for particular pest species makes them especially desirable in biological control programs, in which they are often used. Chemical methods are not restricted to traditional insecticides, but include compounds that attract, rather than kill, insects. **Pheromones** are chemical compounds that are given off by one individual that produce a behavioral response in another individual of the same species. These chemical signals are used by insects for attracting mates, for sending alarms, for finding food (e.g., ant trails), or for aggregating together. Pheromones can be used to attract insects to traps where they can be destroyed by insecticides or physical means. Applications of sex pheromones have been used to disrupt the mating patterns of some insects. But probably the most common use of pheromones is in sampling and detection of pest populations. Insects are extremely sensitive to pheromones, and even if only a few individuals are present in an area, they may travel great distances to a trap containing an appropriate pheromone.

Figure 10-13 The small larvae of the vegetable leafminer tunnel between the upper and lower surfaces of a leaf, producing characteristic serpentine mines or tunnels.

Maintenance of vegetational heterogeneity and diversity is important in managing a number of insect pests. Genetic diversity can be maintained within the crop plant, in the planting of other crop species in the same field (intercropping), in borders of crop fields, or in the landscape of different kinds of crops and natural areas within a region. Vegetational diversity and heterogeneity may benefit insect management in several ways. The noncrop vegetation may serve as a reservoir for predators and parasites of pest species, or pest species may be attracted away from the economic crop to weeds, trap crops, or other vegetation. Tall, densely planted crops such as sugarcane or corn may provide a barrier to the movement of some insects to susceptible crops.

Two important features to consider in the management of insect pests are their mobility and host ranges. Some insects may feed only on one or a few host plants, while others may feed on many different plant species (making it difficult to find resistant or nonhost plants). Because many insects fly and migrate readily, crop rotation is less effective against most insects than against pests like nematodes and soilborne diseases, which tend to be more confined to a site.

The efficacy of some management methods is highly variable. The use of crop nutrition for insect management, particularly N fertilizer, is especially controversial. Addition of N fertilizer benefits plant growth and performance, but there is evidence that N-enriched plants

Table 10-5 Outline of some methods for managing insect pests in agriculture.

Method	Remarks
Biological	
Predators	Least specific biological agent
Parasitoids	Often highly specific
Pathogens	Viruses, bacteria, fungi, protozoa, nematodes
Sterile male release	Regional effort in specific situations
Chemical	
Insecticides	Natural and synthetic products
Acaricides	Target mites
Repellants	Limited to specific situations
Attractants	Pheromones
Physical	
Soil sterilization	Limited to specific situations
Hand-picking	Inefficient; limited to small scale
Traps	Includes flypaper, sticky traps, etc.
Barriers	Limited to specific situations
Heat	Postharvest control of insect pests in stored products
Radiation	Postharvest control of insect pests in stored products
Flooding	Limited to specific situations
Tillage	Affects some kinds of insects
Burning	Destroys pests in crop residues
Preventive	
Quarantines	Regional and government programs
Sanitation:	
Crop residues	Cleanup is important
Overwintering sites	Cleanup is important
Damaged fruit	Cleanup is important
Alternate hosts	Cleanup is important
Pruning	May limit access of some insects to trees
Cultural	
Cropping practices that provide advantages to crop	
Resistant cultivars	Physical or nutritional effects
Crop nutrition	A trade-off; see text
Timing of planting	Useful for some situations
Early harvest	Reduce insect damage
Cropping practices that may reduce insects on crop	
Crop rotation	Useful against insects with narrow host range; limited mobility
Cover crops	Useful against insects with narrow host range; limited mobility
Spatial heterogeneity	Diversity maintained in several areas:
Intercropping	Lowers some insects in main crop
Trap crops	Attract insects away from main crop
Regional diversity	In borders, other fields, landscape

stimulate population growth and development of a number of insect pest species. Therefore, it may be unclear whether any increased plant growth may be offset by increased pest damage. Because specific information is needed for the effective use of many insect management methods, additional details may be found in local extension publications, crop advisory recommendations, and other sources (e.g., Bartlett 1991; Roberts et al. 1991; Romoser and Stoffolano 1998; Stern 1991).

Vertebrates

Although vertebrate pests may be absent from some agroecosystems, a relatively few individuals can cause a considerable amount of damage in others due to their large size compared to some of the more traditional pest groups. Methods for managing vertebrate pests are introduced here briefly, but additional detail on nonchemical methods for managing vertebrate pests can be found elsewhere (Arnold 1991; Davis 1991).

Rodents and Other Mammals

On a worldwide basis, rats and mice cause extensive losses to rice and other grain crops (pre- and postharvest), as well as to sugarcane and various fruit crops. However, a variety of other mammals may be pests in specific situations. Rabbits, prairie dogs, ground squirrels, woodchucks and other rodents, and even deer, can seriously damage crop vegetation. Fruit bats may cause losses to tree fruits in some tropical locations, and various predators may affect domestic livestock. Rodenticides and other toxins are often used to manage these pests, but fences and other barriers can be effective in many cases. Cultivation and cleanup of excessive weeds and hiding places can reduce incidence of some rodents. A viral disease has been used to reduce populations of rabbits in Australia and other locations.

Postharvest losses of grain to rats and mice is a significant concern in many locations. Effective sealing of grain storage containers and the rat-proofing of storage buildings are essential in minimizing these losses.

Birds

Agricultural problems from birds occur most often as a result of their feeding on grain crops such as sorghum, millet, corn, rice, or wheat. In some situations, birds may feed on fruit, nuts, or seedlings. Contact with wild bird species may offer the possibility of disease transmission to poultry.

Chemicals used to manage bird populations include toxins, repellents, and wetting agents that affect the insulating properties of feathers. However, their use may raise environmental concerns about effects on nontarget species. Nonchemical management methods include attempts to scare birds away by use of noise, pyrotechnics, scarecrows, or other means. Several of these methods, as well as site modification by pruning or cleanup, may make roost areas less attractive. Plantings may be timed to avoid flocks of bird species that may damage seedlings or grain. It is also possible that habitat and crop diversity within the landscape may provide a more general food source for birds, so that they do not come to depend on grain monocultures.

Pests of Domestic Animals

Most of the examples given in this chapter involve pests of agricultural crops. However, livestock and other domestic animals are also subject to a vast array of specialized pest and disease problems. These problems may be especially severe in the tropics, where warm, humid climates favor the survival of many different pests and pathogens. The situation is particularly serious when animal breeds originating in Europe and other temperate locations are introduced into the tropics, where they are not adapted to the climate or to local pests and pathogens.

The pest and disease problems of each domestic animal species are often highly specialized. As an example, the various groups of parasites and pathogens that affect domestic cattle are summarized in Figure 10-14. Most of the boxes in this figure could be subdivided into a number of distinct problems. For example, serious viral diseases of cattle include rinderpest, foot-and-mouth disease, and rabies. Examples of bacterial diseases include contagious bovine pleuropneumonia, brucellosis, anthrax, and bovine tuberculosis. Major regional efforts have been undertaken to prevent the spread of some of the more serious viral and bacterial diseases through the use of quarantine, local eradication, and vaccination.

A number of arthropod pests and parasites are debilitating to cattle and other livestock. As a group, mites include ticks and a variety of other types that cause mange, scabies, and other problems. Flies include biting species as well as screwworm larvae, which feed and develop in wounds. But ticks and flies are also important as vectors of serious diseases. Tsetse flies (*Glossina* spp.) are annoying biting flies, but are more important as vectors of trypanosome protozoa, which have limited production in domestic animals in portions of central Africa. Ticks can vector a number of important diseases to cattle, including bovine anaplasmosis, caused by rickettsia (bacteria), and Texas fever or babesiosis, caused by protozoa. The vampire bat of Central and South America is more important as a vector of rabies than as a parasite of cattle.

Several different groups of worms are important animal parasites. A number of different kinds of nematodes may infect the organs and tissues of each species of domestic animal. Flatworms (phylum Platyhelminthes) include flukes (trematodes) and tapeworms (cestodes). Some of these worms have very complex life cycles. For instance, some liver flukes, which parasitize sheep, cattle, and other livestock, may require one or more **intermediate hosts** in addition to the vertebrate animal to complete their life cycle. Certain snails are intermediate hosts of some of these flukes, which can be acquired when an animal drinks water contaminated by the snails. Spiny-headed

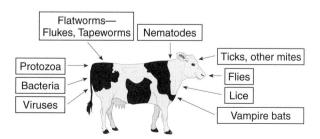

Figure 10-14 Main groups of parasites and pathogens that affect domestic cattle.

worms (phylum Acanthocephala) are a small group of specialized parasites (not shown in Figure 10-14). One kind is an important parasite of swine. **Antihelminthic** is a general term given to drugs used to treat parasitic worm infections in domestic animals.

Although Figure 10-14 and related comments pertain to cattle, similarly specific listings could be developed for swine, sheep, poultry, and other livestock. The diagnosis and treatment of specific animal parasites and diseases are highly specialized. Additional information on this subject can be found in veterinary medicine and parasitology texts, as well as sources in the Bibliography at the end of this chapter. In addition to the pathogens, parasites, and arthropod pests of domestic animals, vertebrate predators of livestock may be important on local and regional scales. The dingo fence is an example of a major regional effort to restrict predatory wild dogs from sheep grazing areas of southeastern Australia.

Summary

Damage and losses in agriculture are caused by weeds, insects, nematodes, vertebrates, and disease-causing pathogens such as fungi, bacteria, and viruses. Various ecological factors, such as temperature, season, moisture, soil type, crop cultivar, and dispersal ability affect the development of these pests and pathogens and their effects on crop plants. Numerous methods are used for managing pests and pathogens, including biological, chemical, physical/mechanical, preventive, and cultural methods. Cultural methods include practices that favor or improve the health of the crop plant and practices aimed at reducing the numbers of pests or pathogens. Distinct kinds of parasites and pathogens affect different types of domestic animals. The identification and management of agricultural problems caused by pests and diseases can be complex, so specialized information may be needed for their effective management.

Topics for Review and Discussion

1. What are some ways in which we can improve the ability of crop plants to compete with weeds?
2. Discuss the importance of vectors in the spread of plant and animal diseases.
3. All items in a list are not equal! Tables in this chapter provide lists of methods for managing weeds, plant diseases, nematodes, and insects. Some of these methods are much more important than others. In your judgment, which methods are likely to be the three most important and the three least important that growers might use to manage weeds? Plant diseases? Nematodes? Insects?
4. Nematodes are often studied in the discipline of plant pathology, along with fungi, bacteria, and other agents that cause plant diseases. Why might it be appropriate to include nematodes with the plant pathogens? What control methods might be applicable both to nematodes and to pathogens like fungi or bacteria?
5. Compare the groups of organisms responsible for the main pest and disease problems of:
    a. The major crop plant in your area; and
    b. The major domestic animal produced in your region.

Literature Cited

Agrios, G. N. 1997. *Plant Pathology*. 4th ed. San Diego, Calif.: Academic Press.
Arnold, K. A. 1991. Environmental control of birds. In *CRC Handbook of Pest Management in Agriculture*, 2d ed., Vol. I, ed. D. Pimentel, 593–600. Boca Raton, Fla.: CRC Press.
Barker, K. R., G. A. Pederson, and G. L. Windham, eds. 1998. *Plant and Nematode Interactions*. Madison, Wisc.: American Society of Agronomy.
Bartlett, A. C. 1991. Insect sterility, insect genetics, and insect control. In *CRC Handbook of Pest Management in Agriculture*, 2d ed., Vol. II, ed. D. Pimentel, 279–287. Boca Raton, Fla.: CRC Press.
Bennett, W. F., ed. 1993. *Nutrient Deficiencies and Toxicities in Crop Plants*. St. Paul, Minn.: The American Phytopathological Society.
Borror, D. J., C. A. Triplehorn, and N. F. Johnson. 1989. *An Introduction to the Study of Insects*. 6th ed. Philadelphia, Pa.: Saunders College Publishing.
Cook, R. J. 1993. Making greater use of introduced microorganisms for biological control of plant pathogens. *Annual Review of Phytopathology* 31: 53–80.
Davis, D. E. 1991. Environmental control of rodents. In *CRC Handbook of Pest Management in Agriculture*, 2d ed., Vol. I, ed. D. Pimentel, 587–592. Boca Raton, Fla.: CRC Press.
Dindal, D. L., ed. 1990. *Soil Biology Guide*. New York: John Wiley and Sons.
DiTomaso, J. M., and D. L. Linscott. 1991. The nature, mode of action, and toxicity of herbicides. In *CRC Handbook of Pest Management in Agriculture*, 2d ed., Vol. II, ed. D. Pimentel, 523–569. Boca Raton, Fla.: CRC Press.
Evans, K., D. L. Trudgill, and J. M. Webster, eds. 1993. *Plant Parasitic Nematodes in Temperate Agriculture*. Wallingford, U.K.: CAB International.
Gill, K. S., M. A. Arshad, and J. R. Moyer. 1997. Cultural control for weeds. In *Techniques for Reducing Pesticide Use*, ed. D. Pimentel, 237–275. New York: John Wiley and Sons.
Gould, A. B., and B. I. Hillman. 1997. Sanitation, eradication, exclusion, and quarantine. In *Environmentally Safe Approaches to Crop Disease Control*, eds. N. A. Rechcigl and J. E. Rechcigl, 223–241. Boca Raton, Fla.: CRC Lewis Publishers.
Luc, M., R. A. Sikora, and J. Bridge, eds. 1990. *Plant Parasitic Nematodes in Subtropical and Tropical Agriculture*. Wallingford, U.K.: CAB International.
Marshall, D. 1997. Cultural controls for crop diseases. In *Techniques for Reducing Pesticide Use*, ed. D. Pimentel, 221–235. New York: John Wiley and Sons.
McSorley, R., and L. W. Duncan. 1995. Economic thresholds and nematode management. In *Advances in Plant Pathology*, Vol. 11, eds. J. H. Andrews and I. C. Tommerup, 147–171. San Diego, Calif.: Academic Press.
Mehrotra, R. S., K. R. Aneja, and A. Aggarwal. 1997. Fungal control agents. In *Environmentally Safe Approaches to Crop Disease Control*, eds. N. A. Rechcigl and J. E. Rechcigl, 111–137. Boca Raton, Fla.: CRC Lewis Publishers.
Mink, G. I. 1991. Control of plant diseases using disease-free stocks. In *CRC Handbook of Pest Management in Agriculture*, 2d ed., Vol. I, ed. D. Pimentel, 363–391. Boca Raton, Fla.: CRC Press.
Minotti, P. L. 1991. Role of crop interference in limiting losses from weeds. In *CRC Handbook of Pest Management in Agriculture*, 2d ed., Vol. II, ed. D. Pimentel, 359–368. Boca Raton, Fla.: CRC Press.
Mundt, C. C. 1990. Disease dynamics in agroecosystems. In *Agroecology*, eds. C. R. Carroll, J. H. Vandermeer, and P. M. Rosset, 263–299. New York: McGraw-Hill.

Radosevich, S., J. Holt, and C. Ghersa. 1997. *Weed Ecology: Implications for Management.* 2d ed. New York: John Wiley and Sons.

Ragsdale, N. N., and H. D. Sisler. 1991. The nature, modes of action, and toxicity of fungicides. In *CRC Handbook of Pest Management in Agriculture,* 2d ed., Vol. II, ed. D. Pimentel, 461–496. Boca Raton, Fla.: CRC Press.

Regnier, E. E., and R. R. Janke. 1990. Evolving strategies for managing weeds. In *Sustainable Agricultural Systems,* eds. C. A. Edwards, R. Lal, P. Madden, R. H. Miller, and G. House, 174–202. Delray Beach, Fla.: St. Lucie Press.

Roberts, D. W., J. R. Fuxa, R. Gaugler, M. Goettel, R. Jaques, and J. Maddox. 1991. Use of pathogens in insect control. In *CRC Handbook of Pest Management in Agriculture,* 2d ed., Vol. II, ed. D. Pimentel, 243–278. Boca Raton, Fla.: CRC Press.

Romoser, W. S., and J. G. Stoffolano, Jr. 1998. *The Science of Entomology.* 4th ed. Boston, Mass.: WCB/McGraw-Hill.

Rush, C. M., G. Piccinni, and R. M. Harveson. 1997. Agronomic measures. In *Environmentally Safe Approaches to Crop Disease Control,* eds. N. A. Rechcigl and J. E. Rechcigl, 243–282. Boca Raton, Fla.: CRC Lewis Publishers.

Shaner, G. 1991. Genetic resistance for control of plant disease. In *CRC Handbook of Pest Management in Agriculture,* 2d ed., Vol. I, ed. D. Pimentel, 495–540. Boca Raton, Fla.: CRC Press.

Slife, F. W. 1991. Environmental control of weeds. In *CRC Handbook of Pest Management in Agriculture,* 2d ed., Vol. I, ed. D. Pimentel, 579–585. Boca Raton, Fla.: CRC Press.

Stern, V. 1991. Environmental control of insects using trap crops, sanitation, prevention, and harvesting. In *CRC Handbook of Pest Management in Agriculture,* 2d ed., Vol. I, ed. D. Pimentel, 157–182. Boca Raton, Fla.: CRC Press.

Stirling, G. R. 1991. *Biological Control of Plant Parasitic Nematodes.* Wallingford, England: CAB International.

Tivy, J. 1992. *Agricultural Ecology.* Harlow, U.K.: Longman Scientific and Technical.

Bibliography

Calnek, B. W., H. J. Barnes, C. W. Beard, W. M. Reid, and H. W. Yoder, Jr. 1991. *Diseases of Poultry.* 9th ed. Ames, Ia.: Iowa State University Press.

Hamilton, C. J. 1991. *Pest Management: A Directory of Information Sources. Volume 1: Crop Protection.* Wallingford, U.K.: CAB International.

———. 1995. *Pest Management: A Directory of Information Sources. Volume 2: Animal Health.* Wallingford, U.K.: CAB International.

Hill, D. S. 1994. *Agricultural Entomology.* Portland, Ore.: Timber Press.

Hill, J. R., and D. W. B. Sainsbury, eds. 1995. *The Health of Pigs.* Harlow, U.K.: Longman Scientific and Technical.

Hillocks, R. J., and J. M. Waller, eds. 1997. *Soilborne Diseases of Tropical Crops.* Wallingford, U.K.: CAB International.

Holm, L., J. Doll, E. Holm, J. Pancho, and J. Herberger. 1997. *World Weeds: Natural Histories and Distribution.* New York: John Wiley and Sons.

Kimberling, C. V. 1988. *Jensen and Swift's Diseases of Sheep.* 3rd ed. Philadelphia, Pa.: Lea and Febiger.

McGovern, R. J., and R. McSorley. 1997. Physical methods of soil sterilization for disease management including soil solarization. In *Environmentally Safe Approaches to Crop Dis-*

ease Control, eds. N. A. Rechcigl and J. E. Rechcigl, 283–313. Boca Raton, Fla.: CRC Lewis Publishers.

Parker, C., and C. R. Riches. 1993. *Parasitic Weeds of the World: Biology and Control.* Wallingford, U.K.: CAB International.

Pimentel, D., ed. 1991. *CRC Handbook of Pest Management in Agriculture.* 2d ed., Vol. II. Boca Raton, Fla.: CRC Press.

Provost, A., and G. Uilenberg. 1990. Constraints to livestock production due to diseases. In *Developing World Agriculture,* ed. A. Speedy, 254–264. London: Grosvenor Press International.

Rechcigl, N. A., and J. E. Rechcigl, eds. 1997. *Environmentally Safe Approaches to Crop Disease Control.* Boca Raton, Fla.: CRC Lewis Publishers.

Smith, A. E., ed. 1995. *Handbook of Weed Management Systems.* New York: Marcel Dekker.

Steelman, C. D. 1991. Environmental control of arthropod pests of livestock. In *CRC Handbook of Pest Management in Agriculture,* 2d ed., Vol. I, ed. D. Pimentel, 603–621. Boca Raton, Fla.: CRC Press.

Chapter 11

Principles of Pest Management

Key Concepts

- Strategies and tactics for managing pests
- Determining when to manage pests
- Ecological basis for pest management
- Integrated pest management programs

Devastating losses to agricultural crops are caused by the pests discussed in the previous chapter. The extent of these losses is difficult to estimate because comprehensive surveys of crop losses are difficult to conduct. Even within a region, a pest may cause crop failure in one field, but leave another field almost untouched. Interpretation can be difficult when losses from multiple pests occur in the same field. There is also the question of whether or not expenditures for pesticides and other control practices should be included in crop loss estimates. Pesticide sales can provide some impression of the magnitude of pest problems, although these data may be biased toward perceived pest problems rather than actual losses. However, data on sales are usually more readily available than actual loss estimates.

World pesticide sales for 1995 were estimated at $37.7 billion (Aspelin 1997). Pesticide sales in the United States that year were estimated at $11.3 billion, of which $7.9 billion was for agricultural uses. Herbicides (55%) and insecticides (32%) made up the largest proportions of U.S. pesticide expenditures. The proportions of various types of pesticides used vary greatly by crop (Figure 11-1). Note the especially high proportion of herbicide use on corn and soybean, two crops with vast acreage in the United States. Atrazine, a herbicide commonly used in corn production, was the most heavily used pesticide in U.S. agricultural crops in 1995 (Aspelin 1997). These examples illustrate the extent and costs of managing agricultural pest problems, particularly weeds.

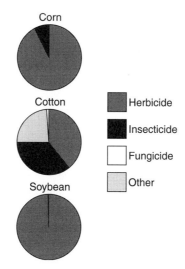

Figure 11-1 Relative proportions of different groups of pesticides used on four major U.S. crops in 1995. The "Other" category includes multipurpose soil fumigants and desiccants.
Based on data presented by the United States Department of Agriculture (1997).

Pest Management Strategies

The various **tactics,** or methods for managing particular types of pests, were summarized in the previous chapter. Specific tactics were arbitrarily designated as biological, chemical, physical/mechanical, preventive, or cultural, although some tactics may fit multiple categories. For example, organic mulches may affect weeds by physically blocking emergence or by releasing decomposition products that can be toxic to weeds.

A **strategy** is the overall plan or approach that we use to manage a particular pest. A management strategy may include one or more tactics. **Integration** is the combined use of multiple tactics in a management plan. For example, a management strategy for a particular foliar disease in a crop might involve combining (integrating) the use of wider spacing between crop rows, reduced overhead irrigation, and optimal timing of planting to reduce disease pressure. Pest management strategies may be of several types:

- *Prevention* or *exclusion*. Preventive strategies are aimed at keeping a pest from entering a region or field in which it is not present. A variety of preventive or sanitation tactics may be helpful for this purpose. Preventive strategies are particularly useful in limiting the spread and introduction of exotic pests across regional and international borders.
- *Eradication*. These strategies attempt to eliminate a pest from an area in which it is already present. When successful, spectacular results can be achieved, as in the elimination of the screwworm from the southern United States or the elimination of various Mediterranean fruit fly (medfly) introductions in Florida. But eradication of

most pests, even from a single field, is an extremely difficult and expensive task. Imagine how difficult it might be to remove *every* aphid or *every* weed from a 1-hectare field. Not only would it be nearly impossible, but it may not even be worthwhile to eliminate *every* weed since very low numbers of weeds may have no impact on crop yield. For these reasons, many people prefer to use the term "management" rather than "control" when referring to reduction of pest populations because, in practice, we rarely control or eliminate any pest.

- *Pest population reduction.* Many pest management strategies focus on reduction of pest numbers. The questions of when to reduce pest numbers and how far to reduce numbers are addressed in more detail later in this chapter. Strategies for pest population reduction may be *responsive,* based on current biological information obtained from a particular field, or they may be *preemptive,* based on calendar or historical expectations. An insecticide spray every Tuesday to avoid buildup of earworms on sweet corn is an example of a preemptive strategy.
- *No action.* This can be an appropriate strategy for many minor pests and for situations when a pest is not expected to be particularly damaging.

Regardless of the strategy chosen and the specific tactics used, two basic management questions may be asked:

1. How should the pest be managed? (Or which tactic[s] should be used?)
2. When should the pest be managed, if at all? (Or is the management plan economical?)

Tactics and Issues in Pest Management

Although various pest management tactics have already been introduced in the previous chapter, several of these tactics are used so frequently that a number of special considerations and issues are associated with their use.

Chemical Pesticides

As indicated at the beginning of this chapter, the amount and dollar value of recent pesticide use in the United States and worldwide remain extremely high. Several issues related to the use of chemical pesticides are discussed here and in Chapter 12. Additional information on this topic may be found in various texts (e.g., Agrios 1997; Radosevich, Holt, and Ghersa 1997; Romoser and Stoffolano 1998) or technical publications about pesticides (e.g., Meister 1995).

Classification of Pesticides. Pesticides can be classified in a variety of ways, depending on the main feature of interest. Some overlap in these classifications may exist.

- *Target pest.* Depending on the target organism, materials can be classified as insecticides, herbicides, nematicides, acaricides, fungicides, bactericides, or rodenticides.
- *Origin.* Materials can be **natural products,** which are produced by or isolated from living organisms, or **synthetic products,** which are artificial compounds produced

in the laboratory. Various degrees of overlap may exist. For instance, some materials are produced from industrial-scale fermentations involving bacteria or fungi, while others are analogues of naturally occurring materials.
- *Chemical structure.* At the simplest level, we can distinguish between inorganic and organic compounds. Many different classes of organic compounds have been used, such as chlorinated hydrocarbons, organophosphates, carbamates, pyrethroids, etc. The chemical structure ultimately determines the toxicity and mode of action of the material. In addition, many environmental effects of pesticides and other agrichemicals depend on chemical structure and related properties, which are discussed in the next chapter.
- *Formulation.* A pesticide product is typically composed of an **active ingredient,** which is the actual toxic compound, and various other materials that may facilitate its delivery, improve its stability, or make it easier and less toxic to handle. Depending on these added materials, the same active ingredient can be formulated in a number of different ways. Pesticide formulations include granules, pellets, dusts, fumigants, liquids or flowables, emulsifiable concentrates, wettable powders, and water-soluble powders or liquids (see Application 11-1). Emulsifiable concentrates and wettable powders are special formulations of nonsoluble pesticides that aid in their suspension in water, so that they can be sprayed. **Adjuvants** are additional materials that are mixed with sprays to enhance their performance. They may be included as part of a formulation or added later by the spray applicator.
- *Mode of action.* Pesticides can act against their targets as toxins, repellents, attractants, desiccants, growth regulators, etc. Various types of poisons can be further classified according to their activity on specific sites, such as the nervous system or the respiratory system. Some pesticides are plant or insect growth regulators or analogs of insect hormones. Of course, these materials are much more specific than are the

Application 11-1 *Names for Chemical Pesticides*

A typical chemical product has three names: a chemical name (which may be cumbersome), a common name, and a trade name (under which it is sold). For the pesticide aldicarb:

Common name: aldicarb

Chemical name: 2-methyl-2-(methylthio) propionaldehyde O-(methylcarbamoyl) oxime

Trade name: Temik® 15G

Temik® 15G is the trade name for a specific formulation of aldicarb. The "G" indicates that the formulation is a granule. The "15" indicates that the amount of aldicarb (the active ingredient) in this formulation is 15%. In specifying an application rate, it is critical to know whether we are referring to the formulated product or the active ingredient. An application of 10 kg/ha of Temik® 15G is the same as 1.5 kg/ha of aldicarb. The aldicarb rate can be written as 1.5 kg a.i./ha, where the "a.i." stands for active ingredient, which avoids any ambiguity.

general purpose fumigants or other toxins. The mode of action is important in determining how selective a compound will be for a particular pest or group of pests.
- *Use or route of entry.* Pesticides can also be classified by how they are delivered to the pest. Herbicides can be soil-applied or foliage-applied, and insecticides can be stomach poisons (ingested by eating), contact poisons (absorbed through cuticle), or fumigants (enter through respiratory system). Many modern pesticides are systemic. A **systemic** pesticide is taken up by the plant and can be translocated to various plant parts. A systemic herbicide applied to the leaves may move through the plant to kill roots as well. A systemic insecticide incorporated into leaves may kill insects feeding on the leaves. Systemic pesticides are regulated, and most governments have established tolerances for the levels of pesticides permitted in harvested food products.

Advantages of Pesticide Use. Of the many pest management tactics available, pesticides provide the most rapid method for responding to an emergency situation. They are fast-acting and often highly effective, resulting in large numbers of dead pests in a relatively short time. They may be the only means of saving a crop in advanced stages of pest infestation. In many situations, pesticide use is economical, and the value of the increased yield returned may be several to many times the cost of the pesticide product and its application. Many pesticides are readily available and accessible, and there may be a variety of products to choose from in managing a particular situation. Finally, while a knowledge of pesticide labeling and application technology is essential for applying pesticides, less detailed knowledge of pest biology and ecology is required for their use, at least in comparison with nonchemical methods.

Disadvantages of Pesticide Use. Many pesticides are toxic to humans and other organisms, although the degree of toxicity depends on the chemical nature of the material. Particular care must be taken to limit exposure to applicators and other workers who are involved in the handling of pesticides. This includes the disposal of pesticide containers and residues as well. Pesticide residues in and on food products are another major safety concern. Most governments attempt to address these human safety and environmental issues through regulation of pesticide use, application rates, protective equipment, residue tolerances in food, harvest intervals, and other aspects involved in pesticide application.

Several environmental issues related to pesticide use are discussed in more detail in Chapter 12, including pollution, groundwater contamination, persistence of residues, and effects on nontarget organisms. Although not usually considered in economic analyses, the environmental costs associated with pesticide application could greatly increase the cost of pesticide use. Of most immediate concern are nontarget effects in and around the application site, including:

- Effects on natural enemies of pests;
- Toxicity to bees and other pollinators;
- Effects on wildlife in and around the treated field;
- Possible phytotoxicity to crop plants (including carryover from herbicide residues); and
- Drift to nontarget crops and residential areas.

The elimination of natural enemies may have long-term effects on pest management in an agroecosystem. A *secondary pest* is one that normally does not cause much economic damage. Some secondary pests may become serious pests when populations of their natural enemies or their competitors are reduced or eliminated by pesticides. For example, several species of mites have become serious pests to various fruit crops when populations of predatory mites were reduced by acaricidal sprays. In some of these situations, the initial reduction of the pest by a chemical pesticide may be followed by long-term increases in pest populations beyond the levels present before any treatment. This phenomenon is called **resurgence.**

Pesticide resistance is reported in many insect pests and in increasing numbers of weeds and fungal pathogens. Frequent exposure to certain pesticides has resulted in selection for genotypes that are resistant or tolerant to these materials. The speed at which pesticide resistance develops is affected by pest genetics, the nature of the pesticide, and various biological and ecological factors. Frequent application of a persistent material maximizes long-term selection pressure on a pest population and increases the likelihood that resistance will develop. Because developmental costs for new pesticides are high, the anticipation and proper management (see Roush 1991) of pesticide resistance is important in prolonging the lifetime of effective chemical products.

Expectations from Pesticide Use. Most people are familiar with the rapid and short-term effects resulting from chemical pesticide use. To some extent, the ready availability of effective short-term solutions to pest problems may have hindered the need to look for and study alternative methods for managing pests. Societal pressures for quick action and the rapid efficacy of pesticides have led to expectations among many people that pests are always "controlled" quickly and effectively by the application of some material (chemical or otherwise). Such expectations are extended to nonchemical methods of pest management as well, even to methods that are not designed for emergency situations, but for long-term management. However, many of the pest management tactics outlined in Chapter 10 will not produce rapid "control" of pest populations. Instead, they may be useful in strategies designed to prevent pest levels from reaching emergency situations in the first place. Considerable patience and biological information may be needed for their effective use. Disappointment may result from their use if immediate relief from a crisis is needed.

Host Plant Resistance

Host plant resistance is used widely for managing pest and disease problems, and examples are discussed in various articles (e.g., Brown 1995; Tingey and Steffens 1991; Young 1998) and in several of the applications at the end of this chapter. Nevertheless, pests may overcome, or "break," resistance in many cases. Continual planting of the same resistant cultivar results in selection toward those pest genotypes that can feed and reproduce (even to a limited extent) on that particular cultivar. As selection continues and pest fitness improves, new pest isolates or populations develop, which are well adapted to the formerly resistant cultivar. Overreliance on the same resistant cultivar should be avoided, and the use of resistant cultivars should be carefully managed over time to ensure that resistance remains stable for a long time. Resistance management can include rotation with other crops or with susceptible cultivars of the same crop to slow the development of new pest isolates.

As examined in Chapter 8, resistant cultivars can be obtained from traditional plant breeding programs or from genetic engineering. The latter approach has provided several applications in pest management. Cultivars of several crop plants such as cotton, corn, and soybean have been developed with resistance or tolerance to certain herbicides. This means that a general purpose herbicide such as glyphosate can be used on glyphosate-tolerant cotton cultivars to manage virtually any kind of weed in the crop. Also, genes that produce the *Bacillus thuringiensis* toxin have been introduced into some crop cultivars. A Bt cotton cultivar produces a toxin that is effective against several caterpillars that are serious pests of cotton. Other transgenic crop cultivars are in development for managing virus diseases and other pest problems. As with any management practice, some risks are associated with the use of transgenic plants (see Fuchs and Gonsalves 1997). Selection toward disease isolates and pest populations that can overcome resistance will proceed whether a resistant cultivar has been developed from genetic engineering or from traditional plant breeding methods. The constant presence of Bt toxin in certain transgenic plants may provide especially strong selection pressure toward Bt-resistant caterpillars, so the use of these cultivars must be managed carefully.

Biological Control

Typically, the biological methods are called biological "control" by most scientists, even though they are a form of pest management. The definition of biological control and the methods that are included in biological control are subjects of some controversy (see Application 11-2).

Application 11-2 *What Is Biological Control and What Is Not?*

Biological control has been variously defined, and people may have very different views about what is or is not included as biological control. For the most part, in this book we consider biological control to be management by predators or parasites, although we have arbitrarily included sterile male techniques here as well (sometimes classified separately as **autocidal control**). Where do you draw the line as to what is included in biological control? Should viruses be included? They are usually studied along with living organisms and they are parasites. Isolates of the bacterium *Bacillus thuringiensis* infect a variety of insect pests, releasing a toxin in the process. Is it still biological control if the toxin is produced in bacterial cultures in the laboratory, isolated, and then sprayed on caterpillars in the field? Should plants that have been genetically engineered to produce Bt toxin be considered biological control agents? Many plants produce alkaloids and other natural products that may be antagonistic to pests. Should antagonistic plants be considered biological control agents? Could cover cropping, host-plant resistance, crop rotation, and related cultural practices be considered types of biological control? Is any use of a natural plant product a form of biological control? For example, if the chemical formula for a natural product is known, it may be possible to synthesize it in a laboratory. The natural source is bypassed, providing a synthetic source of what is normally a natural product. This application is intended to show the difficulty in placing a definition or boundary on a subject like biological control. It also emphasizes the need to recognize that different people may have very different impressions of what is included as biological control.

Classical biological control is the introduction of a natural enemy to manage a nonnative pest. When introduced into a new area, some exotic (nonnative) organisms may become serious pests in their new environment. The introduced pest is free of any predators or parasites that may have kept it under control in the country or region from which it originated. In classical biological control programs, scientists search for the natural enemies of an exotic pest in its native country or region, and then import them to the location where the pest is causing problems. One highly successful example was the introduction of the Vedalia beetle from Australia in the late 1800s to control the cottony-cushion scale insect that was causing serious damage to citrus in California. Success of a planned introduction is hard to predict, since it depends on ecological conditions and how well the introduced predator or parasite adapts to its new environment. Many classical biological control attempts are unsuccessful or have only limited impact. In a few cases, unexpected problems arise. Today, much planning goes into minimizing potential problems from imported natural enemies, which are tested extensively for nontarget effects before they are released.

Natural biological control occurs to some extent in all but the most sterile agroecosystems. Recall, for example, the many different predators and parasites present in soil (Chapter 3), many of which may have some (but usually unmeasured!) impact on pest species. Historically, the vegetable leafminer (see Chapter 10) has not been a serious pest on most crops because it is usually managed by naturally occurring parasitoids and other natural enemies. However, where these biological control agents have been eliminated by insecticidal applications (directed at pests), the vegetable leafminer has become an important secondary pest. The more the leafminer is sprayed with insecticides, the more persistent the problem becomes! Naturally occurring predators, parasites, and pathogens provide various levels of management of potential pests, depending on the situation. Where naturally occurring management is inadequate, additional measures may improve performance. **Augmentation** is the increase in populations of natural enemies in a site. One common example is the addition of ladybird beetles to a field to prey on aphids.

Conservation of natural enemies is important to ensure adequate numbers for pest management. In annual crops, it is particularly important to provide natural enemies with an alternate food source when the key pest is not available. This task is easier with general predators that may persist on a variety of food sources. Maintenance of vegetational diversity in and around the field may help to maintain alternate food supplies, overwintering sites, and hiding places (refugia) for natural enemies. Other practices are also helpful, such as the timing, placement, and choice of materials used in chemical applications to minimize nontarget effects.

Whether or not classical or natural biological control strategies may be more appropriate for a particular pest problem depends on the situation. Among the most important considerations are whether the pest is native to the area or is introduced, and what kinds and levels of natural enemies may already be present and acting on the pest.

Cultural Practices

The effective use of cultural practices for pest management usually requires very specific biological and ecological information. For example, to use crop rotation against a certain fungal disease, you need to know if all stages of the fungus life cycle are confined to soil, if persistent survival stages are included in the life cycle, if candidate rotation crops and cultivars are hosts for the fungal isolate present, if multiple isolates exist, and so on. Some limited trials and testing may be needed to establish expected performance for a particular site. Many cultural practices used for pest management impact other features of crop ecology as well. Crop rotation for disease management could affect soil fertility, soil structure, organic matter, weed distribution, herbicide residues, or soil water. Various cultural practices are discussed in more detail in Chapters 13, 14, and 15.

Integration of Tactics

The integration, or complementary use of multiple tactics, is an important feature of many pest management strategies. For instance, the use of tillage, narrow crop row spacings, and timing of planting may provide better weed management than any of these tactics alone. Use of host plant resistance along with conservation of natural enemies may provide improved insect management. Integration is essential when a planned tactic is not expected to provide adequate management if used alone. Of course, the tactics chosen must be compatible. Applications of nonselective insecticides do not integrate well with biological control by insect predators or parasitoids. Additional examples of integrated pest management strategies may be found in the applications and literature at the end of this chapter.

Deciding When to Manage Pests

Provided that tactics are available to manage a particular pest, the next questions are if and when the pest should be managed. To answer these questions, additional information is needed on:

- Size of the pest population present;
- Estimates of damage expected from this population; and
- Economics of whether the benefit of management is worth the cost.

Sampling of Agricultural Pests

It is virtually impossible to know the exact density, or size, of a pest population in a particular field. This would require the impractical and expensive task of counting every individual pest in the field. Therefore, it is almost always necessary to examine a *sample,* or smaller unit of the field, to *estimate* pest population density. The objective of most sampling plans is to obtain estimates of the mean and standard deviation (see Chapter 1) of the population density of a particular pest. In other cases, data on presence or absence of a pest may be sought.

Accuracy and Methods for Collecting Samples. Accuracy refers to how close our estimate of mean population size is to the actual population size in the field. We will never know the actual population size (unless we count the whole field), but our choice of correct methods and tools for sampling will greatly affect accuracy. For instance, aphids, thrips, whiteflies, and other small insects usually congregate on the undersides of plant leaves. We may find a few if we inspect the upper surfaces of leaves, but our estimate will be highly inaccurate unless we examine the undersides of leaves.

The methods for obtaining the most accurate samples of pest populations vary with pest species and can be highly specialized. Technical information is available on the many tools and methods for estimating specific pests (Crossley et al. 1991; Metcalf and Luckmann 1994;

Southwood 1984). Sampling tools and methods vary in their complexity. In addition to direct observations, we can use nets; various types of traps, baits, and shovels; specialized cores for collecting soil samples; air sampling devices; or other means for removing a sample. Examination of many types of samples requires further analyses in the laboratory or extraction to remove organisms from soil or plant debris. Certain types of fungi and bacteria may need to be cultured on selective media. **Bioassay,** or culture in living plants or animals, may be required in some cases, such as to distinguish certain types of viruses.

Another consideration is the selection of an appropriate **sample unit,** or size of the area or item from which an individual sample is collected. A sample unit may be a single leaf, ten leaves, three whole plants, 10 meters of row, a square meter of land, or whatever is determined to be most appropriate for the pest(s) in question. In many cases, estimates of pest density are reported as numbers of pests per sample unit, such as twelve weeds per 2.0 meters of row, twenty aphids per leaf, fifty nematodes per 100 cm^3 of soil, etc. In other cases, biomass or weight of the sampled object per sample unit may be reported. Sometimes percent infection or coverage may be estimated, such as the percent of animals infected by a particular parasite, the percent of leaf surface covered by a foliar disease, or the percent of ground covered by a weed species. In other instances, estimates of pest damage rather than direct pest counts can be obtained, such as percent defoliation, or number of plants or animals showing a particular symptom. Rating scales can be used instead of numbers or percentages. In all cases, specific literature and local recommendations provide the best information on choosing methods to collect the most accurate sample in a particular situation.

Precision and Spatial Dispersion. Once we know the methods for collecting a sample, we must determine where to collect the sample and how many samples to collect from a field. These are difficult questions that depend on the spatial arrangement of organisms within the field.

In Chapter 7, we learned that pests or other organisms can be distributed or dispersed within a field in regular, random, or aggregated (clumped) patterns (see Figure 7-2). As discussed in that chapter, both the colonization pattern and the biology of the organism determine its pattern of distribution. However, the pattern of dispersion of an organism has important effects on sampling.

Sampling a pest that has a regular dispersion is easy because the pest density in one part of the field is the same as anywhere else. However, regular dispersion is rare in practice. Plant density (number of plants per square meter) in a machine-planted crop approaches regular dispersion. Random dispersion is the ideal case for statistical purposes. Provided that the site is uniform, measurements of many plant characteristics, such as height or yield of individual plants, should approach a random dispersion. Aggregated or clumped dispersion is typical of most agricultural pests. Individual organisms are clumped together in high-density locations, while they may be nearly absent in others. A sample unit (represented by the small box in Figure 11-2) may give a serious over- or underestimate of the mean population density, depending on where it was placed in the field.

We can obtain an impression of the spatial dispersion of an organism by collecting a few samples and determining their mean ($\bar{x}$) and variance (s^2):

$$\text{If } s^2 < \bar{x} \rightarrow \text{regular}$$
$$\text{If } s^2 = \bar{x} \rightarrow \text{random}$$
$$\text{If } s^2 > \bar{x} \rightarrow \text{clumped}$$

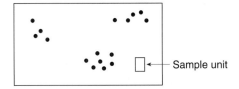

Figure 11-2 An aggregated spatial dispersion of pests (represented by dots) within a field poses problems in sampling. A sample unit (represented by box) could be placed in a high-density or a low-density area, giving very different impressions of the true population density in the field.

For most agricultural pests, clumped patterns will result, for several reasons:

- Dispersal pattern (large numbers of seeds or eggs at one location, limited mobility);
- Microhabitat (preference for certain spots within plant rows or field);
- Physical and environmental differences within the field (due to soil type, topography, etc.); and
- Agricultural practices (cultivation, old row sites, crop history).

Although crop yields, biomass, and various other measurements of plants within a field are expected to be random, they often may be clumped because they are related to soil properties. Soil texture, soil fertility, topography, soil organic matter, moisture retention, and other characteristics are not uniformly distributed, but may be clumped and irregular. Therefore, measurements of crop plants may follow similar patterns.

Unfortunately, clumped patterns of dispersion are typical problems for most agricultural sampling programs. The high variability (high s^2 values) makes it difficult to obtain a precise estimate from a series of samples. *Precision* refers to how close the estimates from individual samples are to one another.

Precision agriculture is an approach for optimizing fertilizer and pest management practices in fields where clumping and gradients in soil characteristics occur. Through the combined use of sampling and global positioning systems (GPS), areas of low fertility can be located within the field, and fertilizer application rates can be adjusted when these areas are reached. Areas of recurrent weed infestations can be located precisely and then spot treated.

Sampling Pattern. Because of clumped dispersion, it is necessary to take multiple samples in estimating average pest density in a field because some samples will invariably be collected from high-density locations and others from low-density locations. Several different patterns can be used for collecting multiple samples from a field (Figure 11-3):

Random. The choice of sampling location can be strictly random, but this can be difficult to set up in practice. Random numbers tables or other statistical schemes are needed to determine the sample locations.

Systematic. Samples are collected at regular intervals throughout the field. Systematic plans are easy to set up and are often used in practice. A **transect** is a series of samples collected along a straight line path.

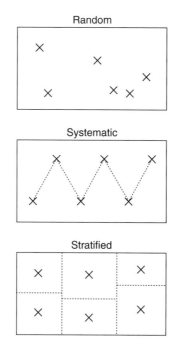

Figure 11-3 Patterns for collecting multiple samples within a field.

Stratified. To use this pattern, the field is divided up into portions, or **strata,** and a sample is collected from each stratum. Sample collection within a stratum can be random (stratified random) or systematic. Strata may be of equal or different sizes. Stratified plans are convenient in fields with multiple soil types or other variations that can be used as strata.

Number of Samples. The question of how many samples to collect has no definite answer! It depends on how much sample error we will tolerate or how much precision we require. Precision and error show opposite trends. A common measure of precision is the coefficient of variation, or CV, which is given by:

$$\text{CV} = (s/\bar{x}) \times 100\% \qquad \text{(Equation 11-1)}$$

The lower the value of CV, the greater the precision. As the number of samples collected increases, error decreases and precision increases (Figure 11-4). Note from Figure 11-4 that taking just a few samples provides a rapid decrease in sampling error. As more samples are collected, error continues to decrease, but the rate diminishes until a point is reached at which we cannot further decrease error (or increase precision). There is always a trade-off between sample precision and the number of samples collected. Much sampling work is required for high precision (low error). In practice, CV and other measures of precision relate to the confidence we have in our estimate of the population mean obtained by sampling,

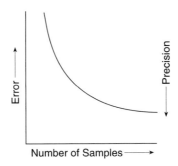

Figure 11-4 Relationship between sample error and number of samples collected.

such as whether the mean is 20 ± 5 or 20 ± 15. The methods and statistics used to develop sampling plans are introduced by other sources (McSorley 1987, 1998; Metcalf and Luckmann 1994; Shelton and Trumble 1991; Southwood 1984) and are the basis for developing sampling plans for specific situations.

Sometimes it becomes obvious when sampling a field that there are so many insects or weeds that we do not need to spend additional time sampling because treatment will be needed anyway. Or we may be finding so few insects or weeds that treatment will not be needed. Sequential sampling plans (see Shelton and Trumble 1991) can provide a shortcut to sampling work in these situations.

Crop Damage and Economic Thresholds

For most pest problems, the severity of crop damage increases with increasing pest population, so that an inverse relationship exists between crop yield and pest population density (Figure 11-5). However, very low numbers of pests, such as one or two weeds in a 100-meter

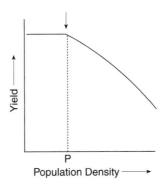

Figure 11-5 General relationship between crop yield and pest population density. Crop damage threshold is at pest population density = P.

row of corn, may not affect crop yield at all. For many crop and pest combinations, we can recognize a **crop damage threshold,** or the pest population level below which no measurable yield loss occurs (represented by population density "P" in Figure 11-5). The crop damage threshold may be called the tolerance limit by some sources. Crop damage thresholds vary greatly, depending on the pest and crop. Some organisms must be present in very high numbers to cause damage, while in other situations, such as with insect vectors of important virus diseases, the crop damage threshold may be very low. In a few cases, feeding or grazing by low numbers of insects or other herbivores may actually stimulate yield to some extent.

Damage to nonmarketable plant parts may be tolerated if it does not affect yield. For example, soybean may tolerate some defoliation (up to 20% or more for some growth stages and cultivars) before any impact on marketable yield is measured. Therefore, some foliage-feeding caterpillars may be tolerated on the crop, but the tolerance for insects like stink bugs, which feed directly on the pods, is low.

Economic Thresholds. Consider the situation depicted in Figure 11-6, which is similar to Figure 11-5 except that a pest population above the crop damage threshold is shown. If the pest population level is at point X, then a yield level B would be expected, which represents a yield loss of Y compared to situations when pest numbers were at or below the crop damage threshold, P.

According to Figure 11-6, if we applied a treatment that could reduce population level X to population level P or below, we would expect to obtain the maximum yield (A) and would have prevented the loss in yield (ΔY). We could evaluate whether this treatment was worthwhile if we knew:

1. The cost of the treatment; and
2. The price we would expect to receive for the crop yield, so that we could calculate the dollar value of ΔY.

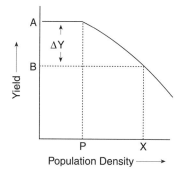

Figure 11-6 General relationship between crop yield and pest population density. A pest population at point X, above the crop damage threshold P, results in a yield loss of ΔY as yield drops from level A to level B.

The **economic injury level** (EIL) is the pest density at which the value of the loss prevented equals the cost of treatment. In Figure 11-6, if point X is the economic injury level (X = EIL), note that:

- If pest density > EIL, then $ value of ΔY > treatment cost, so the treatment is profitable.
- If pest density < EIL, then $ value of ΔY < treatment cost, so the treatment loses money.

We may not choose to wait until the pest population density reaches the EIL before we treat to manage the pest. This is important when there is a time lag between sampling and treatment. If our sampling shows that we are at the EIL (e.g., point X on Figure 11-6), then there is a possibility that the pest population could grow beyond the EIL before we have a chance to treat the field. We may wish to treat the field at some pest density slightly lower than the EIL, so that we do not actually reach or surpass the EIL. The **action threshold** (AT) is the pest density at which treatment is necessary to prevent economic injury (to prevent the population from reaching the EIL). Many authors use the terms "action threshold" and "economic threshold" interchangeably, but sometimes economic threshold is used to mean other things. So we will use the terms action threshold and economic injury level as defined here. The relationship between the economic injury level and the action threshold as pest populations fluctuate over time is shown in Figure 11-7.

Dynamic and Static Management Scenarios. A pest management situation may be considered *dynamic* when pest populations fluctuate during the growing season and treatment decisions may be considered at any time. The dynamic situation shown in Figure 11-7 applies to many common situations, such as insect populations, foliar diseases, or weed infestations developing during the crop season. Because pest populations can change rapidly, it is critical to have detailed, current information and samples about the pest levels and their status in the field. Such information is usually obtained by scouting fields on a regular basis.

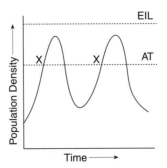

Figure 11-7 Fluctuation of pest population density over time in relation to action threshold (AT) and economic injury level (EIL). Treatment at both X points is needed to prevent the population from reaching EIL.

Some pest management situations are more *static,* such as when pest population fluctuations are not an important issue or when management at various points during the crop season is not possible. Suppose the question is whether to manage a soilborne pest with soil fumigation. The decision must be made before planting. Once the crop is planted it may be impossible to use a treatment like soil fumigation. Future scouting and pest population fluctuations after planting are irrelevant with regard to this treatment decision, as well as with other tactics that must be used prior to planting. In these cases, the data in Figure 11-6 can be applied if yield is related to *preplant* population density rather than *current* pest population density. We must anticipate that a certain preplant pest level will result in a predictable yield loss in the future. The EIL becomes the action threshold and we make treatment decisions based on this preplant pest density. A similar static approach would apply when making the decision to plant a pest-resistant crop cultivar or to use crop rotation.

Cost/Benefit and Uncertainty in Pest Management. To use action thresholds and economic injury levels in making management decisions based on relationships like those shown in Figures 11-6 and 11-7, we need data of several types:

- Estimate of pest population density;
- Expected performance of the treatment tactic;
- Cost of the treatment; and
- Economic forecast of crop value.

Uncertainty and error may be associated with all of these kinds of data. The very nature of the crop yield/pest population density relationship (see Figures 11-5 and 11-6) may be changed by environmental conditions, for example. Environmental conditions may also affect the performance of the treatment method. Fluctuations in market prices can make future crop values hard to predict. Earlier in this chapter, we saw that the estimate of pest population density is always a source of error and uncertainty. Faced with such uncertainty, a grower may opt for the conservative use of a pesticide to err on the side of safety (avoid risk of yield loss). To address this issue realistically, we need more detailed forecasts, using probabilities that certain levels of pest damage will occur (e.g., 80% chance of 10% yield loss, 15% chance of 20% yield loss, 5% chance of 50% yield loss).

At present, many pest management decisions are made based on whether or not the economic injury level is expected to be reached. In these cases, the benefit of treatment is balanced against the cost of control. In the case of pesticide application, this includes not only the cost of the material, but the labor, fuel, protective equipment, and other costs associated with making the treatment application. Rarely are environmental costs (e.g., effects on wildlife, cleanup of pollution, etc.) considered when making on-farm decisions on pesticide use. A cost/benefit ratio favoring economic benefit from pesticide application might change drastically if environmental costs are included. At present, environmental costs are ignored or subsidized by society, so they do not actually enter into the pest management decision-making process.

Pest Management Decision Making

The dynamic process of pest management decision making has several key features (Figure 11-8). Whether or not a farmer or crop manager chooses to implement a particular

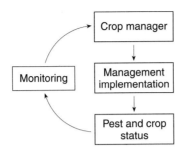

Figure 11-8 Major components of the cycle in dynamic pest (or crop) management.

management tactic will impact the status of the pest and the agroecosystem, which are continually changing over time. Monitoring or scouting of pest populations provides the manager with the most current information needed to make an effective management decision. Thus, feedback is always provided and the process is cyclical. In some cases (small plots, etc.), the farmer may be both the scout and the crop manager. In large operations, scouts may be hired or contracted to perform the monitoring duties. In many cases, scouting has resulted in considerable economic benefit, since the cost(s) of unnecessary pesticide applications may be saved once treatment decisions are based on thresholds and biological information. Additional information and references to specific pest management programs are provided later in this chapter.

Information quality is critical at all stages of pest management decision making (Figure 11-8). The crop manager must rely on biological, ecological, environmental, and management information provided by research, extension agencies, scouts, and other crop consultants. The information base itself is improved as monitoring and scouting provide feedback on effects of tactics that are implemented. Because environment affects all parts of the management cycle shown in Figure 11-8, "monitoring" may include both biological and environmental components. Environmental monitoring may be more important than monitoring or counting pests for some programs. Models predicting severity and damage by several important plant diseases depend directly on weather information.

Although Figure 11-8 was introduced as part of the pest management process, it can be applied to many aspects of crop management as well. Measurements of soil fertility can be monitored on a regular basis and decisions on fertilizer usage can be made based on projected crop needs. A similar approach can be used to make decisions about irrigation or other management practices. Since any of these decisions impacts the crop ecosystem, in the end it may indirectly impact other features of that system as well. A decision to irrigate to benefit the crop may result in germination of a greater weed population. All types of management practices (pest management, water management, fertility management, etc.) are interconnected in the overall process of crop management.

Ecological Basis for Pest Management

From Chapter 6, we recall that populations of many organisms have a tendency to build up until they reach a carrying capacity (K) for a particular agroecosystem and set of environ-

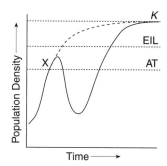

Figure 11-9 Pest population density over time in response to pesticide treatment at point X. K = carrying capacity; EIL = economic injury level; AT = action threshold.

mental conditions. This ecological tendency and our management practices are at cross-purposes when the carrying capacity is greater than the economic injury level.

Reduction in Pest Numbers

Many pest management strategies, including most pesticide-based programs, are aimed at the reduction of pest populations. Unfortunately, many of these reductions are only temporary, when viewed in the context of pest population growth (Figure 11-9). If a pest population is not treated at point X (in Figure 11-9), it will have a tendency to grow to the carrying capacity (dotted line). Even if the pest population is treated and numbers decline drastically, the pest population may eventually recover and grow toward K again if it is not treated. Additional treatments may be needed. In some annual crops, the crop may be harvested before the pest population recovers, so that a single treatment may suffice. A chronic situation can occur in long-term and perennial crops because there is plenty of time for population recovery, resulting in cycles of treatment-recovery-treatment-recovery, etc.

Reduction in Pest Carrying Capacity

An alternative approach to reduction of pest numbers is a strategy aimed at long-term reduction in the carrying capacity of a pest. The agroecosystem is managed in some way that will produce a change in K (Figure 11-10). Such management actions can involve planting a resistant crop cultivar or implementation of biological control or a variety of cultural practices. Many secondary pests, such as certain mites or leafminers, can be maintained at lower carrying capacities if natural enemies are maintained. The objective of these strategies is to maintain a carrying capacity that remains below the economic injury level. Of course, we must be aware of other factors that may result in changes to K over the long term. Since K depends on environmental conditions, fluctuations in these can result in fluctuations in K from year to year.

Changes in plant health can produce changes not only in carrying capacity, but even in the economic injury level itself. Relationships such as that shown in Figure 11-5 can be altered by practices that improve overall crop yield (the whole curve of Figure 11-5 may shift

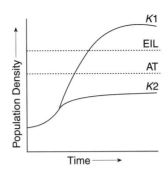

Figure 11-10 Pest population density over time in response to manipulation of carrying capacity. $K1$ = original carrying capacity; EIL = economic injury level; AT = action threshold; $K2$ = new carrying capacity resulting from environmental manipulation.

upward), changing economics and increasing crop value and profits. Thus, approaches that are helpful in the management of various other aspects of plant health may have indirect benefits in the management of plant pests as well.

Multiple Pests

Many of the approaches discussed in this chapter are useful for managing single pest species or groups of pests with very similar biological and ecological characteristics. In practice, however, multiple kinds of pests often occur in the same crop.

Pest Associations and Interactions

Examples of interactions among different types of pest groups are numerous. We have already discussed the importance of insect vectors in the transmission of diseases to plants and animals. When vectors are involved, options may exist to manage the disease in the host, the vector itself, or any alternate hosts of the vector or the pathogen. Soilborne fungal and bacterial diseases are often made more severe by the presence of nematodes, which cause root damage and wounds for colonization by pathogens. An interesting multiple interaction results in annual ryegrass toxicity to livestock in Australia (see McKay and Ophel 1993). A plant-pathogenic bacterium (*Clavibacter toxicus*) is carried from the soil into the seedheads of annual ryegrass (*Lolium rigidum*) and other grasses by foliar nematodes (*Anguina* spp.). The bacteria infect and distort the seedheads, but more importantly, produce a toxin that can cause poisoning in sheep grazing on the infected grass.

The many associations between weeds and other groups of pests can be of particular concern. Should weeds near agricultural fields be managed? After all, they may provide seed sources for crop weeds, alternate hosts for plant diseases and pests, or hiding places for seed-eating rodents. But they may also provide a habitat from which insect natural enemies can enter the field to manage insect pests. The answer depends on the specific situation and the balance between any benefits of biological control and vegetational diversity versus real or potential problems.

Such examples of conflict in pest management decisions often occur. Should we plant crop rows close together to aid in weed suppression or far apart to minimize spread of foliar diseases? In many instances, it may be necessary to focus on a **key pest** that is responsible for major crop damage and ignore or de-emphasize efforts against minor pests. Again, only previous experience and local information about a specific situation can dictate the proper course of action.

Integrated Pest Management

Ideally, it may be possible to integrate complementary tactics for the management of each pest if multiple pests are present. Theoretically, integration in pest management may occur at two levels:

1. Integration of multiple tactics for managing a single pest; and
2. Integration of the management of multiple kinds of pests.

The concept of **integrated pest management** (IPM) was developed in the 1960s as an approach to managing pests only when needed through a variety of methods, with the overall objective of reducing pesticide use and environmental impact. It was anticipated that use of IPM programs would lessen environmental impact, improve consumer health, increase grower profits as costly treatments were reduced, and encourage the development and use of alternative tactics for pest management. Since then, IPM has been variously defined and a wide variety of IPM practices and programs have been developed.

In practice, many "IPM" programs have focused mainly on the decision whether or not to apply an insecticide based on scouting information. Although only partially successful in adopting the original philosophy of IPM (this is essentially insect management with insecticides), these programs have nonetheless been very profitable for growers and have resulted in important reductions in unnecessary insecticide applications.

Efforts to integrate multiple tactics into IPM programs and to develop alternatives to pesticides have varied, depending on regions, pests, and commodities involved. Consumer demand can indirectly influence the pest management tactics used in production. Demand for blemish-free produce with high cosmetic appeal may require heavy use of fungicides and insecticides. Demand for organically grown produce implies use of alternative practices in production.

In some cases, efforts to integrate the management of pests other than insects into IPM programs have proceeded slowly. This may have resulted, in part, from some scientists working only within their own disciplines (e.g., entomology, weed science, etc.). However, as discussed earlier, sometimes the effects of a particular tactic on different classes of pests can be in conflict, and so it can be difficult to build a management program from complementary tactics in such cases.

Despite the challenges involved in building IPM programs, it must be recognized that IPM is only one aspect of crop management (Figure 11-11). Any pest management strategies and tactics proposed must be compatible with other crop management practices as well, which in turn influence the pest management practices. In addition, the individual crop plants are not isolated, but are influenced by other crops and plants in the farming system, as well as in the surrounding landscape. Subsequent chapters examine the effects of landscape (Chapter 12) and cropping systems (Chapter 13) on crop production.

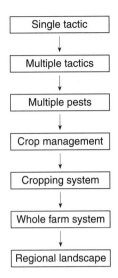

Figure 11-11 Levels of integration progressing from single tactic pest management to integration of multiple crops on a farm and multiple farms into the regional landscape.

Examples and Case Studies

Examples of IPM programs and practices are numerous. A few examples illustrating the uses of various pest management practices are shown in Applications 11-3 to 11-7. Table 11-1 provides literature citations for pest management programs on a variety of crops. Several of these are entire texts or extensive works containing information on the management of different classes of pests.

Application 11-3 *Cotton Pest Management*

Many different pests attack cotton (Figure 11-12) and a variety of tactics are used in their management (Anonymous 1996; El-Zik and Frisbie 1991). Information on economic injury levels at various stages of cotton growth is available for many cotton insects, including key pests such as the bollworm and tobacco budworm (*Helicoverpa* spp., formerly *Heliothis* spp.), lygus bugs (*Lygus* spp.), and the boll weevil (*Anthonomus grandis*). In many cases, insecticide treatments are based on scouting information on these pests. In some states, threshold information from soil samples can be used to make treatment decisions for nematodes or the soilborne fungus, *Verticillium dahliae*. Weather-based models can be used to forecast

(continued)

some foliar diseases. Novel pest management tactics are used in some cases. Transgenic cotton cultivars, which produce Bt toxin, are used against caterpillars of the bollworm/tobacco budworm complex. Other transgenic cultivars are tolerant to glyphosate, a general purpose herbicide that can then be used on the crop.

Through the efforts of an eradication program, the boll weevil has been reduced to an infrequent pest in many locations in the southeast and southwest regions of the United States. Regional coordination of cotton planting and harvest and early destruction of crop residues have limited overwintering sites for this pest. Synchronization of early plantings has improved the detection of early-season populations within or entering a region. Pheromone trapping has improved the sensitivity of detection and the direction of early-season spray programs.

Figure 11-12 Cotton crop ready for harvest.

Application 11-4   *Arthropod Management on Fruit Crops*

A variety of IPM programs have been used for the monitoring and management of insect and mite pests on fruit crops (see Croft and Hoyt 1983; Zalom 1997). Due to the perennial nature of fruit crops, many pest management strategies have focused on the long-term maintenance of low carrying capacities for pest populations. As a result, the use of classical or natural biological control has been emphasized in many systems. Conservation of predators and parasites is considered when applying insecticides or acaricides. For example, decisions to apply acaricides may be based not only on numbers of pest mites, but on relative abundance or ratios of predatory and pest species. Other tactics that conserve biological control agents include the use of pesticides that are less toxic to these organisms, or the release of natural enemies that have been selected in the laboratory for resistance to certain pesticides.

Application 11-5   *Soybean Cyst Nematode*

The soybean cyst nematode (*Heterodera glycines*) is the major nematode pest of soybean in the United States, China, Japan, and other important soybean-producing areas (Riggs and Niblack 1993). The nematode has a relatively narrow host range, so rotation with nonhosts is beneficial in its management. Corn is a convenient rotation crop in the midwestern United States, while cotton or peanut may be used in southern states. Use of resistant soybean cultivars is complicated by the occurrence of different isolates of this nematode. Specific soybean cultivars may be resistant to some isolates but not others, so the decision to plant a particular cultivar depends on which isolate occurs in a particular field. Furthermore, repeated annual plantings of the same resistant cultivar provides sufficient selection pressure for the nematode population to overcome resistance and shift to a new isolate within a few years. If resistant cultivars are used, their proper management is critical. An example rotation for the midwest could be corn in the first year (reduces nematode population), resistant soybean in the second year (reduces population but provides some selection pressure), susceptible soybean in the third year (relieves selection pressure but not much crop damage since population is low after previous two years), and then the sequence is repeated.

Application 11-6   *Late Blight of Potato*

Late blight, which was responsible for the Irish Potato Famine (see Chapter 8), remains one of the most important diseases of potato today (Stevenson 1993; Weingartner 1998). Development of this foliar disease, which is caused by the fungus *Phytophthora infestans,* is favored by cool, rainy weather during the summer and early fall. A number of different isolates, which vary in their aggressiveness, occur throughout the world. An integrated management program is necessary to limit problems from this disease. Use of certified tubers for planting is critical, since the disease may be moved long distances on infected tubers. Other impor-

(continued)

tant sanitation measures include proper and prompt disposal of nonmarketable potato tubers (culls), proper site selection, irrigation management, and avoidance of alternative hosts, mainly certain weeds and vegetables in the same plant family as potato (Solanaceae). In many parts of the United States, regular scouting of fields is used to detect development of late blight. Weather-based forecasting is used to alert scouts and potato growers to conditions under which disease development is likely. Fungicides are used to limit disease development, although there is concern over selection for resistant isolates. Management efforts must continue through harvest to avoid fungal contamination of tubers.

Application 11-7 *Methyl Bromide*

Methyl bromide is a broad-spectrum soil fumigant that is important in the management of various soilborne diseases, nematodes, and weeds. It has been widely used in commercial production of tomato, strawberry, eggplant, and other crops that are especially susceptible to these pests. Use of this volatile fumigant requires covering the soil with a plastic tarp immediately after the methyl bromide is injected into the soil. This application process has been incorporated into the bedded plastic mulch system (Figure 11-13). This production system involves production of crops on raised soil beds that can be easily sealed in plastic. Fertilizer is incorporated into the bed before fumigation, and drip irrigation tubing may be applied with the plastic mulch at the time of fumigation.

Figure 11-13 Plastic mulch covering fumigated beds prior to planting.

(continued)

Over the last few decades of the twentieth century, routine application of methyl bromide in these systems has become commonplace. However, methyl bromide has been implicated in atmospheric ozone depletion (see Ristaino and Thomas 1997). As uses of this material become limited, alternative methods are needed to manage the array of pests in these production systems. Alternative fumigants can be effective against diseases and nematodes, but additional herbicides may be required to manage weeds. With the widespread use and availability of methyl bromide, development of nonchemical alternatives lagged. Use of resistant cultivars, soil solarization, rotation and cover crops, or other practices have potential, but face limitations that require improvement and integration of tactics. It is unlikely, for example, to find a vegetable cultivar that is resistant to such a wide range of different diseases and nematodes, and is highly competitive with weeds as well. This example illustrates the vulnerability of an industry that is dependent on a single pest management tactic, and emphasizes the need for development and integration of a variety of tactics in developing an effective pest management strategy.

Table 11-1 Some examples of pest management programs and practices for selected crops.

Crop	Pest(s) Included[a]					Reference
	Weed	Dis.	Nem.	Ins.	Vert.	
Alfalfa	X	X	—	X	—	Pimentel 1991
Apple	—	X	—	—	—	Sutton and Jones 1991
Apple	—	—	—	X	—	Reissig et al. 1995
Banana	—	X	X	X	—	Gold and Gemmill 1993
Citrus	X	X	X	X	—	Knapp et al. 1996
Corn	—	—	—	X	—	Wintersteen and Higley 1993
Cotton	X	X	X	X	—	El-Zik and Frisbie 1991
Fruits, temperate	—	—	—	X	—	Zalom 1997
Fruits, tropical	—	X	X	—	—	Ploetz et al. 1994
Olive	—	—	—	X	—	Walton 1995
Peanut	—	X	X	X	—	Melouk and Shokes 1995
Pecan	—	X	—	X	—	Harris 1991
Potato	X	X	X	X	—	Rowe 1993
Potato	—	X	X	X	—	Zehnder et al. 1994
Rice	X	X	—	X	X	Grayson, Green, and Copping 1990
Sorghum	—	X	X	X	—	Pimentel 1991
Soybean	X	X	X	X	—	Copping, Green, and Rees 1992
Stored products	—	X	—	X	X	Evans 1987
Strawberry	X	X	X	X	—	Pimentel 1991
Sugarbeet	—	V	—	X	X	Jepson 1987

(continued)

Table 11-1 Continued.

Crop	Pest(s) Included[a]					Reference
	Weed	Dis.	Nem.	Ins.	Vert.	
Sweetpotato	—	V	X	X	—	Jansson and Raman 1991
Tomato	—	V	—	X	—	Schuster, Funderburk, and Stansly 1996
Vegetables	X	X	X	X	—	Pohronezny 1989
Vegetables	—	—	—	X	—	Finch 1987
Wheat	—	—	X	—	—	Brown 1987
Wheat	X	X	—	X	—	Pimentel 1991
Wheat	—	V	—	X	—	Wratten, Elliot, and Farrell 1995

[a]Dis. = plant diseases from fungi, bacteria, viruses, mycoplasmas, abiotic factors, etc; Nem. = nematodes; Ins. = insects, mites, and other arthropods; Vert. = vertebrates
X = information included on this pest group
— = none or limited information on this pest group
V = plant disease information limited to insect-vectored viruses

Summary

Strategies used to manage pests may involve prevention, eradication, reduction of pest populations, improvement of plant health, or no action. Provided that an effective tactic is available for managing a particular pest, a decision must be made about when to manage the pest. Pest management decision making depends on acquisition of specific biological data on the status of pest populations and damage, usually through sampling or monitoring. Damage thresholds and forecasts of projected crop losses may be used to determine whether a decision to treat is economical or not. Some pest management strategies focus on temporary reductions in pest numbers to prevent losses, while others aim for a long-term reduction in the carrying capacity of the pest in the agroecosystem. The simultaneous management of multiple pest problems is probably the most difficult challenge for most pest management programs.

Topics for Review and Discussion

1. Why are agricultural pests usually clumped in their dispersion within a field?
2. List the advantages and disadvantages of using chemical pesticides.
3. What are some ways in which the carrying capacity of a pest population in an agroecosystem can be reduced?
4. What are some of the environmental costs that are not usually included in pest management decision making?
5. Discuss the management of a key pest on a crop grown in your area. What tactics are used in the management of this pest? Is threshold information available for this pest? Does its management involve an organized IPM program and scouting?

Literature Cited

Agrios, G. N. 1997. *Plant Pathology.* 4th ed. San Diego, Calif.: Academic Press.

Anonymous. 1996. *1996 Proceedings Beltwide Cotton Conferences.* Memphis, Tenn.: National Cotton Council of America.

Aspelin, A. L. 1997. *Pesticides Industry Sales and Usage. 1994 and 1995 Market Estimates.* Washington, D.C.: U.S. Environmental Protection Agency.

Brown, J. K. M. 1995. Pathogens' responses to the management of disease resistant genes. *Advances in Plant Pathology* 11: 75–102.

Brown, R. H. 1987. Control strategies in low-value crops. In *Principles and Practice of Nematode Control in Crops,* eds. R. H. Brown and B. R. Kerry, 351–387. Sydney, Aust: Academic Press.

Copping, L. G., M. B. Green, and R. T. Rees, eds. 1992. *Pest Management in Soybean.* London: Elsevier Applied Science.

Croft, B. A., and S. C. Hoyt, eds. 1983. *Integrated Management of Insect Pests of Pome and Stone Fruits.* New York: John Wiley and Sons.

Crossley, D. A., Jr., D. C. Coleman, P. F. Hendrix, W. Cheng, D. H. Wright, M. H. Beare, and C. A. Edwards, eds. 1991. *Modern Techniques in Soil Ecology.* Amsterdam: Elsevier.

El-Zik, K. M., and R. E. Frisbie. 1991. Integrated crop management systems for pest control. In *CRC Handbook of Pest Management in Agriculture,* 2d ed., Volume III, ed. D. Pimentel, 3–104. Boca Raton, Fla.: CRC Press.

Evans, D. E. 1987. Stored products. In *Integrated Pest Management,* eds. A. J. Burn, T. H. Coaker, and P. C. Jepson, 425–461. London: Academic Press.

Finch, S. 1987. Horticultural crops. In *Integrated Pest Management,* eds. A. J. Burn, T. H. Coaker, and P. C. Jepson, 257–293. London: Academic Press.

Fuchs, M., and D. Gonsalves. 1997. Genetic engineering. In *Environmentally Safe Approaches to Crop Disease Control,* eds. N. A. Rechcigl and J. E. Rechcigl, 333–368. Boca Raton, Fla.: CRC Lewis Publishers.

Gold, C. S., and B. Gemmill, eds. 1993. *Biological and Integrated Control of Highland Banana and Plantain Pests and Diseases.* Ibadan, Nigeria: International Institute of Tropical Agriculture.

Grayson, B. T., M. B. Green, and L. G. Copping, eds. 1990. *Pest Management in Rice.* London: Elsevier Applied Science.

Harris, M. K. 1991. Pecan IPM. In *CRC Handbook of Pest Management in Agriculture,* 2d ed., Volume III, ed. D. Pimentel, 691–706. Boca Raton, Fla.: CRC Press.

Jansson, R. K., and V. K. Raman, eds. 1991. *Sweet Potato Pest Management: A Global Perspective.* Boulder, Colo.: Westview Press.

Jepson, P. C. 1987. Sugar beet. In *Integrated Pest Management,* eds. A. J. Burn, T. H. Coaker, and P. C. Jepson, 295–327. London: Academic Press.

Knapp, J. L., J. W. Noling, L. W. Timmer, and D. P. H. Tucker. 1996. Florida citrus IPM. In *Pest Management in the Subtropics,* eds. D. Rosen, F. D. Bennett, and J. L. Capinera, 317–347. Andover, U.K.: Intercept.

McKay, A. C., and K. M. Ophel. 1993. Toxigenic *Clavibacter/Anguina* associations infecting grass seedheads. *Annual Review of Phytopathology* 31: 151–167.

McSorley, R. 1987. Extraction of nematodes and sampling methods. In *Principles and Practice of Nematode Control in Crops,* eds. R. H. Brown and B. R. Kerry, 13–47. Sydney, Aust: Academic Press.

———1998. Population dynamics. In *Plant and Nematode Interactions,* eds. K. R. Barker, G. A. Pederson, and G. L. Windham, 109–133. Madison, Wisc.: American Society of Agronomy.

Meister, R. T., ed. 1995. *Farm Chemicals Handbook '95.* Willoughby, Oh.: Meister Publishing Company.

Melouk, H. A., and F. M. Shokes, eds. 1995. *Peanut Health Management.* St. Paul, Minn.: The American Phytopathological Society.

Metcalf, R. L., and W. H. Luckmann, eds. 1994. *Introduction to Insect Pest Management.* 3d ed. New York: John Wiley and Sons.

Pimentel, D., ed. 1991. *CRC Handbook of Pest Management in Agriculture.* 2d ed., Volume III. Boca Raton, Fla.: CRC Press.

Ploetz, R. C., G. A. Zentmyer, W. T. Nishijima, K. G. Rohrback, and H. D. Orr, eds. 1994. *Compendium of Tropical Fruit Diseases.* St. Paul, Minn.: The American Phytopathological Society.

Pohronezny, K., ed. 1989. The impact of integrated pest management on selected vegetable crops in Florida. Bulletin 875, Agricultural Experiment Station, University of Florida, Gainesville, Fla.

Radosevich, S., J. Holt, and C. Ghersa. 1997. *Weed Ecology: Implications for Management.* 2d ed. New York: John Wiley and Sons.

Reissig, H., A. Agnello, J. Kovach, and J. Nyrop. 1995. Development and implementation of simplified monitoring and forecasting techniques for key arthropod pests in NY apple orchards. In *International Symposium on Agricultural Pest Forecasting and Monitoring,* eds. G. Bourgeois and M. O. Guibord, 143–151. Quebec: Réseau d'avertissements phytosanitaires du Québec.

Riggs, R. D., and T. L. Niblack. 1993. Nematode pests of oilseed and grain legumes. In *Plant Parasitic Nematodes in Temperate Agriculture,* eds. K. Evans, D. L. Trudgill, and J. M. Webster, 209–258. Wallingford, U.K.: CAB International.

Ristaino, J. B., and W. Thomas. 1997. Agriculture, methyl bromide, and the ozone hole: Can we fill the gap? *Plant Disease* 81: 964–977.

Romoser, W. S., and J. G. Stoffolano, Jr. 1998. *The Science of Entomology.* 4th ed. Boston, Mass.: WCB/McGraw-Hill.

Roush, R. T. 1991. Management of pesticide resistance. In *CRC Handbook of Pest Management in Agriculture,* 2nd ed., Volume II, ed. D. Pimentel, 721–740. Boca Raton, Fla.: CRC Press.

Rowe, R. C., ed. 1993. *Potato Health Management.* St. Paul, Minn.: The American Phytopathological Society.

Schuster, D. J., J. E. Funderburk, and P. A. Stansly. 1996. IPM in tomatoes. In *Pest Management in the Subtropics,* eds. D. Rosen, F. D. Bennett, and J. L. Capinera, 387–411. Andover, U.K.: Intercept.

Shelton, A. M., and J. T. Trumble. 1991. Monitoring insect populations. In *CRC Handbook of Pest Management in Agriculture,* 2d ed., Volume II, ed. D. Pimentel, 45–62. Boca Raton, Fla.: CRC Press.

Southwood, T. R. E. 1984. *Ecological Methods.* 2d ed. London: Chapman and Hall.

Stevenson, W. R. 1993. Management of early blight and late blight. In *Potato Health Management,* ed. R. C. Rowe, 141–147. St. Paul, Minn.: The American Phytopathological Society.

Sutton, T. S., and A. L. Jones. 1991. Apple disease management. In *CRC Handbook of Pest Management in Agriculture,* 2d ed., Volume III, ed. D. Pimentel, 117–133. Boca Raton, Fla.: CRC Press.

Tingey, W. M., and J. C. Steffens. 1991. Environmental control of insects using plant resistance. In *CRC Handbook of Pest Management in Agriculture,* 2d ed., Volume II, ed. D. Pimentel, 131–155. Boca Raton, Fla.: CRC Press.

U.S. Dept. of Agriculture. 1997. *Agricultural Statistics 1997.* Washington, D.C.: U.S. Government Printing Office.

Walton, M. P. 1995. Integrated pest management in olives. In *Integrated Pest Management,* ed. D. Dent, 222–240. London: Chapman and Hall.

Weingartner, P. 1998. Late blight: The "new" epidemic. *Citrus and Vegetable Magazine* 62 (7): 32–36.

Wintersteen, W. K., and L. G. Higley. 1993. Advancing IPM systems in corn and soybeans. In *Successful Implementation of Integrated Pest Management for Agricultural Crops,* eds. A. R. Leslie and G. W. Cuperus, 9–32. Boca Raton, Fla.: Lewis Publishers.

Wratten, S. D., N. C. Elliott, and J. A. Farrell. 1995. Integrated pest management in wheat. In *Integrated Pest Management,* ed. D. Dent, 241–279. London: Chapman and Hall.

Young, L. D. 1998. Breeding for nematode resistance and tolerance. In *Plant and Nematode Interactions,* eds. K. R. Barker, G. A. Pederson, and G. L. Windham, 187–207. Madison, Wisc.: American Society of Agronomy.

Zalom, F. G. 1997. IPM practices for reducing insecticide use in U.S. fruit crops. In *Techniques for Reducing Pesticide Use,* ed. D. Pimentel, 317–342. New York: John Wiley and Sons.

Zehnder, G. W., M. L. Powelson, R. K. Jansson, and K. V. Raman, eds. 1994. *Advances in Potato Pest Biology and Management.* St. Paul, Minn.: The American Phytopathological Society.

Bibliography

Abawi, G. S., and J. Chen. 1998. Concomitant pathogen and pest interactions. In *Plant and Nematode Interactions,* eds. K. R. Barker, G. A. Pederson, and G. L. Windham, 135–158. Madison, Wisc.: American Society of Agronomy.

Andow, D. A., and P. M. Rosset. 1990. Integrated pest management. In *Agroecology,* eds. C. R. Carroll, J. H. Vandermeer, and P. Rosset, 413–439. New York: McGraw-Hill.

Cavigelli, M. A., S. R. Deming, K. L. Probyn, and R. R. Harwood, eds. 1998. *Michigan Field Crop Ecology.* Michigan State University Extension Bulletin E-2646, East Lansing, Mich.

Cook, R. J. 1993. Making greater use of introduced microorganisms for biological control of plant pathogens. *Annual Review of Phytopathology* 31: 53–80.

Debach, P., and D. Rosen. 1991. *Biological Control by Natural Enemies.* 2d ed. Cambridge, U.K.: Cambridge University Press.

Kogan, M. 1998. Integrated pest management: Historical perspectives and contemporary developments. *Annual Review of Entomology* 43: 243–270.

Luna, J. M., and G. J. House. 1990. Pest management in sustainable agricultural systems. In *Sustainable Agricultural Systems,* eds. C. A. Edwards, R. Lal, P. Madden, R. H. Miller, and G. House, 157–173. Delray Beach, Fla.: St. Lucie Press.

McFadyen, R. E. C. 1998. Biological control of weeds. *Annual Review of Entomology* 43:369–393.

McSorley, R., and L. W. Duncan. 1995. Economic thresholds and nematode management. In *Advances in Plant Pathology,* Volume II, eds. J. H. Andrews and I. C. Tommerup, 147–171. San Diego, Calif.: Academic Press.

Parry, D. 1990. *Plant Pathology in Agriculture.* Cambridge, U.K.: Cambridge University Press.

Chapter 12

Agriculture in the Landscape

Key Concepts

- Diversity in the agroecosystem
- Interactions of agroecosystems and natural ecosystems
- Environmental effects of fertilizers and pesticides

Whether they are integrated and complementary or not, all crop management practices (fertilizer applications, pest management tactics, irrigation practices, soil management, etc.) interact as components of the crop ecosystem. The individual crop agroecosystem in turn is usually a part of a larger cropping system (see Chapter 13), which in turn is included in a whole farm, which in turn is a part of a regional landscape. A **landscape** is a general term that includes all of the various ecosystems (e.g., agricultural, natural, urban, aquatic, etc.) in a defined geographic region. The individual agroecosystem is influenced by the larger landscape in which it is contained, but it also affects other components of the landscape in various ways. In this chapter, we examine several important aspects of agroecosystem-landscape interactions: maintenance of diversity, interactions of agricultural and natural systems, and agricultural chemicals in the landscape. Interactions between agricultural and urban ecosystems are discussed in Chapter 19.

Diversity in the Agroecosystem and the Landscape

In the 1990s, the term biodiversity came into popular use to refer to all of the species of living organisms in an ecosystem. The term has been variously defined (see Chapter 8 and Hawksworth 1995) in attempts to include variation arising from genetic differences within species and the difficulty in defining what makes up a species across vastly different groups of organisms. Nevertheless, the concept of diversity has been recognized by ecologists for many years. **Diversity** (also called *species diversity* if we are referring to species) consists of two parts: **richness,** or number of species in a community, and **evenness,** or the relative abundance of individuals among the various species present in the community.

Measurement of Diversity

The difference between these two components of diversity can be illustrated by the hypothetical data set presented in Figure 12-1. Communities A and B both have the same number of species, five, and so they have the same richness or biodiversity. However, the structures of these two communities are very different. In community A, most individuals that we find are the same (species 1), while in community B, the distribution of individuals among species is much more balanced. Community B has a greater evenness than does community A. *Dominance* is the opposite of the diversity concept, and refers to the degree that a community is dominated by one or a few very common species. In the example, community A is dominated by species 1.

Diversity can be expressed by a **diversity index,** which is a numerical measure of diversity, or by a *species abundance curve,* which expresses the relative abundance of species graphically. There are many different kinds of diversity indices and species abundance curves (see Magurran 1988). A few of the more commonly used measures are presented here.

Richness. Counting the number of species remains a convenient and simple way of estimating diversity, but as just discussed, it measures only one part of diversity.

Species Abundance Curves. Species abundance curves provide a visual means of quickly comparing the structure of two or more communities (Figure 12-1). In this example, the species are ranked from most to least abundant along the *x*-axis, and the percent abundance of each species is plotted on a log-scale. Greater evenness is indicated by a flatter curve rather than by a steeply sloped curve. Richness is indicated by the length of the curve along the *x*-axis (e.g., seven species instead of five species). Many different versions of species abundance curves are available, each showing different parameters on the axes, and mathematical analyses of certain curve types are possible (Magurran 1988).

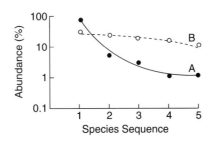

Figure 12-1 Species abundance curves for two communities, A and B, based on data from Table 12-1.

Shannon Index or Shannon-Weaver Index. The Shannon Index, developed by Shannon and Weaver (1949), is an often-used diversity index. It is given by the formula:

$$H' = -\sum_{i=1}^{s} p_i \ln p_i \qquad \text{(Equation 12-1)}$$

where s is the number of species in the community, and p_i is the proportional abundance of species i (= number of species i divided by total numbers in the community). The term $p_i \ln p_i$ is calculated and summed for each species in the community. Application 12-1 provides a sample calculation. In that example, note the higher diversity index for community B, in which individuals are more evenly distributed than in community A. With this index, diversity increases as:

- Species become more evenly distributed in abundance; and
- More species are added to the community.

The maximum value that the Shannon index can reach depends on the number of species in the community (maximum $H' = \ln s$).

Evenness Index. This index is useful for comparing evenness in communities with different numbers of species. The formula for this index is:

$$J = \frac{H'}{\ln s} \qquad \text{(Equation 12-2)}$$

This standardizes evenness on a scale from 0 to 1.

Simpson's Index. The index developed by Simpson (1949) is really an index of dominance rather than diversity. Its formula is:

$$\lambda = \sum_{i=1}^{s} (p_i)^2 \qquad \text{(Equation 12-3)}$$

This index measures dominance on a 0 to 1 scale. If only one species is present in the community, $p_i = \lambda = 1$, the maximum value. Note the high value of this index in the sample calculation (Application 12-1) for community A, which is dominated by species 1.

Reciprocal Simpson Index. Because Simpson's index is a measure of dominance, its reciprocal, $1/\lambda$, is a measure of diversity since it shows an opposite trend. This index is especially convenient when using spreadsheets to manage large data sets. If zeroes are entered in a data set (e.g., if species #4 was absent in community A of Table 12-1), calculations involving λ are unaffected, but the natural logarithm of zero could cause problems in the calculation of H'.

Limitations of Diversity Indices. Various diversity indices provide useful measures of richness and (or) evenness for a community. But this is the only information that they

Application 12-1 *Sample Calculations of Diversity Indices*

Sample calculations are based on data for community A from Table 12-1.

Table 12-1 Numbers of individuals of five different species in two hypothetical communities, A and B.

	Community A	Community B
Species 1	90	30
Species 2	5	25
Species 3	3	20
Species 4	1	15
Species 5	1	10
Total	100	100

Shannon Index

$$H' = -\sum_{i=1}^{s} p_i \ln p_i$$

For community A, $s = 5$ species, and p_i for species 1 is $p_i = 90/100$, so:

$$H' = -\left(\frac{90}{100} \ln \frac{90}{100} + \frac{5}{100} \ln \frac{5}{100} + \frac{3}{100} \ln \frac{3}{100} + \frac{1}{100} \ln \frac{1}{100} + \frac{1}{100} \ln \frac{1}{100}\right)$$
$$= -(0.9[-0.105] + 0.05[-2.996] + 0.03[-3.506] + 0.01[-4.605] + 0.01[-4.605])$$
$$= -(-0.095 - 0.150 - 0.105 - 0.046 - 0.046)$$
$$= -(-0.442) = 0.442$$

Evenness Index

$$J = H'/\ln s = 0.442/\ln s = 0.442/1.609 = 0.275$$

Simpson Index

$$\lambda = \sum_{i=1}^{s} (p_i)^2$$
$$= \left(\frac{90}{100}\right)^2 + \left(\frac{5}{100}\right)^2 + \left(\frac{3}{100}\right)^2 + \left(\frac{1}{100}\right)^2 + \left(\frac{1}{100}\right)^2$$
$$= 0.81 + 0.0025 + 0.0009 + 0.0001 + 0.000$$

(continued)

Reciprocal Simpson Index

$$1/\lambda = 1/0.814 = 1.229$$

For community B in Table 12-1, we can calculate the following:

Shannon index = H′ = 1.544
Evenness = J = 0.960
Simpson index = 0.225
Reciprocal Simpson index = 4.444

---

provide. From Table 12-1, you can calculate that community B has a higher diversity than does community A, but you would not know whether the species in community B were foliage-eating caterpillars or beneficial predators.

One of the greatest limitations to use of diversity indices is the difficulty in recognizing and identifying species. This may not seem like much of a problem in evaluating plants or birds. But in measuring mites, nematodes, fungi, or bacteria, species may become increasingly difficult to recognize, and unidentified species probably occur in most agroecosystems. In these cases, we can substitute genus or family (or even a higher-level taxon) for species in the formulas for diversity indices. Provided we are consistent in their use, patterns in genus diversity are similar to those observed for species diversity.

Because of the uneven recognition of species, genera, etc., across groups of different kinds of organisms, diversity indices are difficult to apply across all members of a community. Most examples show diversity of only plants, birds, or insects for instance, rather than all of these groups together.

Diversity and Stability

Almost by definition, a monoculture is dominated by a single plant species, and therefore represents an extreme example of an agroecosystem with low diversity. Such a system may be susceptible to weather disasters, pest or disease outbreaks, or other catastrophes. A high degree of management and intervention may be required to maintain this type of agroecosystem. In contrast, many natural ecosystems appear to be more stable and less subject to fluctuations in populations of the organisms making up the community.

Are ecosystems with higher diversity more stable? This question has been much debated in ecology. Difficulty in defining terms has been a problem, and we may think of stability as involving two aspects:

- *Resistance,* or the ability of a community to avoid disturbance; and
- *Resilience,* the rate at which a community recovers following disturbance.

Diversity is only one measure of ecosystem complexity. The community of organisms becomes more complex when a larger number of different kinds of organisms are included, when there are more interactions among organisms, and when the strength of these interactions increases. If we think of the food webs from Chapter 3, imagine how much more complex a food web with fifty members is, compared to a food web with only five members. The larger food web has many more potential connections and interactions among members, and many alternative paths of energy and material flow through it. Intuitively, we might suppose that the more complex community is more stable, and much data would support this idea, but many examples and models support the opposite view as well. As an example, the effect of increased crop diversity on insect species is discussed in Application 12-2. As with many questions in ecology, questions of diversity and stability or complexity and stability have no clear answers. Although complexity and stability may appear to be related much of the time, exceptions may be found quite often.

Most agroecosystems are probably less stable than natural ecosystems. After all, disturbance (planting, harvest, tillage, etc.) is a key feature of agroecosystems. Therefore, resilience is a particularly important feature of stability in agroecosystems, especially annual or short-term crop ecosystems. Succession occurs frequently in such systems, and stability may depend on the rates at which crops develop or beneficial organisms recolonize a recently planted site. In a practical sense, a stable agroecosystem might be one in which less management is required. Various tasks, such as some of the nutrient cycling or pest management, may be performed by nonhuman organisms in the agroecosystem. Although some management is required in all agroecosystems, one goal in developing a more stable system is to reduce the need for interventional management in crisis situations.

Maintenance of Diversity

Increased diversity can lead to a number of important benefits to agriculture, including efficient use of resources by crop plants, varied sources of income and reduced risks for farmers, improved nutrient cycling, conservation of soil and water, and improved management of pests, diseases, and weeds. Diversity can be maintained or increased at several levels:

- *Alpha diversity*—within an individual field or habitat;
- *Beta diversity*—among nearby fields or habitats; and
- *Gamma diversity*—on a regional scale.

Diversity in the Field. Within an individual field or agroecosystem, diversity can be maintained or increased by cultivating a variety of plants (polyculture) rather than a single crop species (monoculture). Polyculture includes crop rotation, cover crops, intercropping, and a variety of other cropping systems discussed in more detail in Chapter 13. Advice on developing polycultural systems is provided by several references in the Bibliography at the end of this chapter. Several practices discussed in Chapter 14, such as addition of organic amendments, reduced tillage, and maintenance of crop residues, may also help to increase diversity and complexity within the agroecosystem.

Diversity in the Landscape. Even if a field is a monoculture, diversity can still be preserved in the landscape around the field. Landscapes with mixtures of different types of annual crops, orchards, pastures, old fields, rangeland, forests, aquatic habitats, or other natural ecosystems may provide a degree of water and soil conservation within a region as well as biodiversity. Many of these habitats may confer direct or indirect benefits to a particular agricultural field. However, a neighboring field with the same crop can be a source of some concern. A poorly managed field can be a source of pests, weeds, or diseases for other fields in the vicinity. Furthermore, if the same crop is grown in various locations over a number of months, the season can also be extended for some of the pests. For example,

Application 12-2 *Crop Diversity and Insect Species*

The effect of crop diversity on insect species provides some data on the diversity-stability issue. As mentioned in a previous chapter, increased diversity can be beneficial in managing plant-feeding insects for a number of reasons. One reason is that the more diverse vegetation may benefit natural enemies of plant pests because the more diverse habitat may include favorable microhabitats for their survival, alternative prey items when the pest species is rare, or alternative food sources for species in which larval and adult food preferences differ (adults of parasitoids may feed on pollen or nectar, for example). Many examples are available comparing insect management in monoculture and more diverse polycultural systems (e.g., see Altieri 1994, 1995; Andow 1991; Cromartie 1991). Andow (1991) summarized data from a large number of these studies, and Figure 12-2 summarizes some of his key findings. For example, 59% of the species of natural enemies (predators and parasites) were more abundant in polyculture than in monoculture, while only 9% preferred monoculture. The increased diversity of the polycultural systems would seem to increase the biodiversity of natural enemies and the complexity of the ecosystem, while limiting the abundance of insect herbivores (presumed to be pests in most cases), which have the potential to destabilize a system with wild population fluctuations. At least this would appear to happen most of the time, but note that some insects showed opposite trends, and others were unaffected.

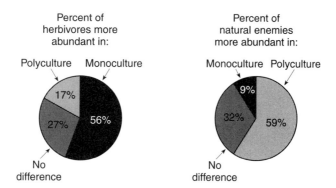

Figure 12-2 Percent of species of arthropod herbivores and natural enemies that were more abundant in monoculture or polyculture.
Calculated from data presented by Andow (1991). "No difference" includes species for which no differences between monoculture and polyculture occurred as well as those with such high variability in numbers that a conclusion could not be reached.

diseased crop residues after an early-season crop may be a source of disease inoculum for later crops.

Border areas of agricultural fields can be a particularly rich source of naturally occurring predators and parasites, pollinators, and other beneficial organisms. An **ecotone** is a transition zone between two different kinds of ecosystems. Ecotones typically contain a greater

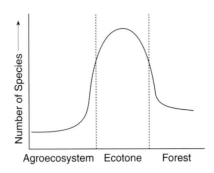

Figure 12-3 General comparison of number of species present in an ecotone and adjacent ecosystems.

diversity of species than that contained in the communities on either side (Figure 12-3). Because an ecotone is usually a narrow border area, species from the community on either side may occur there, as well as species characteristic of a particular ecotone. Near an ecotone, farmers often observe edge effects and organisms that may not be typical on a particular crop. Fence rows, hedge rows, ditches, roadways, alleyways, and windbreaks are common ecotones around agroecosystems. Ecotones may also be created from buffer strips used in soil and water conservation practices.

Agriculture-Environment Interactions

Agricultural and natural ecosystems interact within the landscape in a variety of ways. The effects of agricultural chemicals on the environment have received so much attention that a separate section of this chapter is devoted to those issues. However, any interaction between agricultural and nonagricultural ecosystems (natural or otherwise) depends on several important characteristics of the landscape in which they exist:

Spatial effects—If we consider the landscape as a collection of patches of agricultural fields and natural areas, then the size and proximity of these patches are critical. A single small farm or garden in a vast tract of rain forest may face encroachment from native species. But the survival of those same species could be threatened in a remnant patch of rain forest surrounded by vast acreages of farmland.

Temporal effects—History and timing of events also affect the agriculture-environment interaction. Weather and seasonal events affect timing of events, but human impact (cropping practices, land use history) is just as important in many cases. Through their interactions over time, both the agroecosystem and the natural ecosystem are slowly changed.

Functional effects—Of course, the nature and strength of the interaction are also important. For example, whether agrichemical release from an agroecosystem was minimal or excessive would result in a very different interaction and impact on a natural ecosystem.

Regional Monitoring

Regardless of the structure of the landscape and nature of agriculture-environment interactions, monitoring on a regional or landscape level is very important to agriculture. The most common and familiar form of regional monitoring is weather forecasting. In particular, tracking the movements of fronts or storms through a large region may provide several days' notice to anticipate frost, excessive rains, or other damaging weather events.

Regional monitoring is also important in anticipating some pest problems. Quarantines and pest surveys can detect or prevent the movement of exotic pests into large geographic areas. But surveys can also be helpful in tracking the movement of endemic pests within a region. This type of regional monitoring is important in anticipating movements of migratory locusts in Africa. In the United States, the northward movement of insect pests that overwinter in Florida or the Caribbean can be tracked. Weather forecasting can be used to anticipate the appearance of some plant diseases.

Effects of Natural Ecosystems on Agroecosystems

As discussed previously, the influence of natural ecosystems in increasing diversity in and around agroecosystems is usually beneficial to agriculture. In some cases, weeds, vertebrates, birds, or insects from natural areas can become pests in nearby agroecosystems. However, we should emphasize that most serious agricultural pests are not species that have wandered in from a nearby natural area, but species (often non-native) that are particularly well-adapted to a highly specific crop ecosystem.

Although difficult to demonstrate in controlled scientific experiments, the presence of woodlands, grasslands, and other natural features in the watershed may help in management of erosion and flooding. However, possibly the most important potential benefit of natural ecosystems to agriculture is that they are the sources of future crop germplasm and agricultural products. The development and improvement of future crop cultivars may depend on species preserved in natural ecosystems (see Chapter 8).

Effects of Agroecosystems on Natural Ecosystems

Agricultural activities result in both direct and indirect effects on natural ecosystems. The most direct effect is the conversion of natural areas into farmland, a subject that is examined in more detail in Chapters 18 and 19. **Fragmentation,** or the breakup of natural areas into smaller and more isolated patches, can have serious consequences for natural ecosystems. Large vertebrates and predators that require extensive home ranges are most affected by fragmentation. As patches become smaller and boundaries with agricultural areas increase, the chance for direct impact of agriculture on any species within the patch increases. As patches of natural habitat become more isolated from one another, the dynamics of communities within them tend to resemble those of communities isolated on islands or mountaintops. Dangers of local extinction and colonization by exotic species are increased in these situations.

Agriculture may be a source of pests or weed species that may invade natural areas. Many cultivated plants are exotic or non-native species, and some of these may colonize natural ecosystems. Kudzu, which is grown as a forage legume in some parts of the world, is an exotic species that has colonized many woodlands in the southeastern United States, becoming a serious weed problem. Grazing by cattle and other livestock can also have a great impact on natural systems, changing the structure of plant communities. Other animal species can be affected indirectly by this habitat modification.

Soil erosion resulting from cultivation and other agricultural practices can lead to accumulation of silt, flooding, and other direct impacts to natural areas. Often, however, these effects are subtle, but cumulative, over time. Environmental effects from pesticides and other agricultural chemicals are often indirect (see later in the chapter), but over time, these materials can build up to toxic levels. Certain organisms in the community may be threatened, while others remain unaffected.

The impact of agriculture on natural ecosystems and wildlife is not all negative. It is possible to design agroecosystems that have minimal impact on adjacent natural areas. Some agricultural practices are favorable to wildlife. Orchards, as well as open areas mixed with patches of trees, can provide varied habitat for a variety of species. Some growers deliberately plant millets and other types of grain or leave some residues of grain crops after harvest as food for birds and other wildlife. Also, the ecotones at interfaces of agroecosystems and natural ecosystems may benefit the natural ecosystems as well.

Coexistence of Agricultural and Natural Ecosystems

Within an individual farm, beta level diversity can be increased through design of varied cropping systems (see Chapter 13) or by the creation and proper management of ecotones such as buffer strips, vegetative filter strips, windbreaks, or hedgerows. The design of some of these features can be quite innovative. For example, in Indiana, windbreaks and filter strips composed of trees and shrubs that are valued as fruits or ornamentals may not only improve plant and wildlife diversity, but grower income as well (see Miller et al. 1994).

The maintenance of gamma level diversity can be a more difficult task. On a regional scale, it is possible to confront complicated issues such as fragmentation of natural ecosystems. At this level, it may be possible to design corridors to connect isolated "islands" of natural habitat within a largely agricultural landscape. A typical wildlife corridor may be formed by linking a mixture of various habitats, such as very small patches of natural vegetation, windbreaks, hedgerows, etc. Regional planning and cooperation are needed to devise such systems. Efforts in several European countries to encourage preservation and balance of varied habitats are summarized by Naveh (1994). In the United States, various government set-aside and conservation programs, which encourage cropland to be taken out of production for other reasons, have probably contributed to regional diversity.

Agrochemicals and the Environment

Of the many interactions between agricultural and natural ecosystems, issues relating to pesticides and other agricultural chemicals in the environment have generated the most attention and controversy. We may consider pollution to be unacceptable levels of a substance in the environment, but what is acceptable or unacceptable is a subject of much debate. Standards are constantly changing, based on new evidence and research and on technological improvements resulting in the detection of smaller and smaller concentrations of chemical substances. Detection of concentrations of parts per million (1 ppm = 0.0001%) or parts per billion (1 ppb = 0.0000001%) is common for many chemicals, and measurements of parts per trillion are becoming more frequent.

Whether or not a particular material will persist or become a problem in the environment depends on a variety of physical, chemical, and biological factors, as discussed in the following sections. Among these, one of the most important features is the chemical structure of the substance. Therefore, it is convenient to divide our discussion of agricultural chemicals into three general groups of materials: fertilizer elements, pesticides, and heavy metals.

Fertilizer Elements in the Environment

Although a variety of natural processes and human activities contribute N and P to the environment, agricultural practices are a major contributor, particularly to excess levels of nitrates. Excessive applications of highly soluble inorganic fertilizers are the most common source, but dairies and other concentrated livestock operations may be significant contributors in some locations (see Application 12-3). N and P are of particular concern in aquatic environments.

Application 12-3 *Dairies and Nitrates*

High concentrations of nitrates from animal wastes are of particular concern in high-density or confined animal operations such as dairies, hog pens, poultry houses, and feedlots. When feeds are brought into a small, localized site, much of the nutrients present in the grain or feed (which was produced over a much larger geographic area) are then deposited as manure in that site, resulting in an unusually high nutrient concentration. Removal of large amounts of manure may be a low priority, and it may remain near the original site or stored on-site for an extended period of time, increasing the likelihood and amount of nitrate release.

An important step in reducing this type of nitrate problem is to avoid or reduce the concentrations of animals that result in concentration of waste. Rotational grazing (see Chapter 16), in addition to reducing livestock feeding stress to vegetation, also serves to distribute wastes over a relatively large area. Using this method, the concentration of waste per unit of land is much lower than when animals are restricted to a small, confined area. If this is not practical, the use of concrete floors in structures provides a somewhat expensive method for avoiding contact between wastes and underlying soil, provided that waste removal is timely and disposal is environmentally sound. Site planning may be needed to design optimal plans for construction of drainage ditches, for pumping systems to regulate water (and nutrient) flow in ditches following rainfall, and for settling, drying, or periodic removal of solids. Buffer strips, ponds, and wetlands may be compatible with some design schemes. Nutrient-rich water from ditches can be recycled when used in irrigation of crops. Examples and case studies of various methods for minimizing nutrient enrichment problems from livestock wastes are provided by other sources (e.g., Campbell, Graham, and Bottcher 1994).

Nitrates in Water. Nitrates in water are of concern for several reasons:

- Health concerns occur rarely when nitrates in drinking water are converted to nitrites (see Spedding 1996);
- N enrichment is a contributor to eutrophication (see P section later in chapter) in aquatic systems;
- Nitrates in water represent wasted N and money to growers who intended them to be used by crop plants.

Nitrates are highly soluble in water; thus, excess nitrates move along with water flow from an agroecosystem. The principal routes of nitrate losses are leaching and erosion. Although heavy rains, flooding, and other causes of erosion may move significant amounts of nitrates, leaching is a more consistent and important means by which nitrates move from agroecosystems to aquatic systems. When too much N is applied to a crop, the excess nitrates that the crop could not take up have a tendency to move down through the soil profile by leaching. Eventually they reach groundwater or wells that are sources of drinking water. Nitrate contamination is much more of a problem in shallow wells than in deep wells. The rate and amount of nitrate leaching depends on:

- Amount of N applied;
- Type of N applied;
- Soil type (porosity, drainage, and other features that affect water movement);
- Soil depth (distance to water table);
- Amount of rainfall and irrigation; and
- Season (nitrate leaching is worse in winter when plants do not use nitrates)

A number of practices are helpful in reducing nitrate concentrations in groundwater and other aquatic environments:

- *Optimum fertilizer practices.* Avoiding excess application of nitrates and other highly soluble forms of N is the most important method for minimizing nitrate levels, since nitrates that cannot be used by plants are subject to leaching. The type of N source is also important. Organic N sources and slow-release synthetic formulations spread the release of smaller amounts of nitrates over time, avoiding the high concentrations that may result after the application of an equal amount of nitrate fertilizer. This can be accomplished to some extent with nitrate fertilizers by splitting the dosage into several smaller applications rather than one large application.
- *Timing of fertilizer applications.* Fertilizers should be applied at times that avoid heavy rainfall or irrigation, or applications too late in the season.
- *Buffer strips and other forms of vegetative filtration.* Since nitrates can be used by plants, any vegetation (even weeds) in runoff areas can take up some of this N source, reducing concentrations before they reach other habitats. Trees and other deep-rooted perennials can be helpful in removing nitrates from the soil profile.
- *Cropping and tillage practices.* Methods that favor nutrient management and reduction of erosion should be followed (see Chapters 13, 14).

- *Optimum management of concentrated livestock operations.* (See Application 12-3 and Chapter 16.)

Phosphorus in Aquatic Systems. Much of the P in the agroecosystem is tied up in relatively insoluble compounds (see Chapter 4). As a result, P is not easily leached, and more losses of P occur through erosion and surface runoff. Nutrients and sediments in runoff water are fed into a system of ditches, streams, and gulleys that drain a particular land area (watershed), ultimately reaching a river, lake, or ocean.

The most severe problems result when runoff enters a lake or pond that does not have much outflow. Under these conditions, various stages of eutrophication can occur. **Eutrophication** is the aging process that results from nutrient enrichment of lakes and ponds. This process occurs naturally over long periods of time. **Cultural eutrophication** refers to the acceleration of the process through human activities such as agricultural practices or sewage disposal. Nutrient enrichment leads to a sequence of events beginning with increased growth of algae, phytoplankton, weeds, and other aquatic plants (Figure 12-4). Available P (in dissolved compounds or in plankton) is often limited in lakes, and so addition of P is particularly effective in stimulating production and eutrophication. Cultural eutrophication can also result from addition of N or from erosion. The effects of erosion are twofold: Erosion deposits sediments that may slowly fill up a lake, and some of the eroded sediments and runoff water may contain high concentrations of P or N, which further enrich the lake.

The severity of cultural eutrophication resulting from agricultural fertilizers is affected by a number of factors (Figure 12-5). In the case of P, the most important of these is the distance from the source of P to the lake or pond. Obviously, the potential for problems is much greater if fertilizer application or congregation of livestock occurs on the banks of the

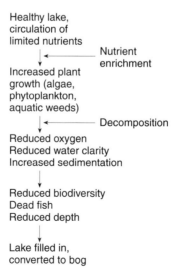

Figure 12-4 Progression of cultural eutrophication following nutrient enrichment of a lake.

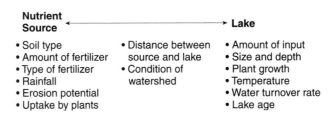

Figure 12-5 Some factors affecting the rate of cultural eutrophication of a lake.

pond than if the source and the pond are several hundred meters apart. Whether cultural eutrophication can be stopped or reversed depends on the size of the lake and how far the situation has progressed. Stopping or reducing the input of nutrients and sediments is a critical first step.

Many of the methods used in reducing nitrates are also effective in reducing the amount of P in the watershed. Optimum rates and timing of fertilizer applications are important. Conservation practices that limit erosion and movement of surface water are especially important in P management. The use of conservation tillage, buffer strips, cover crops, mulches, crop residues, intercrops, or other elements of cropping systems (see Chapters 13 and 14) are examples of helpful practices. If the distance between the P source and the lake, pond, stream, or ditch is sufficient, various types of vegetation (crops, weeds, hedges, etc.) in this zone may be helpful in removing some of the P before it reaches an aquatic ecosystem.

Pesticides in the Environment

Whether or not hazards and problems may result from pesticide use depends on the material used, method of use, and environmental conditions. Despite the best intentions and precautions, accidents can happen (Figure 12-6). Potential problems from pesticide use can be grouped into three general areas:

Human health effects. Many pesticides are toxic to mammals, and therefore may provide occupational hazards to workers and others who come in contact with these materials. Acute poisoning is the most serious consequence of exposure to pesticides. However, potential chronic health effects receive considerable attention because of the large numbers of consumers at potential risk. Pesticide residues occur in vegetables, fruit, milk, and other agricultural products. Although most pesticides used today are less persistent than those used in the past, many more systemic materials are currently used. Governments set tolerance limits for the concentrations of pesticide residues allowed in food products, but government standards, as well as chemicals used, vary in different countries.

Nontarget effects in the agroecosystem. These effects have been discussed in previous chapters. They include phytotoxicity to crop plants or ornamentals from herbicides as well as poisonings of pollinators, domestic animals, or wildlife entering the field. Problems with key pests may become worse and outbreaks of secondary pests may occur

Figure 12-6 Even properly stored pesticides that are isolated in a pesticide storage building may be a source of contamination under unusual circumstances such as a flood.

as a result of reductions in populations of natural enemies. These problems can become more difficult to manage if resistance to pesticides develops.

Nontarget effects in natural ecosystems. Pesticides can directly or indirectly affect a variety of fish, wildlife, and nontarget invertebrates such as earthworms when they enter aquatic and other nonagricultural ecosystems. Many of these effects result from persistent residues, as discussed in the next sections.

Other texts and articles provide additional information on the hazards of pesticide use (Pimental and Greiner 1997) and the potential for environmental contamination from pesticides (Isensee 1991; Radosevich, Holt, and Ghersa 1997; Sharom, McEwen, and Harris 1991). Previous chapters have examined the reduction of pesticide usage resulting from adoption of IPM strategies and alternative methods for pest management. When pesticides must be used, it is essential to follow all labels and other regulations closely to avoid possible hazards, including nontarget effects.

Pesticide Movement from the Agroecosystem. Pesticides can move out of the agroecosystem into air, soil, or water, or can be carried out as residues in harvested material

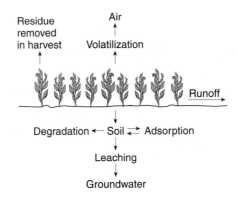

Figure 12-7 Pathways of pesticide movement out of an agroecosystem.

(Figure 12-7). Volatilization is the principal way in which chemicals sprayed on or around crop plants enter the air. Wind, drift, and other events that occur during application contribute heavily to volatilization losses. But probably the most important factor affecting volatilization is the vapor pressure of the chemical involved. The high volatility of fumigants is not surprising, but many insecticides and herbicides have low vapor pressures that allow them to volatilize easily. In the atmosphere, many of these chemicals break down from exposure to light (photolysis). Others may react with atmospheric gases. Methyl bromide and some other halogenated compounds can react with and break down atmospheric ozone (see Chapter 11).

Many pesticides eventually enter the soil ecosystem. These include not only soil-applied pesticides, but also excess material dripping from foliage and pesticide residues in crops or other organic materials. The amount and rate of chemicals entering and moving in soil depend on a variety of factors. Agronomic practices such as application methods, irrigation, or tillage, and meteorological factors such as precipitation, temperature, or wind affect the movement of pesticides into soil. The formulation and chemical properties of the pesticide are also very important, particularly its solubility in water, which is the principal agent that carries pesticides into and through soil. Soil properties that affect rate of chemical movement include soil organic matter, clay content, pH, soil structure and porosity, soil moisture, and soil organisms present. **Adsorption** is the attachment of chemical molecules to the surface of soil particles. Many pesticides are adsorbed easily onto organic matter (particularly humus) and certain types of clays. Adsorption holds the chemicals in soil and restricts their movement until they are released by desorption (= reverse of adsorption). The persistence or degradation of chemical pesticides in soil is discussed in the next section.

Pesticides enter water through *runoff* or by *leaching*. Movement of pesticides in runoff water is affected by rainfall or irrigation soon after application and by soil conditions and erosion potential. In sites where erosion potential is high, runoff, rather than leaching, may be the more important route of pesticides into water. Before pesticides can leach into groundwater, they must pass through soil, where they may be degraded or adsorbed. Adsorption and leaching show opposite trends; the more strongly a chemical is adsorbed onto soil, the less likely that it will be available for leaching. Therefore, leaching of a pesti-

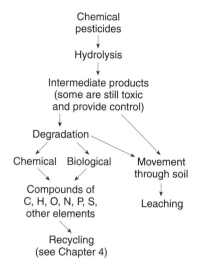

Figure 12-8 Degradation of chemical pesticides in soil.

cide depends on soil properties such as organic matter and clay content, but is also influenced by the water solubility of the material, the persistence of the material, and soil depth.

Persistence and Degradation of Pesticides. Leaching of pesticides into groundwater is limited because most pesticides are broken down, or degraded, in the soil environment. The typical steps in **pesticide degradation** in soil are shown in Figure 12-8. Degradation in soil can be chemical or biological. **Hydrolysis,** or reaction with water, is the most basic type of chemical degradation, and is the initial step in the degradation of most chemical pesticides. In many cases, products from the hydrolysis reaction may still be toxic and provide control. Other intermediate products of varying composition and toxicity can be formed as hydrolysis products react with other chemicals in the soil environment. At any stage, various types of intermediate products may be subject to movement through soil and to leaching. Of course, the specific degradation reactions depend on the chemistries of the individual pesticides involved. However, these chemical reactions can also change with soil type, soil pH, and other soil properties.

Biological degradation, or **biodegradation,** is also an important route of breakdown and inactivation of chemical pesticides. It occurs in most soils along with chemical degradation. Some bacteria and fungi in soil break down pesticides to obtain C, N, and other nutrients. It may seem unusual that organisms use pesticides as food sources, but what is toxic to a mammal or insect may be nothing more than another organic chemical food source to certain kinds of bacteria. Sometimes these organisms are so efficient that they degrade the pesticide before it has reached its target pest (see Application 12-4).

When sufficient biological or chemical degradation has occurred, the toxic properties of the original pesticide are no longer present. Ultimately, the material may be broken down into basic elements such as C, H, O, N, P, or S, which can be recycled in the environment. In reality, most degradation does not proceed to free elements, but to simple compounds

Application 12-4 *Accelerated Biodegradation*

Carbofuran is a soil insecticide that was widely used for management of the corn rootworm in the midwestern United States during the 1970s. By the late 1970s and into the 1980s, it became more and more difficult to control corn rootworms with carbofuran, particularly in fields where carbofuran had been used for a number of years. Many of these insecticide failures resulted not from insect resistance to the material, but from accelerated biodegradation, also called enhanced biodegradation. Years of use of the same chemical had apparently resulted in selection for a soil microflora (especially bacteria) that was particularly efficient in using this chemical as a food source. As time went by, and selection continued, the microflora may have become so rapid and efficient in breaking down the chemical that the material became degraded and ineffective before it ever reached the pest.

Accelerated biodegradation has been observed with a variety of soil-applied pesticides, including a number of insecticides and herbicides. Selection for accelerated biodegradation can be minimized by adopting management strategies that avoid frequent, repeated use of the same chemical material. These include IPM strategies involving treatment when needed rather than on a routine basis, use of nonchemical methods of management where possible, rotations of different chemicals and different classes of chemicals, and avoidance of overdose and excessive application rates.

---

containing them, such as H_2O, NO_3^-, etc. In the process of nutrient recycling (see Chapter 4), it does not matter whether these simple compounds and essential elements originally came from a toxic synthetic pesticide, a fertilizer, or an organic material.

Some chemical pesticides are much more persistent in the soil environment than others. Persistence and potential toxicity in soil depend on a variety of factors including method of use and application (soil-applied pesticides are a particular concern), rate of material applied, soil properties and organic matter, soil organisms present, weather conditions, irrigation, tillage, solubility of the chemical in water and other media, and the half-life of the chemical (how long it takes half of the amount of chemical to break down).

The chemical structure of the pesticide determines its properties, and so it is possible to make some broad generalizations about persistence and degradation of some of the more common classes of pesticides (Figure 12-9). Most types of chemical pesticides, particularly the newer ones, consist only of C, H, O, N, P, or S, which are all essential elements that can be recycled after the pesticide is broken down. Inorganic compounds, represented in Figure 12-9 by copper sulfate, contain metal atoms that cannot be further broken down in most environments. Most are older materials that are not used much anymore. These toxic metals can persist indefinitely. Likewise, toxic levels of halogens, particularly Cl and Br, can build up in ecosystems because the halogen atoms are not degraded and do not lose their integrity. Therefore, even if the pesticide is completely degraded, excessive levels of a toxic element such as Br may remain in the end. Pesticides containing halogens are called organohalides, and those containing Cl in particular are organochlorines. A number of organochlorines, such as DDT, are relatively insoluble in water but are highly soluble in soils or in fatty tissues of animals. Many of the organochlorine pesticides retain their toxic properties for a long time because they are very slow to break down. Organochlorine pesticides were used heavily from the end of World War II into the 1970s, when much of their use was

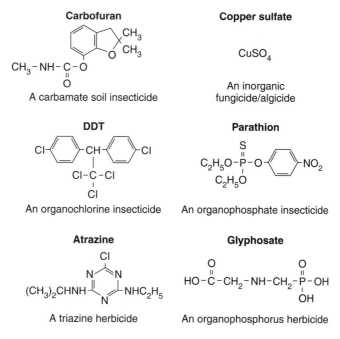

Figure 12-9 Chemical structures of some pesticides.

discontinued due to human health concerns and some of the environmental problems described in the next section (although some are still used in some countries!).

Bioaccumulation and Biomagnification. **Bioaccumulation** is the cumulative buildup of toxic elements or compounds in the body of an organism. When trace amounts of DDT or other organochlorines are ingested, they may be retained in the fatty tissues of the body. Since these materials are not easily eliminated, they can build up over time. **Biomagnification** is the increase in concentration of toxic materials in the higher consumers within a food chain. In Figure 12-10, trace amounts of organochlorine compounds in insects or fragments of insects killed by organochlorines enter the aquatic ecosystem where they are incorporated into the plankton, which form the base of the aquatic food chain. Small fish and invertebrates feeding on these plankton accumulate the organochlorines from all of the plankton they eat. These small fish are eaten by medium-sized fish that accumulate the organochlorines from the smaller fish, and the process continues up to the highest levels of consumers, which receive a disproportionately high dosage of organochlorines from the entire food chain. The concentration of the organochlorines in body tissues is magnified, or increased, through the food chain. Eventually, death, toxicity, or chronic problems may affect the highest members of the food chain as excessive dosages are exceeded. Biomagnification is not confined to aquatic systems. Biomagnification of DDT has been connected with reproductive problems and reduced survival of offspring in some predatory bird species.

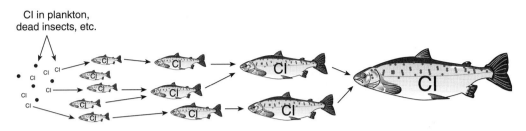

Figure 12-10 Biomagnification of an organochlorine compound in an aquatic food chain. Relative size of "Cl" represents relative concentration of the organochlorine in the organism.

The problem of biomagnification has been recognized for several decades, and as a result, many of the organochlorines responsible are not manufactured or used much anymore (but exceptions can still be found). Instead, most insecticides and other pesticides produced since the 1970s are organophosphates, carbamates, and other materials that are less persistent and break down much more quickly than the organochlorines used in earlier decades. To avoid the problems of accelerated biodegradation (Application 12-4) on one hand and bioaccumulation on the other, an "ideal" pesticide should persist long enough to kill the pest, but break down before it can cause any negative environmental impact.

Heavy Metals and Agriculture

A number of heavy metals can sometimes cause toxicity problems in agriculture (Table 12-2). Other metals may occur rarely, and arsenic (As) and selenium (Se) are metalloids that may be toxic in some living systems. The situation is complicated because a number of these elements, such as Cu, are essential micronutrients to living organisms in trace amounts, but can become toxic at excessive levels. All of the materials listed in Table 12-2 are elements that cannot be degraded further, and tend to be very persistent in the soil environment.

Some of these metals, as well as other potentially toxic elements, are constituents of soils in many parts of the world (see Chapters 5 and 18). The most common source of heavy metal contamination is from industrial and urban sources. But some metals can be added to agroecosystems, where they may accumulate over time. Several of these materials, such as Pb or As, were often used as inorganic pesticides during the first half of the twentieth century. Today, Cu is the heavy metal that is probably added most frequently to agroecosystems because it is a component of some bactericides, fungicides, and algicides. A number of different metals, such as Cu, Zn, Pb, Cd, and Ni, may be added to agricultural sites when sewage sludge, municipal solid waste, or other biosolids of urban origin are used as fertilizer amendments, although this risk is decreasing as tighter restrictions are placed on urban industries and municipalities. However, even if the levels of these elements in a single application of biosolids are low, metals may accumulate over time if repeated applications are made. While the N, C, O, H, P, and S released from breakdown of biosolids are used and recycled, any heavy metals released can persist in the environment.

Table 12-2 Some of the more common heavy metals that may be found in agroecosystems.

Symbol	Metal	Symbol	Metal
Cd	Cadmium	Hg	Mercury
Cr	Chromium	Mo	Molybdenum[a]
Co	Cobalt[a]	Ni	Nickel
Cu	Copper[a]	Sn	Tin
Fe	Iron[a]	V	Vanadium[a]
Pb	Lead	Zn	Zinc[a]
Mn	Manganese[a]		

[a]Essential micronutrient for some organisms.

Summary

Ecological diversity can be maintained within an individual field, in the landscape surrounding the field, or within a larger geographic region. Every agroecosystem affects the surrounding landscape, which in turn affects the agroecosystem as well. Agricultural chemicals such as fertilizers and pesticides are a particular concern in natural ecosystems. Fertilizer elements such as N or P can cause problems such as eutrophication when they enter aquatic systems through leaching or runoff. Pesticides vary in their persistence and potential toxicity in the environment, since degradation depends on chemical structure, soil type and properties, soil organisms, and a variety of environmental and other factors. The potential for bioaccumulation and biomagnification has been lessened through the use of chemical products that are more biodegradable and therefore less persistent.

Topics for Review and Discussion

1. What are some ways in which the movement of N and P from agroecosystems into natural ecosystems can be reduced?
2. Discuss the features of the landscape in your area. Which are the largest components of your landscape—agricultural, natural, or urban ecosystems? Which ecosystem(s) has the most impact on the others?
3. How have the properties of commonly used chemical pesticides changed over the last quarter of the twentieth century in response to environmental concerns?
4. What is a stable agroecosystem? What characteristics should such a system have? Give examples of agroecosystems that are relatively stable versus those that are relatively unstable.
5. What factors would you consider in designing a landscape in which agroecosystems and natural ecosystems could coexist indefinitely?

Literature Cited

Altieri, M. A. 1994. *Biodiversity and Pest Management in Agroecosystems.* New York: Food Products Press.

———. 1995. Biodiversity and biocontrol: Lessons from insect pest management. In *Advances in Plant Pathology,* Vol. II, eds. J. H. Andrews and I. Tommerup, 191–209. London: Academic Press.

Andow, D. A. 1991. Control of arthropods using crop diversity. In *CRC Handbook of Pest Management in Agriculture,* 2d ed., Vol. I, ed. D. Pimentel, 257–284. Boca Raton, Fla.: CRC Press.

Campbell, K. L, W. D. Graham, and A. B. Bottcher. 1994. *Environmentally Sound Agriculture.* St. Joseph, Mich.: American Society of Agricultural Engineers.

Cromartie, W. J., Jr. 1991. The environmental control of insects using crop diversity. In *CRC Handbook of Pest Management in Agriculture,* 2d ed., Vol. I, ed. D. Pimentel, 183–216. Boca Raton, Fla.: CRC Press.

Hawksworth, D. L., ed. 1995. *Biodiversity: Measurement and Estimation.* London: Chapman and Hall.

Isensee, A. R. 1991. Movement of herbicides in terrestrial and aquatic environments. In *CRC Handbook of Pest Management in Agriculture,* 2d ed., Vol. II, ed. D. Pimentel, 651–659. Boca Raton, Fla.: CRC Press.

Magurran, A. E. 1988. *Ecological Diversity and its Measurement.* Princeton, N.J.: Princeton University Press.

Miller, B. K., B. C. Moser, K. D. Johnson, and R. K. Swihart. 1994. Designs for windbreaks and vegetative filterstrips that increase wildlife habitat and provide income. In *Environmentally Sound Agriculture,* eds. K. L. Campbell, W. D. Graham, and A. B. Bottcher, 567–574. St. Joseph, Mich.: American Society of Agricultural Engineers.

Naveh, Z. 1994. Biodiversity and landscape management. In *Biodiversity and Landscapes,* eds. K. C. Kim and R. D. Weaver, 187–207. Cambridge, U.K.: Cambridge University Press.

Pimentel, D., and A. Greiner. 1997. Environmental and socio-economic costs of pesticide use. In *Techniques for Reducing Pesticide Use,* ed. D. Pimentel, 51–78. New York: John Wiley and Sons.

Radosevich, S., J. Holt, and C. Ghersa. 1997. *Weed Ecology.* 2d ed. New York: John Wiley and Sons.

Shannon, C. E., and W. Weaver. 1949. *The Mathematical Theory of Communication.* Urbana, Ill.: University of Illinois Press.

Sharom, M. S., F. L. McEwen, and C. R. Harris. 1991. Movement of insecticides in the environment and biodegradability. In *CRC Handbook of Pest Management in Agriculture,* 2d ed., Vol. II, ed. D. Pimentel, 613–640. Boca Raton, Fla.: CRC Press.

Simpson, E. H. 1949. Measurement of diversity. *Nature* 163:688.

Spedding, C. R. W. 1996. *Agriculture and the Citizen.* London: Chapman and Hall.

Bibliography

Altieri, M. A. 1987. *Agroecology: The Scientific Basis of Alternative Agriculture.* Boulder, Colo.: Westview Press.

Begon, M., J. L. Harper, and C. R. Townsend. 1990. *Ecology.* 2d ed. Boston, Mass.: Blackwell Scientific Publications.

Carroll, C. R. 1990. The interface between natural areas and agroecosystems. In *Agroecology,* eds. C. R. Carroll, J. H. Vandermeer, and P. Rosset, 365–383. New York: McGraw-Hill.

Cavigelli, M. A., S. R. Deming, L. K. Probyn, and R. R. Harwood, eds. 1998. *Michigan Field Crop Ecology.* Michigan State University Extension Bulletin E-2646. Michigan State University, East Lansing, Mich.

Gliessman, S. R. 1998. *Agroecology: Ecological Processes in Sustainable Agriculture.* Chelsea, Mich.: Ann Arbor Press.

Logan, T. J. 1990. Sustainable agriculture and water quality. In *Sustainable Agricultural Systems,* eds. C. A. Edwards, R. Lal, P. Madden, R. H. Miller, and G. House, 582–613. Delray Beach, Fla.: St. Lucie Press.

Ludwig, J. A., and J. F. Reynolds. 1988. *Statistical Ecology.* New York: John Wiley and Sons.

Soule, J., D. Carré, and W. Jackson. 1990. Ecological impact of modern agriculture. In *Agroecology,* eds. C. R. Carroll, J. H. Vandermeer, and P. M. Rosset, 165–188. New York: McGraw-Hill.

Tivy, J. 1992. *Agricultural Ecology.* Harlow, U.K.: Longman Scientific and Technical.

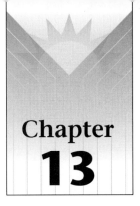

Chapter 13

Cropping Systems

Key Concepts

- Diversity in polycultures
- Multiple cropping systems
- Advantages of crop rotation
- Shifting cultivation

Monocultures and polycultures have different advantages and disadvantages associated with them, and cropping systems may need to be designed to minimize the disadvantages in any given situation. A **monoculture,** which is the repeated production of one crop on a given piece of land over time, results in highly specific production practices and mechanization needs. In contrast, a **polyculture** is the production of several different crops on a given piece of land (temporally or spatially), is often less specialized, and may or may not involve mechanization.

Monoculture versus Polyculture

Diversity, as we have seen, is one of the most important attributes of a stable and sustainable agricultural system. It provides for improved pest management, nutrient cycling, resource use, and yield increases, while also diversifying productivity and decreasing risk of loss. Why is it, then, that large-scale producers in industrialized nations still produce monocultures?

The answer lies in mechanization. Producers in industrialized nations have machinery, specialized equipment, and, often, the knowledge to grow one crop very well. Nevertheless, these systems may not be sustainable. They require intensive management and large inputs of fertilizers and fossil fuels. But they are responsible for producing very large quantities of food products that are exported to feed a hungry world. Should we therefore convert all monocultures to polycultures to add diversity? Not without expecting large drops in productivity, at least initially. Still, we should not just shrug our shoulders and give up on monocultures. Adding diversity is still possible by using rotations, borders, and cover crops.

In many regions of the world, food security is a function of the small farm or family garden rather than the large, highly mechanized farming system. These small farms are much more diversified, and their overall output feeds families rather than nations. Should we therefore convert all polycultures to monocultures to increase productivity? The answer is no, for their purpose is very different from the monoculture. Security is most important where most polycultures are grown. For this reason, we should not attempt to change them, but should instead work to develop a base of scientific information appropriate for these agroecosystems.

As we look at agricultural systems around the world, it becomes obvious that there are two extreme types of major farming systems that are very different from each other. Industrialized agriculture is usually supplemented by large amounts of energy from fossil fuels, and produces large amounts of single types of crops for sale across the nation and to other countries. Subsistence agriculture is supplemented only by human or animal energy, and produces small amounts of many different kinds of crops for the survival of a family or for limited market sales (Table 13-1). In general, industrialized farms produce enormous yields, but can be vulnerable to epidemics and adverse environmental events. Subsistence farms produce lower yields of individual crops, but are more stable and less vulnerable to a disaster.

Problems with Monocultures

Many of the problems associated with monoculture production have already been mentioned elsewhere in this text, but two of these problems deserve to be revisited. Perhaps the most serious danger of growing a monoculture is the susceptibility of that crop to diseases and natural disasters. If only one variety is grown, it will take a hard hit when a natural

Table 13-1 A comparison of industrialized agriculture and subsistence agriculture.

Industrialized Farms	Subsistence Farms
Grow large amounts of one kind of crop for sale	Grow enough for a family and maybe some for market
Plant one variety of seed	Plant many varieties of seed
Use irrigation, fertilization, pesticides	Rely on natural methods of improving soil and fighting pests; irrigation may be done, but only on a small scale
Highly mechanized	Most work done by animals or humans
Farms are vast and are usually flat and easy to cultivate	Farms may be on small parcels of land that are not easy to cultivate
Soils are fertile or maintained by inorganic additions	Soils are relatively infertile, but are improved by organic additions
System is not integrated	System is integrated with family life

disaster strikes or if a pest breaks its resistance. If the crop fails, all is lost. There is no "insurance" available that comes from growing more than one kind of crop. But lack of diversity in an agroecosystem also means that the system has less of an ability to rebound from environmental stresses. Increased diversity allows numerous habitats for natural enemies, so pest populations are less likely to cause serious problems. When a monoculture is grown, a pest can destroy the crop in very little time, or the crop may require expensive pesticide applications to protect it.

Principles of Multiple Cropping Systems

Multiple cropping systems (polycultural systems) allow diversity that can bring some inherent stability to the system. Planting a variety of crops in the same field provides diverse habitats for natural enemies of crops, allows for unique niche exploitation, and allows the grower to guide ecological succession in the field, rather than fight it. In addition, multiple cropping systems provide "insurance" to the farmer, so that if one crop fails as a result of some environmental stress, a variety of other crops will likely remain available to feed his or her family. Growing a variety of crops also ensures a good diet to the small farmer who grows primarily for his or her family.

Multiple cropping systems also allow for a more integrated farming system, with some crops used to support and provide for others. For instance, deep-rooted crops can gather nutrients from deeper in the soil profile. As the leaves of these crops fall to the soil surface and decompose, nutrients are then provided for the more shallow-rooted crops. Cover crops or green manures can provide organic material and nutrients for companion crops. Legumes can symbiotically fix nitrogen that will then be available for the subsequent crop. Corn and other upright plants can serve as "stakes" for other crops that are planted later. Branches honed from trees can also serve as stakes. The most significant advantage of multiple cropping systems is that everything can be cycled and reused, and resource use efficiency within a system can be maximized.

Examples of Multiple Cropping Systems

Intercropping

Intercropping is quite uncommon in modern U.S. agriculture, but it is a very common system in most tropical regions of the world. Intercropping systems make up at least 80% of the agriculture in some areas, and is perhaps most common in Africa, Asia, and Latin America. In Africa, almost all cowpeas are grown as intercrops, and in Latin America more than half of all corn is grown in association with other crops. Intercropping is most prevalent in areas where farms are small and farmers lack the capital to buy fertilizers, pesticides, or equipment, but it should not be considered to be confined to such regions. Intercropping may also be used on larger, highly mechanized systems.

The basic definition of **intercropping** is the planting of two or more crops simultaneously in the same unit of land. Still, there are many different kinds of intercropping systems, some of which are more common than others. Crops can be planted either as mixtures in a field (no rows), or as alternate rows or strips (Figure 13-1). Relay cropping and agroforestry are specific examples of intercropping systems, and will be described in detail.

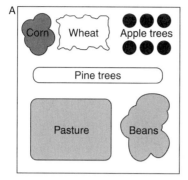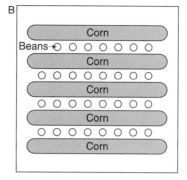

Figure 13-1 Intercropping systems can be planted as a mixture of several crops in different sections of a field (A), or in alternating rows within a field (B).

Advantages of Intercropping. The ecological and economic advantages of intercropping over monocropping are many, and some seem almost counterintuitive at first glance. Because agriculture in industrialized nations is heavily weighted toward monocropping, we tend to think of that system as being the most efficient and most productive. The opposite is often true, however. Perhaps one of the most significant advantages of intercropping over monocropping is that intercropping suitable crops with each other almost always increases the total yield per hectare, due to *decreased* competition!

It seems very odd to think that planting two crops in the same area can actually increase yield. However, recall that competition between plants is most severe when the resource requirements of plants are very similar (see Chapter 7). Intraspecific competition (between two plants of the same species) is almost always more intense than interspecific competition (between two plants of different species). Therefore, intercropping should generally decrease competitive interaction between plants in a field. This is particularly true when the two plants fill two radically different niches. Increased total yield per hectare is especially important when farms are small and limited in size by how much land can be worked with only human or animal labor.

Because of the different niches filled by each plant, intercrops allow for quite efficient resource use. If the crops grown together differ in the way they use resources, they complement each other and can make better use of resources when combined than when grown alone. Plants with the greatest differences in resource use are generally most compatible in intercrops. Combinations of perennials and annuals are particularly efficient because minerals lost by annuals are commonly taken up by perennials and stay in the system, rather than being lost by leaching or erosion.

What are some other good intercrop combinations?

- C3 and C4 plants differ in their photosynthetic pathways. An efficient intercrop would place C4 plants in full sunlight, at the top of mixed canopies, and C3 plants underneath the C4 canopy in the shaded understory. Both plants will

photosynthesize at maximum rates, and more of the total sunlight hitting the field will be captured and converted to organic compounds.
- Leaves with different growth patterns and different maturation times can intercept more sunlight over the course of the growing season. This, in turn, will prevent sunlight from hitting the ground, so less water is lost to evaporation, leaving more available for plant uptake.
- Plants with complementary rooting patterns can exploit a greater volume of soil and have more access to nutrients like phosphorus (which is immobile). For this reason, intercropping plants that have deep tap roots with plants that have shallow feeder roots will use more of the available soil moisture for production of plant biomass. Such a system will also take up more nutrients as they become available through mineralization, so fewer nutrients will be lost over time to leaching.
- Legumes that use N_2 as a nitrogen source can be intercropped with nonlegumes that use NO_3^- to maximize nitrogen assimilation.

While efficient resource use is an important advantage of intercropping systems, there are many other advantages that are perhaps more readily perceived or articulated by a grower. We have already discussed yield stability and protection against risk, environmental extremes, and natural disasters, but these are important enough to warrant further discussion. Yield stability simply refers to the benefits provided by growing a greater variety of food crops in a small unit area. This may be increasingly important in the future if we begin to assess the success of intercrops based on such criteria as caloric or protein production per hectare per day, which is, after all, a good indicator of what farmers actually look for when planning their systems to provide a nutritious diet for their families. For farmers who send some of their produce to market, economic return from their diverse farms may vary from year to year, depending on the market prices of the crops. In some cases, the farmer will be particularly interested in yield from one crop, but will then add other species for risk protection, improvement of soil fertility, erosion control, or for other minor household uses.

Reducing risk of crop failure is probably as important to farmers as both the nutritional value of their crops and potential economic returns. Risk protection is absolutely critical when a family relies on the farm for all or most of its food needs, and polyculture is much better than monoculture at producing enough of a subsistence diet for a family to survive. If one crop fails, something will likely remain to eat or sell. Crop diversity is particularly important in areas that are susceptible to rapid environmental changes, and sometimes may even provide *yield compensation*. Yield compensation occurs when, if one crop fails due to drought, insect attack, or some other environmental stress, another crop in the system actually increases in yield, partially "compensating" the farmer for the failed crop.

Modification of the environment, or *facilitation*, occurs in an intercrop when one of the crop species grown in combination with other plants gains access to resources that would not be available to it if it were grown alone. It can also result from enhanced resource utilization by one plant species that is made possible by improvements in microhabitats by another. The two most common examples of facilitation are improved soil quality and improved pest management.

There are many examples of reduced pest incidence in intercrops as opposed to monocrops. For example, Gahukar (1989) found that there was a significantly lower number of stem borer larvae in fields with a very high ratio of millet:cowpea plants. As the ratio decreased, stem borer populations rose. Such reductions in pest incidence are generally attributed to one of three hypotheses:

- *Disruptive-crop hypothesis:* In this case, the presence of a second plant species interferes with the ability of the pest to find the first plant. In general, insect pests that have narrow host ranges may have a harder time finding plants to feed on when the patches of suitable food plants are small and dispersed among other plants, rather than when food plants occur in large aggregations.
- *Trap-crop hypothesis:* When a vulnerable host crop is intercropped with a trap crop, the trap crop attracts the pest, keeping it away from the more vulnerable crop. This can have the disadvantage of actually increasing the number of pests in the system, however.
- *Natural enemies hypothesis:* This hypothesis states simply that more predators and parasites are present in an intercrop than in a monocrop. An intercrop has a higher variety and number of food sources, so the habitat is generally more suitable for the growth and reproduction of natural enemies (see Chapter 11). Intercrops are better for generalists than for specialized parasites or parasitoids because specialists require very specific food sources (see Chapter 9).

Improved soil quality in intercrops often occurs as result of including legumes in the system. Legumes have symbiotic relationships with nitrogen-fixing bacteria, so they add fixed nitrogen to the system, which is then available for uptake by other plants as well. Cereal/legume mixtures are quite common because fixed nitrogen from the legumes is available for the cereal, and the cereal usually complements the legumes in amino acids, creating a quality diet for the farmer. Intercrops also minimize erosion in a field because they provide good soil coverage, thus preventing runoff.

Other advantages of intercropping include maintenance of genetic diversity, physical support for vine crops, and weed suppression. Shading helps to suppress weeds; a well-chosen combination of crops can intercept 90% or more of light. In addition, intercropping has the socioeconomic advantages of improving marketing opportunities without the need for storage, spreading labor costs evenly through the growing season, and improving local diets.

Determining the Intercrop Advantage. Whether or not an intercropping system will be preferable to a monoculture varies with each specific situation and depends on such factors as compatibility of crop species or cultivars used, planting density of each species, arrangement of crops (in rows, strips, blocks, etc.), and the timing of planting of each crop. If more than one plant species pursue resources in the same niche at the same time, there may be no advantage to intercropping systems. In fact, intercropping can dramatically decrease yield in this case.

Intercrop advantage occurs if interspecific competition (between species) is less than intraspecific competition (between plants of the same species). In this case, increased yield is likely to result. How much of an advantage that intercropping has over monocropping can be determined by calculating the land equivalent ratio (LER) for the system.

The LER was initially proposed by scientists at the International Rice Research Institute (IRRI) in the Philippines. It is calculated as the amount of monoculture land area needed to produce the same amount of polyculture yield. In other words, how much land is needed to plant two crops grown in monoculture in order to get the same yield produced when the crops are planted together? If the LER is greater than 1, the intercrop outyields the monoculture. If it is less than 1, the monoculture is more efficient. LER is calculated by adding up the ratios of yields of intercropped systems to monocropped systems:

$$\text{LER} = \left(\frac{y_i}{y_m}\right)_1 + \left(\frac{y_i}{y_m}\right)_2 + \cdots + \left(\frac{y_i}{y_m}\right)_n \qquad \text{(Equation 13-1)}$$

where each term represents a different crop, and y is the yield (quantity per area) of the crop grown as an intercrop (y_i) or in a monoculture (y_m). This ratio is calculated for crop 1, crop 2, etc., up to n crops in the system.

Suppose that you wish to grow both corn and beans, but are interested in knowing whether to grow them as monocultures or as an intercrop. You plant fields to all possibilities, and examine yield (kg/ha) at the end of the growing season:

Cropping System	Yield (kg/ha) Corn	Yield (kg/ha) Bean	Total Yield (kg/ha)
Monocultures			
Corn	2000		2000
Bean		900	900
Intercrop	1600	270	1870

After examining the data, you note that your corn yield was fairly similar in the intercrop, but your bean yield decreased fairly significantly, so it is not an easy call.

Using equation 13.1:

$$\text{LER} = \left(\frac{y_i}{y_m}\right)_1 + \left(\frac{y_i}{y_m}\right)_2 + \cdots + \left(\frac{y_i}{y_m}\right)_n$$

For crop 1 (corn):

y_i = yield of corn grown as an intercrop = 1600 kg/ha
y_m = yield of corn grown as a monoculture = 2000 kg/ha

For crop 2 (bean):

y_i = yield of bean grown as an intercrop = 270 kg/ha
y_m = yield of bean grown as a monoculture = 900 kg/ha

$$\text{LER} = \left(\frac{1600}{2000}\right) + \left(\frac{270}{900}\right) = 0.8 + 0.3 = 1.1$$

Therefore, although individual yields are less (1600 kg/ha of corn from the intercrop compared to 2000 kg/ha from the monoculture), the overall land use efficiency is greater for the intercrop than for the monoculture (LER > 1). If this seems counterintuitive (after all, the overall monoculture yield was greater than the overall intercrop yield), remember that the monocultures were each grown on 1 hectare of land, so 2 hectares were used to produce those yields, while only 1 hectare was used to produce the yields seen in the intercrop.

Agroforestry

Agroforestry is a specific intercropping system that combines the cultivation of trees with other trees, livestock, and/or annual and perennial crops. The trees in the system shield smaller, shade-tolerant crops and vines while stabilizing soil temperatures, slowing decomposition of organic matter, and supplying nutrients to companion crops via leaf litter.

While agroforestry systems share many of the same advantages as intercropping systems of annual crops, there are some advantages that are unique to tree-based systems. For example, the addition of trees to the system actually stabilizes the system, providing permanent above- and below-ground structure to the cropping system. This, in turn, reduces water and wind movement through the system, reduces soil erosion, moderates temperature extremes, and increases soil fertility. Higher insect diversity is encouraged because the trees serve as permanent reservoirs for predators and parasites. Resources are more efficiently shared because light, moisture, and nutrients are distributed horizontally and vertically within the system. In other words, tap roots of trees can harvest moisture and nutrients from deep in the soil profile while annual crops planted between the trees can harvest these resources from the soil surface. This leads to more nutrient uptake in the forested areas and also serves to conserve nutrients so they are not leached from the system. Light is also used more efficiently because trees intercept direct light, and understory crops harvest diffuse light that passes through the canopy.

There are several socioeconomic advantages of agroforestry systems as well. An integrated and diverse system can dramatically increase year-round productivity per unit land area, which is particularly important for subsistence farmers. Because trees are long-lived, they are fairly low maintenance, so the management needed in this system is of the shorter-lived crops that are planted below or adjacent to the trees. Because little management is needed for tree or perennial growth, the ratio of energy input to output tends to be fairly low, and is thus quite efficient. Finally, products of one component of the system can be used as inputs for others. Branches from trees can serve as stakes for crops. Senescing leaves from the trees may be nutrient rich, and are good additions of organic matter to the system.

Nevertheless, despite all of the obvious advantages of agroforestry systems, there are several potential problems associated with them. It is very important to choose carefully the plants that are to be included in an agroforestry system since not all trees are adapted to all systems, and competition may therefore occur. In some cases, it may be difficult to set up an agroforestry system because it often requires substantial investment to get started. Trees can be expensive, and because they take some time before production reaches an optimal level, return may be more long-term than short-term. It may be hard to convince farmers concerned about feeding their families now that a fairly expensive long-term investment will pay off in the future. And in some cases, land tenure may become an issue. If a grower is not the owner of the land, anything that he or she plants on it will become the property of the owner in the future. Growers are seldom willing to purchase trees with their own money that will immediately become the property of someone else once the trees are established in the ground.

Still, despite these problems, agroforestry systems are very attractive for growers willing to look to the future. Productive trees can be very easily integrated with annuals or perennials, or with grasses and animals, or any combination of these. Depending on the needs of the grower or the region, the system can be managed to serve a particular function. For instance, in areas where soil erosion is a serious problem, increasing the density of trees can add protection to the soil. If firewood is needed (a serious issue in many African nations), fast-growing trees such as *Leucaena* can be planted. If food for the farmer's family is the most pressing need, the ratio of sustenance crops to trees can be increased, or the number of fruit and seed trees can be increased. If production of an export crop is important, adding coffee or cacao to the system can increase the market value of the land. Trees can be planted around borders and boundaries to serve as natural fences or windbreaks, or can be planted to provide fodder for livestock. These systems are easily adaptable to nearly every situation, provided that needs are considered carefully during initial planning stages.

A good working example of an agroforestry system is found in Nigeria, where *Leucaena* trees are grown with maize or sorghum. *Leucaena* (Figure 13-2) is a fast-growing leguminous tree, so nitrogen-fixing bacteria are associated with the tree in a symbiotic relationship. The leaves of *Leucaena* are commonly used as a natural organic amendment, providing nitrogen to the crops associated with it. *Leucaena* residues decompose rapidly, and nitrogen in these residues is more readily recovered from soil than from other plant species (Vanlauwe, Sanginga, and Merckx 1998a). This is largely due to the low C:N ratio of *Leucaena* residues (less than 15:1) (Vanlauwe, Sanginga, and Merckx 1998b). Nutrient recycling in these systems is greatly increased if earthworms are present in the soil, since their nutrient-rich casts are brought to the soil surface, contributing greatly to the uptake of nutrients by annual crops (Hauser 1993).

In Nigeria, *Leucaena* is commonly planted in parallel hedges that are 10–20 meters apart. Annuals (maize or sorghum) are planted between the rows of *Leucaena*. Throughout the growing season, the *Leucaena* is periodically cut and the leaves and stems are added to the annual crops to add nutrients. The woody limbs can be used as fuelwood. In addition to adding nutrients to the annual crops, the *Leucaena* serves to stabilize and improve the soil, and also provides a useful habitat for building up populations of beneficial insects, which then keep pest populations low on the associated crops. An agroforestry system of this type is known as **alley cropping.**

Relay Cropping

Relay cropping is a form of intercropping in which the second crop is planted during the life cycle of the first crop (Figure 13-3). In other words, a second crop species is planted between plants or rows of an already established crop during the growing period of the first planted crop(s).

The advantages of relay cropping are similar to those of other intercropping systems, but relay cropping has the added advantage of minimizing competition between plants because the major demands on resources from crops often occur at different times. In these systems, one crop is often used as a physical support for crops planted later. Perhaps one of the most common relay cropping systems in the world is the bean/corn system of Latin America. Corn is planted into the field first. Beans are planted later, in close association with the corn. By the time the beans are starting to climb the corn plants, the corn has been harvested, and

Figure 13-2 *Leucaena* trees are fast-growing leguminous trees, so they can be used in a variety of farming situations to improve soil while providing fuel and fodder to the household.

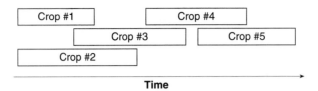

Figure 13-3 Relay cropping cycles can include more than one crop in a field at a time. Because planting dates differ, most crops are in different stages of development at any given time.

only the stalks remain in the field as a support system for the beans. There is very little competition for resources in this system because the beans grow at a time when the corn takes up very few resources, and the corn is well-established before the beans are planted in the field.

Rotation

Crop rotation, or sequential cropping, is not normally thought of as a multiple cropping system. In most cases, particularly in the United States, rotation involves replacing one monoculture by another over time. Nevertheless, crop rotation does involve growing more than one crop on the same piece of land, but the crops are separated temporally.

Crop rotation is the multiple cropping system most commonly used in industrialized nations. Most farms in the developed world use some form of rotation, either true crop rotation (very diverse crops planted over time) or varietal rotation (land planted to different varieties of one crop over time).

There are many advantages of crop rotation. Some are quite different from the benefits derived from an intercropping system, and some are quite similar. Like intercropping systems, rotation allows for a diversification of production over time. If legumes are used in the rotation, they can usually add enough nitrogen to the soil to produce a good yield of the crop following in the rotation (and can provide a high-quality forage for livestock as well). For example, Caporali and Onnis (1992) reported that a crop of alfalfa grown prior to a crop of sunflower provided 130 kg N/ha to the soil. Perhaps most importantly, rotation can break disease cycles of pests that are not mobile. Rotations are frequently used for the management of such pest problems as soil nematodes, fungi, and soil-dwelling insects. If a resistant plant is rotated with a good host plant, the pests will be deprived of a food source when the resistant plant is in the ground. Pests that cannot move in search of another host may starve to death, causing enough of a decline in their population that the good host plant can often be planted again in the field with little economic damage. Crop rotation has also been shown to reduce weed densities in many cases (Liebman and Dyck 1993). In addition to these ecological benefits, crop rotation should broaden the economic base of the farm on which it is used because it spreads labor out over the entire year, and yet still allows for the intensified production of high value crops.

Choosing a Crop Sequence. Choosing a particular sequence of crops can be a lot harder than it sounds. Crops used in different regions of the world vary with climate, tradition, economics, labor availability, and other factors. Choices of crops are situation-specific, and should be based on a variety of information gathered both from the scientific literature and from the field itself.

For example, some crops will show increased or decreased yield, depending on the crop that they follow. Legumes tend to increase yields of subsequent crops because they provide nitrogen to the soil, although in some semi-arid regions they may deplete soil water. Sorghum, however, is a very difficult crop to follow. Sorghum residues have a very high C:N ratio, so their decomposition stimulates growth of microorganisms, which then seek nitrogen from other sources. This leaves very little nitrogen available for the next crop. However, such a situation could be beneficial if nitrate pollution in a field is a concern. In some crop rotations, allelochemicals exuded from one crop (e.g., wheat) may play a role in the success of the subsequent crop. Herbicide residues left on the field may carry over into the next crop. For example, if a broad-leaf herbicide is applied while corn is in the ground, it may still be around when soybean is planted next, and, thus, will inhibit growth.

To better illustrate this point, consider a rotation with potato, legume, and sorghum (Figure 13-4). Which crop would you grow before the other crops, and why? You might say that you should plant the legume before the potato, since the legume would add nitrogen to the soil, and therefore increase potato yield. However, if you have a field infested with root-knot nematode, the legume may allow the nematode populations to build up, causing serious problems for the potato. In that case, you may choose to plant sorghum first. But sorghum planted before the potato may decrease nitrogen levels and decrease potato yield. And if wireworms are present in the system, planting sorghum before potato may increase damage from these insects to the potato! Crop rotation, like any multiple cropping system, is never simple.

Crop Rotation in Industrialized Nations. While rotation is still a common practice in both developing and industrialized nations, fewer farmers currently use rotation than in the past, largely because synthetic fertilizers are now widely used, so there is less of a need to increase fertility by this method. Continuous cropping of only one crop became more popular in the mid-1900s as farmers began to specialize. Pesticides and fertilizers were used in large quantities to keep production high, but if prices of these synthetic products

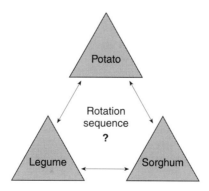

Figure 13-4 Choosing the best order for a crop rotation can often be difficult.

increase, rotation may become more popular again. Rotation is still widely used in many areas to manage certain pests and to adjust crop production for changes in climate.

In cool-temperature systems, only one crop is commonly grown per growing season, and farmers may choose to rotate a different crop each year, or to concentrate on one "specialty" crop. An example of a common crop rotation system is found in the midwestern United States, where farmers grow corn and soybean in alternate years. Winter crops, where available, are always different from crops grown during the main growing season. They are often good for pest management, and tend to improve soil fertility and reduce erosion as well.

In warm-temperature systems, more than one cash crop can be grown per year, so continuous crop sequences are common. Economics usually determines the crop sequence, and cover crops are occasionally used in the rotation to manage pests or improve fertility. These cover crops are not cash crops, but may be used as forage for livestock associated with the system. Occasional unintentional rotations can occur as well, such as when weeds take over an area or erosion causes a section of land to go fallow for a period of time.

Cover Crops

Cover crops are lower-value crops that are usually grown in seasons that are less favorable for cash crop production. They are not usually grown as cash crops themselves, but may be used for forage, animal feed, green manure, or erosion control. They can be either annuals or perennials, and are often legumes, cereals, or a mixture of the two. They can also be grown as a ground cover in vineyards and orchards (Figure 13-5).

The primary purpose of growing cover crops is to improve the health of the agroecosystem. They serve to protect soil from erosion, improve soil structure, enhance soil fertility, and suppress pests and pathogens. While they may be used as a forage or feed crop, cover crops are not normally grown for harvest, but are instead grown to cover the soil for some part of the year when the land is not planted to cash crops and would otherwise remain bare. In the northern regions of the globe, most cover crops (rye, vetch, clover, alfalfa) are grown during cooler seasons. In the tropics, cover crops are normally grown during the dry season. If legumes are used, they are planted in the rainy season, but are left in the field throughout the dry season.

How do cover crops improve the health of the system? First, they improve soil structure and water penetration. The addition of organic matter and roots increases aeration, and the plants themselves intercept rainfall and reduce the force of the drops on the soil surface. This in turn decreases soil erosion. The use of cover crops also reduces the need for tillage (see Chapter 14), which reduces soil compaction and prevents the formation of a hard pan that can result from tillage. The addition of organic matter to the soil also improves soil nitrogen fertility, particularly if the cover crop is a legume.

Some insect control may occur as a result of using cover crops. Pest insects may decline on some cover crops or the cover crops can provide a niche for beneficial insects. The cover crop may provide alternate prey to support a larger population of natural enemies. Because predators may leave the area when their food source begins to decline, providing them with an alternate food source throughout the year may allow enough of a population to stay in the field to provide adequate control when pests on cash crops begin to multiply.

Other benefits of cover crops include the moderation of soil temperatures (temperature fluctuations of bare soil are much greater than soil with plant cover), the utilization of excess fertilizer, which can help to reduce pollution resulting from fertilizer runoff, and potential control of weeds in the field.

Figure 13-5 Cover crops are useful additions to many agricultural fields because they can minimize soil erosion while improving the soil. They are often used in vineyards and orchards to prevent weed growth. This photograph is a summer cover crop of cowpea in north Florida.

Cover Crop Management. For maximum benefit to the soil, the cover crop must decompose on the land on which it is planted. Decomposition improves when conditions are moist, so the crops should be turned fairly deeply into the soil at the end of their growth season. Of course, if cover crops are used in an orchard or vineyard, special care must be taken to prevent damaging the roots of those perennial crops.

When choosing a cover crop to plant, several things should be kept in mind. First, because of their extensive root system, grasses are good at minimizing soil erosion and improving the penetration of rainfall. Legumes are not as good for water infiltration, but they add nitrogen to the soil and their residues break down more rapidly than do grass residues (see Chapter 14 for additional information on using legume cover crops to supply nitrogen). While clovers are occasionally used, they are slow growers, poor competitors with weeds, and generally require more management (e.g., mowing).

Forage legumes are the most common cover crops used in all areas of the world, and are often chosen based on their ability to perform well under the environmental conditions of the particular region (cool weather in the northern regions, minimal irrigation in tropical regions, etc.). Examples of cover crops grown in cool temperate, warm temperate, and tropical regions are listed in Table 13-2.

Table 13-2 Common cover crops in temperate and tropical regions of the world.

Region	Cover Crops
Cool temperate	Alfalfa (*Medicago sativa*)
	Pea (*Pisum sativum*)
	Red clover (*Trifolium pratense*)
	White clover (*Trifolium repens*)
Warm temperate	
Winter cover crops	Crimson clover (*Trifolium incarnatum*)
	Hairy vetch (*Vicia villosa*)
	Lupine (*Lupinus* spp.)
	ªRye (*Secale cereale*)
Summer cover crops	Alyceclover (*Alysicarpus* spp.)
	ªSudax (*Sorghum bicolor* × *Sorghum sudanense*)
Tropical	Cowpea (*Vigna unguiculata*)
	Desmodium (*Desmodium* spp.)
	Jointvetch (*Aeschynomene* spp.)
	Pigeonpea (*Cajanus cajan*)
	Velvetbean (*Mucuna pruriens*)

ªNonlegumes

Living Mulches. Living mulches are cover crops, usually legumes, that are grown in association with a cash crop, and are usually used for weed control. The goal is for the cover crop to establish quickly enough to outcompete weeds, but not so tenaciously as to compete with the annual cash crop. The intercrop competes for resources that the weeds would use, or it may suppress weed growth with allelopathy, significantly reducing overall weed biomass in the system.

Shifting Cultivation

Shifting cultivation, sometimes referred to as **slash-and-burn agriculture,** is one of most important adaptations of multiple cropping systems found in the world today. Its use extends throughout the tropics and subtropics, and is perhaps most important in Africa. It is a source of life for millions of people. It is also known as slash–burn agriculture, one of the most controversial agricultural systems that exists, and is blamed for much of the deforestation that is occurring in tropical nations the world over.

In shifting cultivation, a natural system is cleared, planted for a few years with annuals or short-term perennials, and then abandoned and allowed to remain fallow for a much longer period than it was cropped in order to regenerate soil fertility. It is a common subsistence method of agriculture: Most cultivation is done by hand, without the use of machinery, and once the crops are in the ground, little management is needed. While livestock may play a role in the system, they are commonly used for fertilizing the soil (manure) rather than for much animal labor. The size of the plots cropped by each farmer depends largely on availability of labor and on population density.

How Does Shifting Cultivation Work?

Land to be cropped is initially cleared, either with an axe or a hoe (depending on vegetation present), and then burned to release nutrients from the vegetation into the soil and to kill weeds and insects. Although some nitrogen is volatilized by burning, soil fertility is essentially maintained with the ash from the burned area and from some residues of the vegetation (Figure 13-6). Farmers may choose to retain and protect useful tree species and perhaps some bushes to stake plants to. They are trimmed, however, and the trimmed leaves and twigs are burned.

Seeds are then planted directly into the ground, usually in some pattern of intercropping. The system used varies with farmer, region, and tradition. The land is then cropped for a period of several years, until yields begin to significantly decrease or until competition from weeds gets too high. During these first years after burning, yield levels often approach 1000 kg/ha, which is quite high, considering that input levels tend to be very low. After several years, when the nutrient level of the system is depleted, the land is abandoned, left to fallow, and later reclaimed. The length of the fallow determines the extent of recovery. Since the recovery follows general succession patterns, grasses appear first (in less than five years), followed by bushes and shrubs (six to ten years), and finally trees (twenty to twenty-five years). If such a long fallow is allowed, the system may recover fairly well, although the

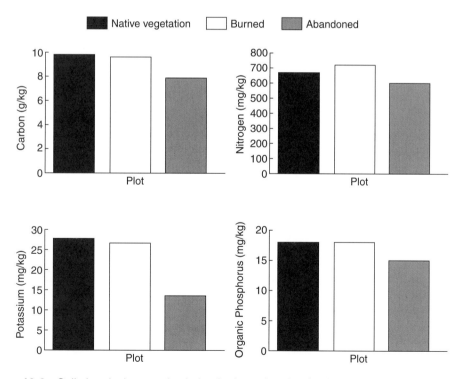

Figure 13-6 Soil chemical properties in land prior to burning (native vegetation), recently burned, and abandoned after five years of cultivation.
Data from Salcedo, Tiessen, and Sampaio 1997

composition of tree species present may be quite different from that which occurred before burning. Most of the remarks in these next sections pertain to sites where shifting cultivation has been practiced in the past. Biodiversity is expected to decrease when a tropical forest site is cut for the first time.

Why Does Shifting Cultivation Have Such a Bad Reputation?

Shifting cultivation has a bad reputation because it is no longer sustainable in most areas where it is still used. As originally practiced, shifting cultivation was a useful food production system. Recycled nutrients were locked up in the biomass, released to crops, and then restored as the land was allowed to recover. Biodiversity was enhanced. Pest populations were kept low. Productivity per input was high, ecological damage was low, and the system was sustainable.

However, the system is only sustainable if the fallow is sufficiently long (to allow for nutrient recycling, to break pest cycles, to restore organic matter, etc.). During the fallow, nutrients are stored up in plant biomass, and are then released when the land is again cleared and burned. But the time of fallow is important. If the fallow lasts longer than necessary for the regeneration of the soil, the system is inherently stable (like having money in the bank). If the fallow period is just long enough to restore soil fertility to its original level, the system is stable, but there is no margin for error (like living from paycheck to paycheck). If the fallow period is shortened, the fallow is not long enough to restore soil productivity (like being in debt and continually adding to it). In this case, yield/hectare falls dramatically, and the land is continuously degraded. This is the problem in most places where shifting cultivation is practiced. Fallow periods are now much too short: Many pieces of land are left alone for only five years compared to the most beneficial twenty to thirty years. Why?

The answer is found in population growth. Shifting cultivation is only sustainable when population levels are low. As more and more people require land to grow their subsistence crops, there is more demand for farmland. Farmers are simply unable to leave land fallow for such long periods of time. In most places where shifting cultivation is now practiced, population levels are too high for the land to be allowed to recover sufficiently. Because nutrient levels do not have time to build up on the land, most pieces of land can be farmed only for a year or two before they are abandoned to be burned again within a few years. It is a vicious cycle of nutrient depletion that cannot be easily broken.

Can Shifting Cultivation Work Under Some Conditions?

Shifting cultivation can work when population levels are low, but this is not the case in most areas where shifting cultivation is common. Can the system be improved to prevent some of the degradation currently underway? The answer is somewhat ambiguous, and depends largely on region. Still, there are some steps that can be taken to improve productivity and sustainability of the system. Some of these steps can actually lead the grower away from a shifting cultivation system to an agroforestry system, which may be more sustainable in the long run:

- Intensify the fallow by planting crops that will improve the soil and the nutrient level of the land. Plant legumes to add nitrogen to the soil. Use fast-cycle trees like *Eucalyptus* or *Leucaena* to harvest nutrients and water from deep in the profile.

- Reduce the need to fallow. Use a living mulch to add nitrogen and to prevent weed competition. Keep the ground covered by multiple cropping to reduce weeds and to keep diversity high. This will allow for a good population of natural enemies and will prevent pests from getting out of hand.
- Use tree crops or multiple-story crops to intercrop with food crops. This will allow the exploitation of many different niches, and thus ensure better resource use. Nutrients will be held within the system, rather than leaching out of the soil.
- In some cases, apply lime and phosphorus to increase pH and to provide nutrients commonly lacking in these systems. The problem with this step is, of course, availability and expense.

Multiple Cropping Systems as an Integrated Farming Method

While each of the methods just described has been discussed separately, it is important to remember that these methods do not need to be separate and self-contained. Any combination of them can work in different situations. Intercropping systems can be rotated. Cover crops can be included in a relay-cropping system. Agroforestry systems can include rotation as part of their management. The most successful multiple cropping systems are those that have been well-matched to the needs of the farmer, the environment, and the market. Many of these methods also include animals as a part of their nutrient cycling and subsistence needs which we discuss in Chapter 16.

Problems with Polycultures

Multiple cropping systems tend to be advantageous in almost all respects. The increased diversity and increased yield will almost always make these systems attractive to most subsistence farmers. Nevertheless, there are some problems that arise as a result of polycultures, and while these can often be solved by choice of plants, they deserve some special attention.

While we have shown that competition in multiple cropping systems is often less than that in monocultures, it can occur, and it can seriously decrease yields, resulting in systems that are economically unprofitable (Powers et al. 1994). For this reason, it is important to be sure that crops are compatible before placing them together. Crops with very different niches are most likely to be compatible, while those with very similar niches will compete. Proper plant spacing is particularly critical. While two plants may have different needs for light, they may have similar needs for water. Placing these two plants close together may simply exacerbate the competition for moisture.

Allelopathy is another serious potential problem of polycultures. Currently, this is a highly debated area of agroecology, with some scientists warning that allelopathy can cause serious yield reductions, and others saying that it seldom causes any significant problems. Nevertheless, the consensus is that it can occur, so it is a good idea to be sure that crops do not inhibit each other before planting large areas with them.

We have already defined allelopathy as any effect of one plant on another through chemical compounds that escape into the environment. Practical considerations for multiple cropping systems include:

- Choose species that show tolerance to allelopathic effects or that have negative effects on weeds. For instance, mulches that consist of residues of other plants that

are allelopathic to weeds can be incorporated and used as weed control. Alternatively, prior to planting a cash crop, establish a cover crop that is allelopathic to the weeds that compete with that cash crop.
- Select combinations of species that are not allelopathic to each other. Some cover crops may inhibit subsequent cash crops, and some legumes are allelopathic to both crop plants and weeds.

Perhaps the most frequent objection to the use of multiple cropping systems—difficulty in mechanization—comes from farmers in industrialized nations. Most farm equipment is designed for monocultures, and is not easily adaptable to intercropping situations. If multiple cropping and the benefits it can give to an agroecosystem are to take root in industrialized nations, appropriate technology and crop varieties must be developed and implemented. Applications of multiple cropping in United States agriculture are increasing (see examples in Campbell, Graham, and Bottcher 1994, Gallaher and McSorley 1997, Keisling 1998).

Summary

A monoculture that has optimized the production system for a particular crop often provides a maximum yield for that crop. In general, however, diverse agricultural systems exhibit improved nutrient cycling and use, pest management, and resource use, and are less vulnerable to catastrophic loss than are monocultures. Diversity in a field may occur over time, as in crop rotation and relay cropping; or space, as in intercropping, inclusion of cover crops, or agroforestry systems. While crop diversity within a field is common on small farms worldwide in order to provide for all of the food needs of a family, it is less common in technologically advanced countries where mechanization encourages monocultural production.

Topics for Review and Discussion

1. What are some of the benefits of having more than one crop in a field at the same time? What are some of the risks?
2. Given the following data, determine whether these crops are better grown together in an intercrop or separately in monocultures:

Cropping System	Yield (kg/ha)		Total Yield (kg/ha)
	Sorghum	Millet	
Monocultures			
Sorghum	246		246
Millet		248	248
Intercrop	189	107	296

Data from Finney 1990

3. Describe in detail what needs to be considered when determining a good crop rotation sequence.
4. Many people feel that shifting cultivation is a major reason why CO_2 levels are increasing in the atmosphere, and that further clearing of rain forests for agricultural use should be prevented. How would you respond to this assertion?
5. What do you think are the major reasons why monoculture is so prevalent in technologically advanced countries? What changes need to be made before polyculture becomes the accepted way to farm in countries like the United States?

Literature Cited

Campbell, K. L., W. D. Graham, and A. B. Bottcher, eds. 1994. *Environmentally Sound Agriculture*. St. Joseph, Mich.: American Society of Agricultural Engineers.

Caporali, F., and A. Onnis. 1992. Validity of rotation as an effective agroecological principle for a sustainable agriculture. *Agriculture, Ecosystems and Environment* 41: 101–113.

Finney, D. J. 1990. Intercropping experiments, statistical analysis, and agricultural practice. *Experimental Agriculture* 26: 73–81.

Gahukar, R. T. 1989. Pest and disease incidence in pearl millet under different plant density and intercropping patterns. *Agriculture, Ecosystems, and Environment* 26: 69–74.

Gallaher, R. N., and R. McSorley, eds. 1997. *Proceedings of the 20th Annual Southern Conservation Tillage Conference for Sustainable Agriculture*. Special Series SS-AGR-60. Gainesville, Fla.: University of Florida.

Hauser, S. 1993. Distribution and activity of earthworms and contribution to nutrient recycling in alley cropping. *Biology and Fertility of Soils* 15: 16–20.

Keisling, T. C., ed. 1998. *Proceedings of the 21st Annual Southern Conservation Tillage Conference for Sustainable Agriculture*. Special Report 186. Fayetteville, Ark.: Arkansas Agricultural Experiment Station.

Liebman, M., and E. Dyck. 1993. Crop rotation and intercropping strategies for weed management. *Ecological Applications* 3: 92–122.

Powers, L. E., R. McSorley, R. A. Dunn, and A. Montes. 1994. The agroecology of a cucurbit-based intercropping system in the Yeguare Valley of Honduras. *Agriculture, Ecosystems and Environment* 48: 139–147.

Salcedo, I. H., H. Tiessen, and E. V. S. B. Sampaio. 1997. Nutrientilability in soil samples from shifting cultivation sites in the semi-arid Caatinga of NE Brazil. *Agriculture, Ecosystems and Environment* 65: 177–186.

Vanlauwe, B., N. Sanginga, and R. Merckx. 1998a. Recovery of *Leucaena* and *Dactyladenia* residue nitrogen-15 in alley cropping systems. *Soil Science Society of America Journal* 62: 454–460.

Vanlauwe, B., N. Sanginga, and R. Merckx. 1998b. Soil organic matter dynamics after addition of nitrogen-15-labeled *Leucaena* and *Dactyladenia* residues. *Soil Science Society of America Journal* 62: 461–466.

Bibliography

Altieri, M. A. 1993. Ethnoscience and biodiversity: Key elements in the design of sustainable pest management systems for small farmers in developing countries. *Agriculture, Ecosystems and Environment* 46: 257–272.

———. 1995. *Agroecology: The Science of Sustainable Agriculture*. 2d. ed. Boulder, Colo.: Westview Press.

Brummer, E. C. 1998. Diversity, stability, and sustainable American agriculture. *Agronomy Journal* 90: 1–2.

Edwards, C. A., R. Lal, P. Madden, R. H. Miller, and G. House, eds. 1990. *Sustainable Agricultural Systems*. Delray Beach, Fla.: St. Lucie Press.

Risch, S. J., D. Andow, and M. A. Altieri. 1983. Agroecosystem diversity and pest control: Data, tentative conclusions, and new research directions. *Environmental Entomology* 12: 625–629.

Soule, J. D., and J. K. Piper. 1992. *Farming in Nature's Image: An Ecological Approach to Agriculture*. Washington, D.C.: Island Press.

Willey, R. W. 1979. Intercropping—Its importance and research needs. Part 1. Competition and yield advantages. *Field Crop Abstracts* 32: 1–10.

———. 1979. Intercropping—Its importance and research needs. Part 2. Agronomy and research approaches. *Field Crop Abstracts* 32: 73–85.

Chapter 14

Conservation of Nutrients

Key Concepts

- Tillage and crop residues for soil conservation
- Benefits of mulches and surface residue
- Organic amendments and green manures as sources of N

From previous chapters, it is evident that most of the reactions and transformations that supply nutrients to crop plants take place in the soil. Therefore, agricultural practices that conserve soil are often important in nutrient conservation as well. This is true of methods designed to limit soil erosion, including conservation tillage practices; as well as terracing, strip cropping, contour plowing and planting; or use of cover crops, intercropping, living mulches, buffer strips, or windbreaks. This chapter focuses on the role of conservation tillage, mulches, and organic amendments in soil and nutrient conservation.

Conservation Tillage

The benefits of tillage and cultivation are well recognized for seedbed formation, for providing a favorable zone for root development, and for managing weeds. Clean cultivation can provide a "neat" and "orderly" appearance to the agricultural field, and is particularly convenient when establishing a monoculture. But the loose, exposed soil of clean-cultivated fields is subject to water and wind erosion. Thus, the decision to cultivate or not to cultivate is something of a trade-off, balancing the benefits of favorable crop establishment versus the risks of erosion.

Conventional tillage refers to practices in which crop residues are buried and the soil surface is cleaned and smoothed prior to planting. **Conservation tillage** is a general term that refers to a variety of practices (see next section) in which tillage operations are reduced or eliminated. Conservation tillage is characterized by a rough soil surface and plant residues that are left on the soil. Some definitions of conservation tillage require that at least 30%, or some other specified minimum amount, of the soil surface must be covered by crop residues. The objectives of these tillage programs are to reduce soil loss and conserve soil moisture.

The use of conservation tillage in the United States increased greatly during the last two decades of the twentieth century. A major reason for this increase in the United States and throughout the world was the widespread use of broad-spectrum herbicides, which provided an alternative to cultivation for general weed control. The development and improvement of equipment adapted to conservation tillage practices have accelerated this process, as well as legislation that encouraged conservation tillage as a soil conservation practice. Use of conservation tillage may also have resulted in savings of energy, fuel, and labor in some situations, such as when fewer cultivation trips over a field were required. However, some of these savings may have been invested in herbicide applications, although, in general, tillage and cultivation are much more time-consuming than herbicide application.

Definitions and Types of Conservation Tillage

A variety of terms, some synonymous, are used to describe conservation tillage practices. Individual practices vary in their intensity and degree of soil disturbance, on a continuum from no disturbance at all to systems that approach conventional tillage. Some sources classify the various conservation tillage practices based on the amount of crop residue left on the surface. Following are some of the more common practices used, beginning with those involving the least soil disturbance:

No tillage. The soil and surface residues are left undisturbed, except for seed placement. Disturbance may be restricted to a thin slit, with nearly all (>90%) of the soil surface covered by crop residue. No tillage is also referred to as direct drilling, direct seeding, no till, or zero tillage.

Ridge tillage. Tillage is used to form ridges on which crops are planted. No-tillage practices are then used to maintain subsequent crops, although the ridges may be reformed periodically.

Strip tillage. This practice involves tilling a narrow strip where the crop is to be planted, leaving the rest of the soil surface undisturbed (Figure 14-1). Some of the advantages of tillage for seedbed preparation are provided, while maintaining residue cover over most of the site. Strip tillage is also called strip till or zone tillage.

Mulch tillage. Soil is tilled and a portion of the crop residue is incorporated, but a mulch of remaining residue is left on the surface. Mulch-tillage practices include shallow, surface tillage designed to leave residue on the surface, as well as practices that are intermediate between minimum tillage and the less-intensive practices just listed.

Minimum tillage. The amount of tillage is reduced in comparison with conventional tillage, and some crop residue is left on the surface. Some definitions specify a minimum amount of residue that must be present on the soil surface. Minimum tillage is frequently called reduced tillage.

These types of conservation tillage practices may involve various degrees of subsoil or subsurface tillage, as well as different methods for managing residues (mowing, chopping, green versus dry, etc.). In recognition of the crop residues that are typical of conservation tillage practices (Figure 14-2), these practices are also referred to as crop residue management or residue farming.

Figure 14-1 In this strip-tillage system, seedlings have emerged in tilled strips surrounded by residue from a previous rye crop. Some residue from a corn crop before rye is also evident.

The types of conservation tillage practices defined here are not always clearly distinct from one another. Transition and overlap between various practices are common. Each field lies at some point on a spectrum from 0% to 100% of the soil surface covered with residue, depending on the degree of cultivation. Considering the possible ranges in intensity of cultivation and maintenance of residue cover, as well as possible combinations with cropping practices and subsurface tillage practices, a nearly infinite array of unique systems is possible, many of which are considered conservation tillage systems.

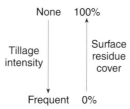

Figure 14-2 Tillage intensity and surface residue accumulation are inversely related. As the frequency of tillage decreases, the amount of crop residue covering the soil surface increases.

Benefits and Risks of Conservation Tillage

Conservation tillage practices affect a variety of different factors in the agroecosystem. Depending on the particular system involved, some of these effects may be beneficial while others may be detrimental. The magnitude of these effects is often related to the extent and intensity of the conservation tillage practices, depending on the amount of residue cover retained and the frequency of tillage operations. The remainder of this section provides comparisons of the effects of conventional versus conservation tillage systems on several specific factors.

Soil Erosion. Conservation tillage was developed as a soil conservation practice. As such, it can be especially beneficial on slopes and other types of highly erodible land. Two features of conservation tillage are particularly important in erosion reduction. First, the presence of mulch and residue means that less area of the soil surface is exposed to the activities of water and wind. These crop residues impede and disrupt the movement and flow of these agents along the soil surface, and trap some of the soil that is eroded. In addition, reduction in tillage minimizes the presence of loose, easily eroded soil on the surface. Reduced tillage practices also promote several other features of soil structure that can contribute to reduced erosion, including increased soil organic matter and reduced soil crusting. Conservation tillage systems can include the use of cover crops, which can further benefit erosion control.

Maintenance of surface residue can reduce soil erosion up to 90% or more in some cases. The benefits of increasing residue cover in reducing erosion from water are shown in Table 14-1. In comparison with clean cultivation, relatively small increases in surface residues can provide substantial reductions in soil erosion. At the 30% surface coverage required by some conservation tillage programs, erosion can be reduced to less than half of the level seen in conventional tillage systems. Maintenance of surface residue usually provides a similar level of protection against erosion from water or wind.

Of course, the amount of residue cover varies seasonally, depending on crops grown, decomposition rates, and management practices. A corn crop generally provides more residue than does a soybean crop. In addition, the higher C:N ratio typical of corn compared to soybean (see Chapter 4) suggests that residue from a corn crop is more resistant to decomposition. The highest levels of residue are achieved when a crop is planted directly into a living mulch or cover crop, and the lowest residue levels result when the previous crop is removed for silage or other purposes.

Reduction of soil erosion through conservation tillage has other environmental benefits as well. Runoff losses of nutrients and pesticides from the agroecosystem may be reduced. The quality of surface water is improved as sediment and other contaminants are reduced.

Table 14-1 Relationship between residue cover and surface erosion from water.

Percent (%) of Surface Covered by Residue	Percent (%) Reduction in Erosion
10	30
30	65
50	83
70	91

From data presented by Schertz (1994)

Soil Moisture. The presence of surface residue reduces evaporation and increases infiltration of water. Thus, soil moisture levels in soils under conservation tillage tend to be higher than those under conventional tillage. The effects of increasing levels of residue on evaporation from the soil surface are shown in Table 14-2. The use of conservation tillage practices can help to stabilize soil moisture fluctuations, conserve water, and reduce irrigation demand on crops grown on well-drained soils during hot, dry growing seasons. However, these benefits of conservation tillage can be disadvantages on poorly drained or waterlogged soils. In cooler climates, wet soils may be slow to dry in spring if a residue layer is present.

Soil Temperature. Crop residues shade the upper layers of soil and result in less fluctuation in soil temperature in conservation tillage systems than in systems without cover. Temperatures in the upper 10 cm of soil may be slightly warmer in covered than in uncovered soil during the winter. But during the sunny summer growing season, soil temperatures are usually lower under conservation tillage than under conventional tillage systems. This temperature difference can range from about 1–5°C, with the greatest differences occurring in the daily maximum temperature values, and lesser differences in daily mean or minimum temperatures.

Whether this cooling of soil temperature during the growing season is an advantage or a disadvantage depends on climate and geographic location. In warm or tropical climates, the reduction of high surface temperatures is beneficial and slows decomposition of organic matter. But in cool climates, the warming of soil in the spring can be delayed by the insulating effect of crop residues, resulting in slower seed germination and growth and in a shorter growing season.

Soil Structure. Although tillage can lead to changes in soil structure such as formation of plow pans and compaction, it is also the most convenient means of managing soil compaction. Avoidance of tillage on soils with a tendency toward compaction can lead to reduced root systems. This problem can be alleviated by subsoiling or other forms of subsurface tillage, while still maintaining much of the surface residue intact. Another option is the planting of deep-rooted cover crops such as grasses or pigeon pea (Figure 14-3), which can improve soil porosity and structure slowly over time.

Soil Fertility. Soil organic matter slowly decreases over time in conventionally cultivated systems, but can be maintained in conservation tillage systems. Differences in soil organic

Table 14-2 Relationship between residue cover and surface evaporation.

Percent (%) of Surface Covered by Residue	Relative Percent (%) of Evaporation[a]
0	100
20	78
40	67
60	61
80	58

[a]Expressed as a percent of the evaporation level with no residue present
From data presented by Shertz (1994)

Figure 14-3 Pigeon pea, a deep-rooted tropical legume.

matter between conventional and conservation tillage systems are most evident in the upper 5–10 cm of soil, and less evident in deeper layers of the soil profile. Increases in organic matter are reported in some conservation tillage systems. Shertz (1994) gives an example of an Illinois farm on which soil organic matter increased from 1.87% to 4.0% during fifteen years of no tillage.

Immobilization of N is a particular concern whenever crop residues with high C:N ratios are present. Thus, the placement of N fertilizer can be critical in some conservation tillage systems. Opportunities to incorporate fertilizers may be limited. If fertilizers are applied in a narrow slit in too close contact with seed, it is possible that a burn to the seedlings may occur. On the other hand, if fertilizers are applied to the soil surface away from the seed, the risks of N immobilization and soil acidification increase. Precise placement of fertilizer can be achieved through use of specialized equipment or strip tillage systems.

Placement of other fertilizer elements can also be more difficult in conservation tillage systems. Typically, P is mixed into the upper soil layers by tillage. This is particularly important for an immobile element like P, which may have difficulty reaching the roots if it is placed on the soil surface. Therefore, deeper placement, such as the banding of P near the seed at planting, is critical in conservation tillage systems.

Because soil moisture levels are typically higher in conservation tillage than in conventional tillage systems, it is expected that NO_3^-, Ca^{2+}, and other water-soluble ions would be more easily leached in conservation tillage systems. Leached Ca^{2+} may be replaced by H^+, resulting in a gradual increase in soil acidity over time. Another factor contributing to increased soil acidification is the release of H$^+$ ions from NH$_4^+$ that has been applied as fertilizer or released from decomposition of residues. This acidification from NH$_4^+$ can be localized near the soil surface if fertilizer is applied on or near the surface and is not incorporated. Thus, soil acidification may proceed more rapidly in conservation tillage than in conventional tillage systems. However, since acidification in conservation tillage tends to be localized in the uppermost soil layers, a surface application of lime may be sufficient to correct the problem. As soil becomes more acidic, nutrient availability and performance of herbicides may be affected.

Some factors (e.g., organic matter) that affect soil fertility are improved by conservation tillage systems, but others (e.g., nutrient availability) are hindered to various degrees. The proper placement of N and other fertilizer elements is probably the most critical of these soil fertility issues. Over a long period of time, however, conservation tillage practices preserve soil fertility better than do conventional tillage practices because of their effectiveness in limiting soil erosion.

Weed Management. Weed management is a critical issue in conservation tillage systems, which usually support higher weed densities than systems that are plowed and cultivated regularly. The increased adoption of conservation tillage practices in the United States and Europe has depended on the development and availability of broad-spectrum herbicides for weed control. Herbicide usage is usually much greater in conservation tillage systems than in conventional tillage systems. Johnson (1994) correlated an increase in acreage under no tillage in the United States since 1988, with an increase in pesticide sales since that year. Much of this increase was due to herbicide sales and use.

Conservation tillage practices can be beneficial in reducing surface runoff and restricting pesticide movement. However, these benefits can be counteracted to some extent by the increased levels of herbicide used and the relative lack of incorporation of herbicides. Nevertheless, although much of the herbicide used in no-tillage operations remains on the soil surface, it is still less subject to runoff than that seen in tillage systems where sediment losses are high.

Efforts are underway to reduce herbicide application rates and to seek alternative methods for weed management in conservation tillage systems. Much herbicide is used to kill cover crops and crop residues in conservation tillage systems. In some cases, it may be possible to coordinate planting of a cover crop so that it has completed its cycle and died naturally before a subsequent crop is planted. New weed problems may emerge in sites that have been under conservation tillage for a number of years. Perennial weeds and others that are better K strategists may replace the r strategists that typically colonize cultivated fields. In addition, frequent herbicide use may provide increased opportunities for the development of herbicide-resistant weeds.

Disease Management. The relative incidence of disease problems in conventional and conservation tillage depends on the particular pathogen involved, the crop, and specific environmental conditions. A number of factors can lead to increased problems with plant pathogens in conservation tillage systems:

- More stable soil environmental conditions and higher soil moisture can favor the survival of some pathogens, particularly soil and rhizosphere-inhabiting fungi like *Rhizoctonia* and *Pythium*.

- Some pathogens may grow on the residues of previous crops.
- There is less opportunity for burial of pathogens by plowing, which can cause increased mortality to some pathogens.

On the other hand, the stable soil environment under conservation tillage may favor the plant more than the disease organism. In one case, water conservation and temperature moderation under no tillage reduced stress to sorghum plants, so that infection by stalk rot was reduced (Kommedahl and Todd 1991). In other cases, by-products from residue decomposition can be toxic or may inhibit certain pathogens.

Two situations are of particular concern in the management of diseases in conservation tillage systems. Planting a crop into a residue of the same crop species is risky because any pathogens growing on the residues may attack the new crop. This type of problem can be minimized by crop rotation, so that a new crop is planted into the residue of a different, preferably unrelated, crop. Fungi that are adapted to cool, moist conditions can cause serious disease problems in soils that are slow to warm and dry out in the spring. The slow germination and seedling development at low soil temperatures can prolong the exposure of vulnerable growth stages to infection.

Insects and Other Animal Pests. Some insect pests tend to be more common in conservation tillage systems than in tilled sites. Wireworms and white grubs are most frequent following sod and other grasses that might provide residue or living mulch for the current crop. Surface debris provides hiding places for cutworms and other pests, but also for carabid beetles, which are beneficial predators. Armyworms and the lesser cornstalk borer may be more common in reduced tillage systems. Because several of these insect pests attack seedlings, early-season insect management is especially important in some crops when conservation tillage is used.

Surface residue increases the amount of cover available for mice and some other kinds of pests as well. Slugs are especially favored by the increased moisture and cover provided by no-tillage systems, and cause increased damage in some regions under these conditions. The occurrence and damage of plant-parasitic nematodes depends much more on the previous crop than on tillage practices.

Energy. The reduction in cultivation and field operations results in a considerable savings in fuel consumption for conservation tillage systems. Less time and labor are required for tractor drivers. However, much energy is required to produce the herbicides that are used heavily in conservation tillage systems. When this energy input is considered, the energy saved through reduced fuel consumption may be nearly balanced in many reduced tillage systems. Lal et al. (1990) reported savings in total energy use of 3–7% when switching from conventional to conservation tillage systems, and concluded that saving energy did not provide a principal advantage for most conservation tillage systems.

Yield. Due to the limitation of erosion and conservation of topsoil, conservation tillage practices should provide a long-term (many years!) yield advantage over conventional tillage systems. Results are mixed for short-term yield differences. Several sources (e.g., Johnson 1994; Sprague and Triplett 1986) have summarized short-term yield comparisons between conservation tillage and conventional tillage systems across a number of locations. Conservation tillage sometimes results in a yield advantage, but other times results in a disadvantage or no difference. Part of this inconsistency results from the varied nature and types of conservation tillage systems, which can include a range of intensity from minimum tillage to no tillage. In addition, the various types of conservation tillage are practiced across a range of soil types, climates, and other environmental conditions. All of these factors can introduce variability into yield comparisons.

A few patterns have emerged from yield comparisons between conservation tillage and conventional tillage systems. Conservation tillage can provide a yield advantage on well-drained soils, particularly to crops exposed to warm temperatures, drought, or moisture stress. Conventional tillage provides a more consistent yield advantage on poorly drained soils, particularly in cool climates. Often, the yield disadvantage of conservation tillage in cool, wet soils can be traced to difficulties in seedbed preparation and stand establishment. Stands can be reduced from losses to seedling diseases or soil insects, and from inability of some plants to emerge through thick layers of residue. Yield of established plants can be affected by the shortening of the growing season if germination and emergence have been delayed by cooler temperatures.

Adoption of Conservation Tillage Practices

The degree to which conservation tillage practices can be applied to a specific situation depends on a variety of factors, as discussed in the previous section. The relative advantages and disadvantages in managing erosion, moisture, temperature, soil structure, fertility, weeds, diseases, pests, and energy use must be considered and weighed in each situation. Additional factors may be important as well. Conservation tillage is advantageous in reducing the turnaround time between successive crops in multiple-cropping systems. It may also be helpful in conserving the beneficial soil fauna, especially larger invertebrates such as earthworms, which may be disrupted by frequent tillage. Of course, the time that growers must spend cultivating and preparing fields is reduced in conservation tillage systems. On the other hand, specialized equipment may be needed to use conservation tillage practices most efficiently (see examples in Sprague and Triplett 1986), although their availability is increasing. The risk of crop failure can be increased with conservation tillage, due to seedling problems in some situations. Also, conservation tillage eliminates the "neat," level appearance of fields that many people have grown accustomed to when conventional tillage is used.

Although future considerations of long-term soil fertility and herbicide costs may affect the economics of conventional versus conservation tillage decisions, in many cases the choice of a tillage system is probably based on short-term economic considerations. If the savings in fuel costs and reduced field operations are balanced by the increased herbicide costs typical of conservation tillage systems, then the economics of conservation and conventional tillage systems may be similar. Barring any major differences in equipment costs, net economic returns may likely depend on the relative yields of the different tillage systems. So we might expect that the use of conservation tillage may result in an advantage on well-drained soils in warm, temperate climates but a disadvantage on cool, poorly-drained soils. Additional information and specific examples of conservation tillage practices can be found in the Bibliography at the end of this chapter.

Mulches

A **mulch** is a material applied to the soil surface. An **amendment** is a material mixed into soil, often to improve soil fertility or structure. Thus, the same organic material can be considered a mulch or an amendment, depending on its use. In minimum tillage and other systems that include some limited tillage, a portion of the previous crop residue can remain on the surface acting as a mulch, while another portion can be buried and subject to decomposition as an amendment. (Figure 14-4).

In time, organic mulches eventually break down and release nutrients. Ultimately, their effects may be similar to the effects of amendments in improving soil fertility. But in the short term, mulches can be used to reduce erosion, to conserve soil moisture, to moderate soil temperature, to suppress weeds, or to improve the quality of surface water. In many cases, the benefits of mulching are identical to those of maintaining crop residues on the soil surface, as described earlier in this chapter. Living plants can also provide many of these benefits. Living mulches can be used as intercrops with other crop plants, as ground covers in orchards, or as a relatively nitrogen-rich or weed-free substrate into which another crop can be planted.

Synthetic mulches include paper, fabric, or plastic materials. Some of these, such as paper, are biodegradable and break down over time just like any other organic material. However, trace amounts of metals from newsprint or other inks can persist in the soil environment.

Plastic mulches are not biodegradable and may be costly in terms of economics and energy used in their manufacture. The use of clear plastic in soil solarization for managing soilborne pests, weeds, and diseases was discussed in Chapter 10. Opaque mulches, which can be white, gray, black, or other colors, are used in some agroecoystems. Although opaque mulches can absorb some solar heat, the soil beneath them does not reach the high temperatures achieved under clear plastic, and so the use of opaque mulches is compatible with plant growth. Plastic mulch is relatively uniform and is easily adapted to mechanized application. It can be applied after soil fumigation to seal in volatile fumigants. Fertilizers and drip irrigation tubing can be added prior to tarping with plastic. The plastic mulch limits leaching of fertilizer placed beneath it, although heavy rainfall on the plastic surface may wash into uncovered areas (such as areas between beds), increasing runoff there. Plastic mulch may suppress some weeds, but emerging nutsedge often punctures plastic and grows through it. Cleanup and disposal of plastic mulch that remains on an old field can pose a problem (Figure 14-5).

Organic Amendments

Numerous materials have been used as organic amendments in agriculture (Table 14-3). Nearly any biodegradable material can be included on this list. Compost is a general term that can include any one material or combinations of the other materials at various stages of decomposition (see Chapter 4; Application 4-5). Use of a material often depends on availability and convenience in transporting large amounts of material. Nevertheless, crop residues and animal manures are by far the most frequently used amendments in agriculture.

Uses of Organic Amendments

Organic amendments are applied to agricultural soils for a variety of reasons. Most improve soil structure and organic matter content, and, depending on the material, may provide some benefits to soil and crop fertility as well.

Conservation of Nutrients ■ 297

Figure 14-4 Young squash plants mulched with rye straw provided from a previous cover crop in the same site.

Figure 14-5 An old vegetable field in which the crop had been grown on gray plastic mulch.

Organic Amendments as Sources of Nitrogen. The utility of an amendment as a source of N depends not only on the amount of N it contains, but also on the rate at which the N is released to plants through decomposition and mineralization. As discussed in Chapter 4, a number of factors affect decomposition rate—soil, season, tillage, and C:N ratio of the amendment. If many of these environmental factors are constant, then the choice of the amendment and its C:N ratio will be particularly important in determining N

Table 14-3 A partial list of materials that have been used as organic amendments.

Crop residues, green manures
Animal manures
Municipal solid wastes (paper, food waste, yard waste)
Biosolids, sewage sludge
Marine products (seaweed, fish emulsion, crab chitin)
Processing by-products (food waste, oil cakes)
Animal by-products (bone meal)
Cellulose (wood chips, sawdust, paper)
Ashes
Peat moss
Compost

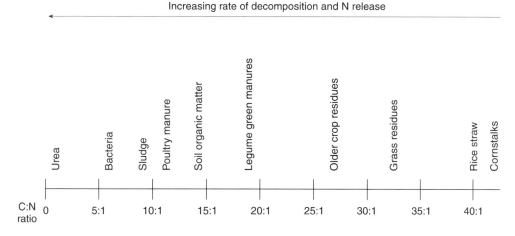

Figure 14-6 Relative C:N ratios of some common organic amendments.

availability. Common organic amendments differ greatly in their C:N ratios and relative speed of N release (Figure 14-6). With C:N ratios of 20:1 or less, most green manures are particularly good sources of organic N for crop plants. But N release from amendments with C:N ratios of 30:1 and higher may be too slow to benefit most crop plants. The implications of using amendments with very different C:N ratios as N sources are examined in more detail in Application 14-1.

The relative concentrations of mineralized forms of N (NO_3^-, NH_4^+) that are available to crop plants are expected to increase following the application of an N-rich organic amendment. A possible sequence of events is outlined in Figure 14-7. Suppose that the concentration of NO_3^- in soil is at some equilibrium level before an organic amendment is added. Immediately after the amendment is added, soil NO_3^- levels may decrease as soil bacteria use this free NO_3^- for growth stimulated by addition of a *carbon* source (these bacteria use the amendment for its C rather than for its N). Although this period of N-rob, or immobilization, may be minimal following many organic amendments, its length increases as the C:N ratio of the amendment increases. Eventually, the organic N from the amendment is mineralized, and we see the expected increase in soil NO_3^-. Levels of NO_3^- in soil increase over time, but eventually decline and return to the equilibrium level as NO_3^- is reduced through plant uptake, leaching, denitrification, and other causes.

Assume that we apply a suitable amendment that soon provides an increase in NO_3^- levels in soil (Figure 14-8). Amendment application and planting of the crop must be timed to make optimum use of the N released. If the amendment is allowed to decompose for some time before planting at time B in Figure 14-8, then some of the NO_3^- may be released and lost before the crop can use it. This can be an important loss if rapidly decomposing amendments are used in warm climates. Ideally, the crop should be planted at time A in Figure 14-8, before the NO_3^- is released, so that the roots of the germinating crop will be ready to take up NO_3^- as soon as it is available. A growing crop has a characteristic demand curve for NO_3^- over the course of the season (Figure 14-9). In an ideal situation, the supply curve

Application 14-1 Meeting Crop Demand from Organic N Sources

Suppose that a particular crop requires 100 kg of N/ha. Theoretically, this need could be met by applying an appropriate amount of inorganic or organic fertilizer. But the composition of potential organic fertilizer sources can vary widely. Examples of several N sources are shown in Table 14-4.

Suppose that one of the amendments in this table is to be used as a source of N, applied as an organic fertilizer source near planting time of a two- to three-month vegetable crop. The amount of amendment needed to deliver the required 100 kg N/ha is shown in the right-hand column. Some of these amounts are large and may be difficult to handle and apply, but there are more serious considerations in evaluating these amendments as potential N sources. Matching N release with crop demand is a critical issue. The decomposition rate of the legume hay might be about right, so that N was released and available during much of the crop season. With an extremely low C:N ratio, urea should be broken down and mineralized quickly. This rapid release may even cause some phytotoxicity to young seedlings. Even if that did not occur, most of the N is still released very early during the life of the crop, and may not be available later on when N demand by the crop may be substantial. The opposite problem would occur with the composted wood chips or the rye straw. These materials would decompose so slowly that they may not release significant amounts of N for crop growth. Even if the amount of rye straw is doubled to 40,000 kg/ha, the equivalent of 200 kg N/ha is immobilized in the straw residues, and is not available to a growing crop. Clearly, rye straw or wood chips could not be considered as N fertilizer sources for a short-lived crop.

Table 14-4 Composition of some amendments as potential N fertilizer sources.

Amendment (N source)	C:N Ratio	Percent (%) N	Amount Needed for 100 kg N/ha
Urea	0.43:1	45	220 kg
Legume hay	20:1	2.5	4000 kg
Composted wood chips	40:1	0.9	11,000 kg
Rye straw	60:1	0.5	20,000 kg

of NO_3^- released from the amendment (Figure 14-8) should closely match the demand curve of the plant (Figure 14-9). Synchronizing N supply and demand is an important research goal if organic N sources are to be used efficiently.

Organic Amendments as Sources of Other Elements. Many organic amendments contain P, K, and other nutrient elements in addition to N. Guano and other animal manures may contain significant levels of P and K. Bone meal is an especially rich source of P and Ca. Ashes may be a source of K. Seaweed and other marine products may be significant sources of P, K, Ca, S, and various micronutrients and trace elements.

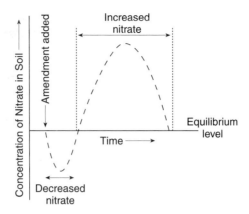

Figure 14-7 Relative concentration of nitrate available in soil following addition of an organic amendment.

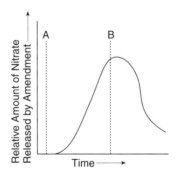

Figure 14-8 Supply of nitrate available in soil following decomposition of an organic amendment.

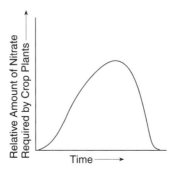

Figure 14-9 Demand of crop plants for nitrate over time during the growing season.

Although much of the discussion in this chapter focuses on the conservation and recycling of N from cover crops and intercrops, similar opportunities may also exist for the recycling of other nutrient elements. Some deep-rooted crops may be helpful in recovering K that has been leached into deeper soil layers. Cover crops rich in P and their associated mycorrhizae could be of use in some cropping systems.

Not all elements received from organic amendments are desirable. Municipal solid wastes and biosolids may contain variable amounts of heavy metals such as Cd, Cu, Pb, Ni, and Zn. Although they may be present in trace amounts in a single application, they can accumulate over time following multiple applications of metal-rich amendments. In addition to metals, other relatively inert materials such as plastic, glass, and rubber may accumulate from applications of some municipal solid wastes.

Organic Amendments and Pest Management. Decomposing residues from organic amendments affect pests and diseases in various ways. Decomposition can release ammonia, organic acids, alkaloids, or other by-products, which, if present in sufficient concentration, can affect pest organisms. It is also possible that the increased soil fertility and organic matter resulting from amendment addition could provide an environment in which plants are better able to tolerate certain pests and diseases, or an environment in which competitors or natural enemies of pests and pathogens are enhanced. Interpretation of experiments involving organic amendments for pest management can be difficult, since the organic amendments may stimulate plant growth on their own, whether pests and pathogens are present or not. Appropriate data are required to demonstrate whether a significant level of pest management has actually occurred.

Results of experiments with organic amendments for pest management are often variable and highly dependent on environmental conditions. Nevertheless, there is increasing evidence that residues of cabbage, rapeseed, and other cruciferous crops may be suppressive to certain plant pathogens. Effects against plant-parasitic nematodes have been less consistent. Residues of the neem tree *(Azadirachta indica)* have shown some potential as amendments for nematode management.

Organic amendments may result in an environment that is favorable to plant diseases or may even serve as inoculum sources of some diseases. Residues and amendments that are closely related to the economic crop are the most likely to harbor pathogens of similar host range that could affect the crop. The use of any diseased plant material is risky, even if it has been composted. Although the high temperatures of composting should kill most plant pathogens, diseased material at the edge of a compost pile or in immature compost may not have reached these lethal temperatures.

Green Manures as Organic Amendments

Sustaining the N fertility of agroecosystems through decomposition of organic amendments can be difficult because large quantities of N are periodically removed by harvest. Robertson (1997) estimated that 100–200 kg N/ha is removed during the harvest of most crops. The average rates of synthetic N fertilizer used to produce three important grain crops are summarized in Table 14-5. In general, these rates of N input are somewhat less than the estimates of N removal by harvest. Of course, some of this N deficit can be made up from organic and inorganic forms of N already present in soil. Overall, inputs of N to the agroecosystem from fertilizers, organic matter, and precipitation are balanced against losses from harvest, leaching, erosion, denitrification, and volatilization. The net result can be a deficit,

Table 14-5 Average rates of N fertilizer applied in various geographical regions.

Crop	Range in Average Rates Across Regions (kg N/ha)	Region with Highest Average Rates	Average Rate for North America (kg N/ha)
Wheat	23–146	East Asia	56
Rice	37–127	East Asia	117
Corn	59–173	Western Europe	157

Summarized from data presented by People, Herridge, and Ladha 1995

with net losses of N estimated at 24 kg/ha/year for Canada, 112 kg/ha/year for Kenya, and 20–70 kg/ha/year for various tropical regions (Giller and Cadisch 1995; Peoples, Herridge, and Ladha 1995).

Green Manures and N Fixation. The return of harvested by-products to the agricultural site for recycling is one way in which N can be conserved and these deficits reduced. However, N fixation provides a way in which significant amounts of new N can be added to the agroecosystem. Under optimum conditions, some legumes can obtain more than 80% of their N through biological N fixation. In many cases, 50–200 kg N/ha can be fixed by a legume crop. However, the amount of N fixed can range widely above and below these amounts, depending on conditions. Rates of N fixation are affected by plant species and cultivar, plant health, *Rhizobium* species present, soil aeration, temperature, water, pH, and a variety of other factors. Nitrogen fixation is decreased by increased NO_3^- levels in soil, as may result from application of inorganic N fertilizer. However, under conditions that promote N fixation, an addition of 100 kg N/ha through N fixation might be possible for many legume crops.

Legume green manures are a particularly attractive source of biologically fixed N. If the crop is used while it is still green, N content and moisture content will be high and the C:N ratio will be relatively low. These factors will aid in the rapid decomposition of the residue and release of organic N. Also, there may be less temptation for removal of harvest from a green, immature crop. If used entirely as green manure, either grain or forage legumes can provide similar levels of N enrichment. However, if a grain crop is harvested from a legume, it is possible that the N removed by harvest may exceed that added by fixation. This limits the effectiveness of the legume residue as a fertilizer source for other crops.

Green Manures as Fertilizers for Subsequent Crops. Fresh green manure can decompose rapidly, and so its use on-site must proceed quickly. Giller and Wilson (1991) point out that more than 40% of the N in some legume green manures may be released within two weeks after the amendment is added to soil. Decomposition can be especially rapid in tropical climates. Here, it may be necessary to plant the succeeding crop immediately after the green manure is turned in, or as a relay crop into a living mulch of the green manure. Some green manures can be cut and stored as hay for later use or for use on another site.

Since the N in a green manure is in organic form, its availability will depend on decomposition. Although fresh green manures can release large amounts of N rather quickly, release of some of the organic N may be so slow that it is not available to the following crop.

In Minnesota, 18–70% of the N in legume residues was used by a succeeding crop of corn (Power 1990). For tropical systems, 12–74% of the legume N was used by the next crop (Peoples, Herridge, and Ladha 1995). These are wide ranges, and they depend on the nature of the material used, the decomposition rate, and environmental conditions.

Many experiments have been conducted in which a green manure was used to supply some or all of the N fertilizer needs of a subsequent crop. The relative performances and yields of a crop fertilized with green manure or with synthetic N fertilizer are compared often. A few examples are summarized in Table 14-6. In some cases, the use of the legume residue produced a benefit in the next crop equivalent to using 95 or 112 kg N/ha of synthetic fertilizer. Some of the additional N needs of the crop can be made up from N reserves present in the soil. These reserves can be retained or even increased over time by practices (such as addition of green manures or other organic amendments) that favor the buildup and maintenance of soil organic matter.

Future Outlook for Green Manure Use. It is difficult to predict whether use of a particular green manure will satisfy the N needs of a subsequent crop. The ranges of N uptake shown in the previous section are evidence of the variable nature of release of immobile organic N. Although N from organic sources is less subject to loss and leaching, its uptake by plants is less predictable than uptake of N from inorganic sources. However, research has shown that significant amounts of biologically fixed N can be added to many cropping systems.

Numerous legumes have potential use as green manures and cover crops. Several of these have been introduced in this chapter and in Chapter 13, but many other possibilities are

Table 14-6 Some situations in which green manures or cover crops were used as an N source to produce a subsequent crop.

Green Manure or Cover Crop	Crop Produced	N Benefit, if Estimated (kg N/ha)[a]	Location	Reference
Hairy vetch	Tomato	Met needs of crop	Maryland, U.S.A.	Abdul-Baki and Teasdale 1994
Hairy vetch	Cotton	68	Alabama, U.S.A.	Power 1990
Crimson clover	Cotton	68	Alabama, U.S.A.	Power 1990
Hairy vetch	Corn	95	Kentucky, U.S.A.	Power 1990
Crimson clover	Corn	83	Kentucky, U.S.A.	Power 1990
Alfalfa/sweet clover	Corn	45–112[b]	Midwest U.S.A.	Power 1990
Peanut	Corn	60	Ghana	Giller and Wilson 1991
Cowpea	Corn	60	Ghana	Giller and Wilson 1991
Pigeonpea	Corn	38–49	India	Giller and Wilson 1991
Hairy vetch	Sorghum	90	Georgia, U.S.A.	Power 1990
Cowpea	Rice	28% of cowpea N used	Indonesia	Peoples, Herridge, and Ladha 1995
Various crops	Rice	30–80	China, India	Rao 1993
Pasture legumes	Grazing animals	—	Latin America	Thomas 1995

[a] Equivalent amount of N fertilizer needed to obtain yield response equal to that from green manure
[b] Data from average soils

available (see examples in Peoples, Herridge, and Ladha 1995; Power 1990; Thomas 1995). Specific examples of legumes and green manures used in intercropping systems in Central America are discussed by Thurston et al. (1994). Intercropped legumes may also be used to supply N to graminaceous crops in mixed croppings of legumes and cereals or in mixed pastures. Clippings or leaf-fall from leguminous trees or shrubs like *Leucaena* spp. can supply organic N in some agroforestry systems. *Azolla* spp., although not legumes, are important providers of biologically fixed N to some rice production systems.

Despite the advantages of green manures as a N source, their use is declining, while the use of synthetic N sources continues to increase. In some systems, there may be good reasons for *not* using cover crops and green manures. In semiarid regions with low rainfall and high evapotranspiration rates, cover crops, particularly deep-rooted ones, can further deplete reserves of soil water needed for crop growth. In these cases, the depletion of soil moisture can affect yield of the subsequent crop more than the gain of any N provided by a previous cover crop. However, the worldwide decline in use of green manures is not restricted to semiarid locations. Preference for synthetic N has increased in most regions and types of agroecosystems. Crop uptake of synthetic N is more predictable than uptake of organic N from crop residues. The production of green manures is labor-intensive compared to application of synthetic N. Land resources, which may be limited, are tied up during the production of the green manure crop. But probably the most important consideration is the fact that, in most cases, synthetic N fertilizers are widely available and are very cheap in comparison to the costs incurred by growers in producing green manures.

Summary

Conservation tillage includes a variety of related practices involving a reduction in the frequency of traditional tillage practices and the maintenance of crop residues on the soil surface. These practices are beneficial in limiting erosion and conserving soil fertility and soil moisture, but alternative methods for weed management may be required if the frequency of cultivation is reduced. Mulches can provide a variety of benefits to a cropping system, including reduction of erosion, suppression of weeds, moderation of soil temperature, and conservation of soil moisture and nutrients. Organic amendments are used mainly to improve soil structure and organic matter content, but some types, particularly green manures, can be used as N sources. When supplying N to a future crop from green manures or organic amendments, it is critical to match the supply of N released by the amendment with the crop demand for N over time.

Topics for Review and Discussion

1. List the advantages and disadvantages of conservation tillage.
2. Based on local soil and climatic conditions, do you think that conservation tillage would be a benefit to crop production in your area?
3. Discuss ways in which we could maximize the amount of N supplied to a crop by a green manure.
4. Describe how tillage practices, multiple cropping, organic amendments, and cover crops can be interrelated in the same cropping system.
5. Effectiveness of organic amendments for pest management can be difficult to demonstrate because increases in crop yield may be due to release of N or other nutrients from the decomposing amendment. How would you design an experiment

to show that a beneficial effect from an organic amendment resulted from management of a pest rather than from release of nutrients?

Literature Cited

Abdul-Baki, A. A., and J. R. Teasdale. 1994. *Sustainable Production of Fresh-Market Tomatoes With Organic Mulches.* Farmers' Bulletin FB-2279. Washington, D.C.: USDA.

Giller, K. E., and G. Cadisch. 1995. Future benefits from biological nitrogen fixation: An ecological approach to agriculture. *Plant and Soil* 174: 255–277.

Giller, K. E., and K. J. Wilson. 1991. *Nitrogen Fixation in Tropical Cropping Systems.* Wallingford, U.K.: CAB International.

Johnson, R. J. 1994. Influence of no-till on soybean cultural practices. In *Proceedings of the 1994 Southern Conservation Tillage Conference for Sustainable Agriculture,* eds. P. J. Bauer and W. J. Busscher, 12–22. Florence, S.C.: USDA-ARS Coastal Plains Soil, Water, and Plant Research Center.

Kommedahl, T., and L. R. Todd. 1991. Environmental control of plant pathogens using eradiction. In *CRC Handbook of Pest Management in Agriculture,* 2d ed., Vol. I, ed. D. Pimentel, 339–361. Boca Raton, Fla.: CRC Press.

Lal, R., D. J. Eckert, N. R. Fausey, and W. M. Edwards. 1990. Conservation tillage in sustainable agriculture. In *Sustainable Agricultural Systems,* eds. C. A. Edwards, R. Lal, P. Madden, R. H. Miller, and G. House, 203–225. Delray Beach, Fla.: St. Lucie Press.

Peoples, M. B., D. F. Herridge, and J. K. Ladha. 1995. Biological nitrogen fixation: An efficient source of nitrogen for sustainable agricultural production? *Plant and Soil* 174: 3–28.

Power, J. F. 1990. Legumes and crop rotations. In *Sustainable Agriculture in Temperate Zones,* eds. C. A. Francis, C. B. Flora, and L. D. King, 178–204. New York: John Wiley and Sons.

Rao, N. S. S. 1993. *Biofertilizers in Agriculture and Forestry.* 3rd ed. New York: International Science Publisher.

Robertson, G. P. 1997. Nitrogen use efficiency in row-crop agriculture: Crop nitrogen use and soil nitrogen loss. In *Ecology in Agriculture,* ed. L. E. Jackson, 347–365. San Diego, Calif.: Academic Press.

Schertz, D. L. 1994. Conservation tillage—A national perspective. In *Proceedings of the 1994 Southern Conservation Tillage Conference for Sustainable Agriculture,* eds. P. J. Bauer and W. J. Busscher, 1–5. Florence, S.C.: USDA-ARS Coastal Plains Soil, Water, and Plant Research Center.

Sprague, M. A., and G. B. Triplett, eds. 1986. *No-Tillage and Surface-Tillage Agriculture.* New York: John Wiley and Sons.

Thomas, R. J. 1995. Role of legumes in providing N for sustainable tropical pasture systems. *Plant and Soil* 174: 103–118.

Thurston, H. D., M. Smith, G. Abawi, and S. Kearl, eds. 1994. *Tapado Slash/Mulch: How Farmers Use It and What Researchers Know About It.* Ithaca, N.Y.: Cornell International Institute for Food, Agriculture and Development.

Bibliography

All, J. N., and G. J. Musick. 1986. Management of vertebrate and invertebrate pests. In *No-Tillage and Surface-Tillage Agriculture,* eds. M. A. Sprague and G. B. Triplett, 347–387. New York: John Wiley and Sons.

Bacon, P. E., ed. 1995. *Nitrogen Fertilization in the Environment.* New York: Marcel Dekker.

Baker, C. J., K. E. Saxton, and W. R. Ritchie. 1996. *No-Tillage Seeding.* Wallingford, U.K.: CAB International.

Boosalis, M. G., B. Doupnik, and G. N. Odvody. 1991. Conservation tillage in relation to plant diseases. In *CRC Handbook of Pest Management In Agriculture,* 2d ed., Vol. I, ed. D. Pimentel, 541–568. Boca Raton, Fla.: CRC Press.

Carter, M. R., ed. 1994. *Conservation Tillage in Temperate Agroecosystems.* Boca Raton, Fla.: Lewis Publishers.

Gallaher, R. N., and R. McSorley, eds. 1997. *Proceedings of the 20th Annual Southern Conservation Tillage Conference for Sustainable Agriculture.* Special Series SS-AGR-60. Gainesville, Fla.: University of Florida.

Hauck, R. D., ed. 1984. *Nitrogen in Crop Production.* Madison, Wisc.: American Society of Agronomy, Crop Science Society of America, and Soil Science of America.

Huber, D. M. 1991. The use of fertilizers and organic amendments in the control of plant disease. In *CRC Handbook of Pest Management in Agriculture,* 2d ed., Vol. I, ed. D. Pimentel, 405–494. Boca Raton, Fla.: CRC Press.

Keisling, T. C., ed. 1998. *Proceedings of the 21st Annual Southern Conservation Tillage Conference for Sustainable Agriculture.* Special Report 186. Fayetteville, Ark.: Arkansas Agricultural Experiment Station.

Ozores-Hampton, M., T. A. Obreza, and G. Hochmuth. 1998. Using composted wastes on Florida vegetable crops. *HortTechnology* 8: 130–137.

Chapter 15

Water Conservation and the Politics of Irrigation

Key Concepts

- Sources of fresh water
- Water conservation methods
- Salinity and sodicity
- Irrigation and drainage

Of all of the resources on this earth, water is perhaps the most precious. While energy can be harvested from the sun, from wind, or from the movement of water, water itself has no alternatives. The decisions that we make concerning water use now will affect agricultural production not only in this decade, but in many decades to come.

There would be more than enough water in this world to provide for all agricultural production *if* it were evenly distributed. Unfortunately, it is not. South America and the Caribbean receive about one-third of all of the precipitation that falls to the earth. Surprisingly, Africa receives more rainfall than Europe, but distribution within the continent ensures that only the Congo basin in the west central region gets plenty of rainfall, while the rest of the continent is consistently quite dry. Timing of rainfall is also an issue. In many parts of the world, much of a country's annual rainfall comes at one time of the year (perhaps in monsoon or hurricane season) while the rest of the year can be very dry. This, of course, has very serious implications for water use in both agricultural systems and in urban areas.

It is hard to imagine that water shortages can be a problem. We look at a globe and see that over two-thirds of the earth's surface is covered with water. There are 1.4 billion cubic kilometers of water on earth. So why do we concern ourselves with water conservation? The reason is that so little of the water on this earth is usable for drinking, agricultural use, or most industrial uses. Approximately 97% of all of the earth's water is saline, leaving only a remainder of 3% that is fresh water. Of that, much is tied up in glaciers or ice caps of mountains, some is too deep in the earth to extract without considerable cost, some is held tightly by the soil, and some is too polluted for use. In fact, only 0.003% of all of the earth's water is available for consumption and agricultural production (Miller 1990). Annually renewable freshwater supplies on land account for only 0.000008% of all water on earth (Postel 1996). Nevertheless, even that small percentage of available water turns out to be enough for each person on earth, provided that we do not overuse the water or pollute it. How can we overuse water? By using it faster than it can be replenished by the hydrological cycle of the earth.

Sources of Fresh Water

The fresh water that we use for agriculture and human consumption comes from one of two sources: surface water runoff and groundwater (Figure 15-1). The surface water is rainfall that does not penetrate the ground or return to the atmosphere via evaporation. It simply runs off the land and flows into rivers, streams, lakes, ponds, and reservoirs. The area of land that collects the water and diverts it into these fresh bodies of water is known as the **drainage basin,** or **watershed.** Fortunately, many surface sources of fresh water tend to be rather quickly replenished by precipitation, provided that rainfall levels are adequate.

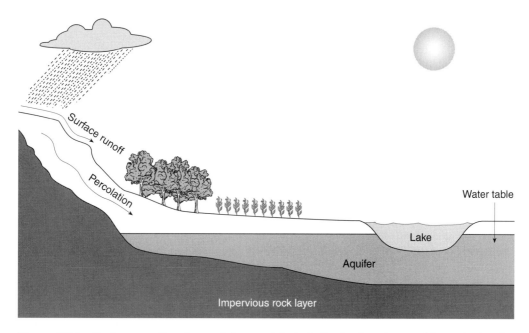

Figure 15-1 Surface runoff and percolation contribute to the fresh water used for agricultural and domestic purposes. While percolation serves to recharge underground aquifers, surface runoff does not penetrate the ground, but accumulates in bodies of water such as rivers, streams, lakes, and ponds.

Groundwater is rainfall that does penetrate the ground. It slowly moves downward through the soil and accumulates in deep layers of water within underground **aquifers.** The top of the water that collects in the aquifer is known as the **water table.** The land that the water filters through to reach the aquifer is known as a **recharge area,** since it essentially recharges or refills the aquifer. Unfortunately, the process of recharging an aquifer usually takes a very long time (hundreds of years), so when we pump out groundwater for agricultural use, it is not readily replenished. This kind of water use is not sustainable over the long term, and can eventually lead to serious problems in water shortages all over the globe.

In the United States, about 75% of the fresh water used each year comes from surface runoff into lakes, rivers, and reservoirs. The remaining 25% comes from groundwater stores. These numbers vary from country to country; many European nations rely exclusively on surface runoff supplies, while many Asian nations are depleting groundwater supplies at alarming rates (see Application 15-1).

Worldwide, about 65% of all water removed from lakes, rivers, and reservoirs is used for agriculture. Another 22% is used for industries, and 7% is required by households and municipalities. Evaporative losses from storage reservoirs account for the remaining 6% (Postel, Daily, and Ehrlich 1996). Most of the water used for industry and household use is returned to the environment (in some cases, polluted), but water used in agricultural systems is usually not recoverable. It is either incorporated into plant biomass or lost to the atmosphere through evapotranspiration.

Agricultural Problems Associated With Water

Most agricultural problems associated with water are related to having too little water (drought), too much water (flooding), or contaminated water. We now look at each of these in turn.

A *drought* occurs because precipitation rates are too low, temperatures are too high (which increases evaporation), or both. We commonly hear of drought conditions occurring in Africa and the Middle East, as a result of both low rainfall and high temperatures. This drought, of course, can lead to famine because seed germination will decline dramatically (see Delouche 1980), and crops that are able to establish simply cannot be productive without adequate moisture (see Dornbos, Mullen, and Shibles 1989). In many regions, the dry conditions are intensified by poor land management. Desertification (see Chapter 16 for a more in-depth discussion) can be a serious problem in many arid regions of the world. If plants are unable to grow, water is less likely to infiltrate the soil and will run off the surface, becoming unavailable for plant needs. If the water is not available, plants cannot grow, and the cycle continues.

Flooding occurs when too much rain falls on an area, or when runoff from other areas collects at one site. When a large volume of rain falls over a short period of time, it cannot be absorbed by the soil, so it will run off the surface and cause flooding. Soil pores fill with water, displacing air, and essentially "drowning" the plant roots. Because roots of plants need oxygen to respire and need to be able to give off CO_2, too much water in the soil inhibits respiration and can kill the plants. We will discuss this more when we talk about waterlogging, which occurs when humans apply too much irrigation water in an area.

Contaminated water is a huge problem in some areas of the world, and it affects both agricultural and human use. When water drains off of an agricultural field, it often contains a variety of organic and inorganic chemicals, including nitrogen, phosphorus, and potassium

Application 15-1 *Water Tensions Within a Nation: The Growing Demand for Water in Northern China*

As population continues to increase worldwide, water demand increases accordingly, both to meet domestic needs and to irrigate crops to feed a growing number of people. In China, where population growth is expected to continue despite strict regulations on family size, water tensions within the nation are already beginning.

In the last fifty years, China's population increased by nearly 700 million people, and is expected to grow by another 300 million in the next thirty years (Brown 1995). Most of this population growth is occurring in areas around the major rivers in China, where water is plentiful enough to meet domestic demands. Growth into other areas of the nation is constrained by desert land and mountainous terrain, where few people are able to live and work. Because growth outward is constrained, the majority of the Chinese population, along with most industry and farmland, is located primarily in one section of the country.

Evidence of Water Stress

Perhaps the most obvious sign that water demand is beginning to exceed supply in China is the seasonal drying of the Yellow River, which flows through the northern territory of the country and empties into the sea. For the first time in China's history, the Yellow River disappeared before reaching the sea in 1972, and has run dry every year since 1985. Each year, the dry period is longer, and in the drought year of 1997, the river ran dry for a total of 226 days. The significance of this shortage in terms of agricultural productivity is obvious. When the river runs dry, water does not reach the Shandong Province, the coastal province closest to the sea where a great deal of corn and wheat is grown. These agricultural fields count on irrigation water from the Yellow River. As the river runs dry earlier and for longer periods of time, productivity in these areas will drop considerably.

Other rivers in China are also showing signs of stress. Hundreds of local lakes and streams are going dry and disappearing each year. The Haui River was drained dry in 1997, and the Fen River, which was heavily used by industry in the area, has dried completely and no longer exists. Water tables throughout the country have fallen, and many wells are pumped dry, leaving no available water for agricultural irrigation. In 1997, when drought was prevalent and rivers were drying, irrigation wells in the Shandong Province were unable to supply irrigation water for farms in that area. China's groundwater is being depleted so quickly that the water table under the North China Plain is falling at a rate of about 1.5 meters per year. Over the last forty years, the water table under Beijing dropped about 37 meters, to a current depth of more than 50 meters. At this rate, the pumping of the groundwater aquifers is greatly exceeding the rate of natural recharge. Once the aquifers are depleted, the only water that will be available from these aquifers will be that which is added to them during recharge (a minimal amount), drastically dropping the potential supply of water for domestic and agricultural use.

The Growing Demand for Water

As the Chinese population continues to grow, water needs continue to grow. According to the Worldwatch Institute (Brown and Halweil 1998), even if demand per person remains the

(continued)

same, water needs of the nation will increase to 25% above the current need. However, demand per person is growing. As the Chinese population grows more affluent, many individuals are seeking to diversify their diets by adding animal products. As we will explore in Chapter 16, it requires more grain to produce a diet of meat products than that needed to produce a vegetarian diet, and more grain will require more irrigation water. In addition, the urbanization of the Chinese population is leading to more residential water demand with the addition of more showers, flush toilets, and other plumbing fixtures. As standard of living increases, water needs increase accordingly. Industry is also contributing to the high water demand in the nation. As the economy of the nation grows, more businesses and factories are built, requiring more water and polluting much of the water they use.

As industry and residential demands on water increase, farms in the area are losing out. Industry can better compete for water supplies economically, since the same amount of water produces much more of an economic output in industry than in agriculture. Urban residents also have a great deal of influence on water use decisions, which means that farmers may be denied access to water reservoirs if urban demand increases. As water becomes less available, farmers may be forced to depend primarily on rainfall to meet the water needs of their crops, which is likely to drop yields by as much as one-half to two-thirds (Brown and Halweil 1998).

Can Anything Be Done to Prevent Disaster?

Currently, water deficits in China are most prevalent in the north, where two-thirds of its agricultural fields lie. In the south, land is fed by the Yangtze River, which never runs dry, ensuring a constant water supply to that region. Unfortunately, the major Chinese cities of Beijing and Tianjin are in the northern part of the country, where water shortages are most serious.

Several solutions have been suggested to alleviate the growing water problems of the nation. China could simply divert water from agricultural use to domestic and industrial use, and let the farmers shift to dryland, rainfed agriculture. However, the population of China is so large that its grain needs would quickly overwhelm the world market, pushing prices up and causing many poorer nations to go hungry. Using more water-efficient crops and discouraging the consumption of meat could also help to reduce water demand for crop irrigation. Farmers could also be encouraged to shift to irrigation systems with higher water use efficiencies by pricing water high enough to motivate serious conservation efforts.

On the domestic front, increasing water use efficiency in homes and industry could decrease water demand. Brown and Halweil (1998) also suggest that China could intentionally keep industrial wastewater separate from domestic wastewater, so that the domestic water could be readily reused as irrigation water without extensive treatment. Reducing pollution that enters freshwater systems could be of great benefit in increasing the supply of usable and safe irrigation water.

More drastic solutions that have been suggested involve the direct diversion of water from the south to the north. While this seems like an ideal solution, it likely would not provide enough water to meet the deficits in the north, and more water would be lost to evaporation in the process. In addition, it would be an enormous and expensive project. Brown and Halweil (1998) calculated that the diversion of water from the Yangtze River to Beijing

(continued)

would be equivalent to diverting the Mississippi River in the United States to supply water to Washington D.C.!

It is obvious that no one solution will solve the water deficit problems in China, and that numerous approaches could be taken to alleviate the problem. As population continues to grow, and affluent people continue to diversify their diets and improve their standard of living, it is likely that water stress will occur in many other regions of the world. China's problem is likely to become a problem faced by most nations in the future.

---

from fertilizers and manures, or potentially toxic chemicals from pesticides. On the farm itself, this can result in a depletion of nutrients from the soil, perhaps necessitating more frequent fertilizer applications (which can then end up in the drainage water again). Off site, the chemicals can accumulate in lakes, ponds, or other bodies of fresh water, and can stimulate the production of algae and other organisms, leading to eutrophication (see Chapter 12). They may also end up in the groundwater, making that water unsafe for human consumption.

Irrigation and Water Conservation

The amount of irrigated land in the United States and around the world has continually grown since the early 1960s (Figure 15-2), but that expansion is currently slowing (Brown, Kane, and Roodman 1994). It is unlikely that the amount of irrigated land will be able to keep pace with population growth in the future, for several reasons. First, the cost of adding new irrigation systems to dry lands in many areas of the world is growing to prohibitively high levels, so new investments in irrigation are beginning to slow. In other areas, public

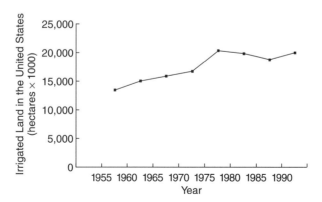

Figure 15-2 Increase in irrigated farmland (in thousands of hectares) in the United States over time.
Data from Brown et al. 1994

concern over environmental and social issues has slowed investments in dams, which often provide the main source of irrigation waters. In addition, irrigated land in many arid regions of the world is starting to lose productivity as a result of poor management of irrigation. In some cases, the application of too much irrigation water has resulted in a condition known as **waterlogging.** Waterlogging occurs in poorly drained lands when seepage from unlined canals and overirrigation of fields raise the water table to the point that the soils are too wet to farm. In other cases, poor drainage combined with rapid evaporation of surface water can lead to accumulation of salts in the soil, gradually reducing yields and ultimately ruining the land for agriculture. This process is called **salinization.**

It is estimated that between 20–30 million hectares in the world (8–12% of the world's irrigated area) suffer from serious salinization, and another 60–80 million are moderately affected (Brown, Kane, and Roodman 1994). Salinization is spreading at a rate of about 1.5–2 million hectares/year, and is widespread in many important agricultural regions, including the United States, China, Pakistan, India, and Mexico.

Another problem with irrigation systems that suggests that our current use of water is not sustainable is the evidence that many nations are overpumping groundwater. Recall that groundwater reserves take hundreds of years to recharge. When removal of this groundwater exceeds the recharge rate, water tables drop. Many nations are currently pumping groundwater reserves at rates that are much more rapid than recharge rates. This is occurring in areas such as northern China, the western United States, North Africa, the Middle East, and some portions of India. Because this use of water is not sustainable, food produced using this groundwater is not a reliable part of the world food supply over the long term. Overuse of groundwater has also led to problems with subsidence in some areas (Figure 15-3) because removal of the water weakens the soil structure.

In spite of the decrease in irrigation expansion over much of the globe, it is obvious that water needs will continue to expand as world population continues to grow. We will need more water for human consumption, as well as for agricultural uses to produce even more food to supply to more people. As this occurs, we will see increasing competition for water supplies, both among countries, and within countries. Among countries, water rights and fair use will become major issues. Within countries, competition will increase between farmers and urbanites, as well as between farmers and environmentalists. In the United States, many federal and state initiatives now legislate the use of water to restore fisheries and wetlands across the nation, and most of that needed water comes from agriculture.

Desalinization is often proposed as the answer to water shortage problems in the world. With so much water readily available from the ocean, the ability to remove salt from that water would provide us with all of the water we could ever need. Unfortunately, the process of desalinization is very expensive. Current desalinization technology requires vast amounts of fossil fuels, so it is not a sustainable means to obtain fresh water. The expense and need for fossil fuels has made the process of desalinization available to the oil-rich Arab states, which currently use about half of the desalinated water produced in the world. For now at least, it is not an option for most large-scale agricultural systems.

Irrigation Efficiency

Worldwide, about 65% of the water used in irrigation is wasted (evaporated or transpired), meaning that only about 35% of applied irrigation water actually contributes to crop growth

Figure 15-3 Subsidence can result from overuse of groundwater when removal of the underground water actually weakens the soil structure, causing collapse. This can lead to the formation of sinkholes, such as the one shown here.

(Miller 1990). Why is so much of this water wasted, and what can we do to prevent this waste of water? The answer is not an easy one because it involves a wealth of ecological, sociological, and economic factors.

When faced with the question of irrigation efficiency, we must first take a look at the different irrigation systems used to apply water in different parts of the world. The systems can be divided into three categories: surface irrigation, overhead irrigation, and micro-irrigation systems.

Surface irrigation, or flood irrigation, is the oldest means of irrigation in the world. With this system, water is brought to a field through canals or ditches. The field may be either completely flooded (common when growing grains such as wheat or rye), or the water may be diverted along furrows throughout a field (common for row crops). The water then moves into the soil and through the soil pores by both gravitational pull downward and capillary movement, which allows for some horizontal movement. The advantages of such a system are that it is quite inexpensive to operate and very little labor is required once the system is set up. Initial costs, however, may actually be quite high because the land should be leveled to allow for even irrigation of the land. Unfortunately, this method is very inefficient in terms of water use. A great volume of water is exposed to the air for a long period of time, and evaporation rates of the water are quite high. Some may also be lost from the field because water is difficult to control. And, despite many farmers' best efforts at leveling their fields, surface irrigation seldom results in even irrigation of a field, so one part of the field usually gets more water than another.

Overhead irrigation systems usually involve a sprinkler system that sends water outward from central points (Figure 15-4). Some systems consist of self-propelled sprinklers that travel the length of the field and sprinkle water on crops or soil as the sprinklers move. A center-pivot system travels in a circular path around a well or other water source. Overhead irrigation systems are used on more than 35% of irrigated land in the United States and are used on both level lands and on lands that are hilly or contoured. The equipment costs for overhead irrigation are higher than those for surface irrigation, but water use efficiency tends to be higher as well. Nevertheless, evaporation costs are great. Much less than half the water applied is actually used by the crop for growth, and the rest is lost to the atmosphere through evapotranspiration. New precision-application systems aim the spray lower, so the water falls through less air before it hits the ground, and therefore evaporation losses are a bit less. However, all overhead irrigation systems wet the crop foliage, which can make conditions ideal for many plant diseases.

Figure 15-4 An overhead irrigation system.

Micro-irrigation, or drip irrigation systems, are the most water-efficient irrigation systems available (Figure 15-5). With these systems, water is dripped out of pipes, tubes, or specialized emitters (or allowed to flow out of buried perforated tubes) directly into the growth region of the plant, so the soil between plants is not watered. This allows the crops to obtain water, but far less is lost to evaporation from the soil surface. Disease incidence is also reduced because the foliage does not get wet. Why, then, are not all farmers all over the world using this system? Unfortunately, the initial costs of installing a drip irrigation system are very high, often prohibitively so, and upkeep can be expensive too. In addition, the system requires maintenance. Output must be carefully measured, and the tubes and pipes must be carefully maintained because they can be easily clogged. And it is not practical for all cropping systems (delivering water to every wheat plant in a field would require quite an extensive network of tubes and pipes!). Nevertheless, the water use efficiency is very high for these micro-irrigation systems, and they are easily the best system for arid regions, particularly for the prevention of salinization.

While micro-irrigation may be the best system for water conservation and, ultimately, for land management, it is not a widely used system, even in industrialized nations like the United States. The reason for this is primarily economic. As long as farmers can get very inexpensive water, they will not invest in these systems.

How Can We Encourage Water Conservation?

The most straightforward means of encouraging water conservation is to charge more for the use of water. Currently, water prices are simply too low in most regions to encourage conservation. In truth, there are very good reasons to subsidize irrigation for farmers, especially those who are poor, but the degree of subsidization that occurs in the United States today does not provide any incentive for farmers to conserve. When water is essentially free, why would people make any attempt to conserve it?

The Broadview water district of California, with over 4000 hectares of irrigated farmland, provides us with an example of how water conservation can work (Postel 1996). In the late 1980s, the district established a tiered water pricing structure to reduce water use on farms. This tiered system was based on an average volume of water per hectare that had been used in the district over a two-year period. This volume was charged at a subsidized base rate. Any amount of water that was used in excess of that base volume was then charged at a much higher rate: 2.5 times higher! In 1991, only seven out of forty-seven fields in the district used water in excess of the base volume, and the average amount of water applied on all irrigated land dropped by almost 20% from the previous year. In truth, the farmers in the district were still paying much less for the water than it was really worth, but the economics of using more water gave them the incentive to conserve (Postel 1996).

Using treated wastewater from urban areas to irrigate crops is another means by which water may be reused, and thus conserved (see Application 15-2). While more expensive than other conservation options, the treatment of wastewater for agricultural production may actually cost less than developing new sources of water (Postel 1996). An additional advantage of using wastewater for irrigation is that the water itself contains nutrients such as nitrogen and phosphorus, which are important for crop growth and may reduce fertilizer needs of a field. Moreover, with urban areas expanding rapidly, more water will be diverted to cities for personal use. This will result in less water available for agriculture, but will also

Figure 15-5 A drip irrigation system on tomatoes. Note tube supplying water to the tomato plants grown on a bed of soil covered by plastic mulch. Drip irrigation systems work very well in combination with other water-conserving materials, such as plastic mulch, because they can deliver water directly to the plant roots while much of the soil remains covered by the mulch.

Application 15-2 *Using Wastewater for Agricultural Irrigation Systems*

The rapid growth of urban areas in the world is leading to increased water demand for domestic purposes, and also to an increased supply of wastewater that can potentially be used for agricultural purposes. The advantages of recycling wastewater can be numerous. It can reduce the need for pumping water from underground reservoirs, alleviating serious water shortages where urban and agricultural needs are in direct conflict, and allowing long-term recharge of these aquifers. In addition, wastewater may contain nutrients that can reduce the need for fertilizer use while increasing agricultural productivity. As water becomes more scarce, prices of water and agricultural products requiring irrigation will be forced higher. By recycling wastewater from urban systems, demand on freshwater decreases, allowing prices of water and food products to remain lower.

Nevertheless, there are some costs associated with the use of wastewater in agricultural systems. Water that contains sewage, pollutants, salts, and potentially harmful microorganisms must be treated before it is applied to agricultural fields as irrigation water. If the costs of treatment are too high, it may not be economically advantageous to reuse wastewater, at least until water prices are driven higher than the costs of treatment. Cost of storage and transport from urban systems to wastewater treatment centers to farmers may also be prohibitively expensive.

Some nations already rely on agricultural reuse of wastewater. Israel uses treated wastewater for crop irrigation, recycling approximately 1900 million cubic meters annually (260 m^3/person). It is estimated that within the next forty years, most agriculture in Israel will rely heavily on treated wastewater from urban areas (Haruvy, 1997). It is quite likely that we will see increased use of wastewater for irrigation of agricultural fields in the future, particularly in arid regions, as urban areas continue to grow and farmers strive to increase productivity to feed a growing world population.

---

result in a steady supply of urban wastewater that can then be recycled for use by crops (Postel 1996).

Another important means of reducing irrigation losses and conserving water is to use a system that accurately determines soil moisture, and thus tells a grower exactly when to irrigate. Many farmers irrigate according to a schedule. Perhaps they irrigate their fields every three days, or every Monday, Wednesday, and Friday. Such irrigation is not based on the need of their crops for moisture, but on the calendar. If the field does not need to be irrigated, this schedule can lead to waste of water, runoff, or leaching. In addition, increasing irrigation efficiency by adjusting irrigation schedules to the seasonal uptake of water by a crop can save a significant amount of water over the course of a growing season (Diez et al. 1997).

Determination of soil moisture is a relatively simple process. One commonly used method is to bury plaster-of-Paris (gypsum) blocks in the soil. Wire electrodes lead from the blocks to the surface, and the electrical resistance between the electrodes is measured. As the soil gains or loses moisture, the electrical resistance changes. When it reaches a certain level, the farmer knows that irrigation is necessary. This prevents overwatering and helps the farmer to deliver water to his or her fields when the crops most need it. Of course, a knowledge of the water needs of a particular crop is fundamental in this type of irrigation decision making.

Conservation Tillage and Water Conservation

Many water conservation methods are designed to decrease evaporative water loss from a field. Frequent tillage, on the other hand, exposes the soil surface to evaporation and runoff. Conservation tillage is an important and frequently used practice for conserving moisture because it leaves crop residues on the surface of the field, thus reducing water loss (Kovar et al. 1992). The beneficial effects of conservation tillage in reducing evaporation from the soil surface and stabilizing fluctuations in soil moisture were discussed in Chapter 14.

Any practice that covers a portion of the soil surface can be beneficial in reducing evaporative water loss. Mulches are commonly used both to prevent surface evaporation by covering the field with an organic (sawdust, manure, leaves, straw, crop residues) or synthetic (paper or plastic) substance, and to reduce weed growth by preventing light from reaching germinating weed plants. Unwanted weeds can draw water from the soil profile and transpire it into the atmosphere, preventing that water from reaching crop plants in the same field. Mulches are most practical for high-value crops or small farms. An occasional drawback with the use of some plastic mulches is that heavy rainfall cannot penetrate the mulch, resulting in heavy runoff into nonmulched areas.

Other Methods of Water Conservation

Slope of land can seriously influence the amount of water that runs off and is lost from the agricultural system, as well as the amount of soil that the runoff water takes with it. As slope increases, the amount of runoff and soil loss increases. Water and soil loss, however, can be reduced with vegetative cover. Land under cultivation can reduce soil and water loss by more than 50% over land left fallow, regardless of slope (El-Hassanin, Labib, and Gaber 1993). For this reason, keeping land in vegetation, or leaving crop residues on the surface, can improve water infiltration, and thus the efficiency of water use.

Timing of irrigation can also affect the percentage of water that is lost to evaporation. Watering in the heat of the midday causes significantly more water to be lost than watering in the very early morning (watering at dusk is not generally recommended because standing water encourages the growth of harmful fungi and bacteria on plants and on the soil surface). When intercropping, choosing plants with complementary water use strategies improves overall water uptake.

Salinity and Sodicity: The Chemistry of Land in Trouble

Saline and sodic soils are most commonly found in arid and semiarid regions of the world, where more than half of the world's agricultural lands are located. In these dry areas, agricultural productivity is significantly increased when they are irrigated, but this same irrigation has led to serious problems with both soil salinity and **sodicity** (accumulation of Na⁺ ions in soil).

Salinity

We have defined salinization as the condition that occurs when salts accumulate in the soil, often ruining lands for agricultural production. This accumulation can be either natural or irrigation-induced. The natural accumulation of salts occurs when cations such as Ca^{2+}, K^+, Mg^{2+}, and Na^+ are slowly released from the weathering of rocks or minerals or brought in by rainfall. They form soluble salts like $CaCl_2$, KCl, $MgCl_2$, and $NaCl$, which cannot be leached from the soils because there is not enough rainfall in the region. Irrigation-induced accumulation results from the use of irrigation water that has significant amounts of soluble salts. Irrigation waters, whether from freshwater resources or pumped from groundwater, contain some level of dissolved soluble salts. If irrigated land is well-drained, enough water can move through the soil to leach those salts out of the upper layers of soil and to prevent excessive buildup of salts. However, in poorly drained soils or in rainfall-deficient locations where leaching is limited, the salts do not readily move downward and out of the upper soil layers, but instead accumulate over time. As water evaporates from the soil surface, even more salts are brought upward, leading to significant salt buildup and a saline soil.

In addition, the soil itself contains soluble salts or added fertilizer salts that dissolve into the standing water, and can make the drainage water quite saline. This drainage water feeds into local lakes and streams, which are then tapped as sources of irrigation water, cycling the salts through the systems again. The buildup of salts in the soil that results from the continued addition and accumulation of soluble salts from irrigation water can damage both the soil itself and the plants that grow in it. The percentage base saturation of these soils tends to be quite high and pH values above 7 are common.

Salts vary in their toxicity to crop plants. Salts of Na^+ are often the most problematic, while K^+ is of little concern. Most plants are relatively intolerant to saline soils. High pH can lead to low micronutrient availability. More importantly, however, is the difference in osmotic potentials that develops between the soil solution and water in root cells. Water from the plant root cells is actually attracted to the salts in the soil solution, and is therefore pulled out of the cells by osmosis. This leads to the collapse of the root cells and the relatively rapid death of the plant. Some plants, including barley, cotton, sugar beet, some wheat grasses and wild ryes, can actually tolerate higher salinities. Recent breeding efforts in this area have started to produce plants with greater tolerance to salinity.

It is much easier to prevent salinization than to "cure" it. The selection of efficient irrigation systems that minimize both the amount of water added to the system and evaporation from the soil surface can help to slow salt buildup. The use of crop residues or mulches reduces surface evaporation and also reduces soil temperature, which can reduce amount of water that is lost from the soil surface and thereby increase the amount of water available to plants. Finally, good drainage systems are indispensable in arid and semiarid regions because poor drainage is probably the most important factor leading to salinization. We discuss drainage in more detail in the following section.

Sodicity

The problem of saline soils can be much worse if Na^+ ions begin to saturate the soil complex, a condition known as sodicity. The accumulation of Na^+ ions results not only in extremely alkaline soils, but can also damage the actual physical structure of soils. Good drainage in a soil profile is largely dependent on good aggregation of soil particles, creating an assortment of pore spaces for water to move through. When Na$^+$ ions start to build up in soils, they cause the dispersal of these aggregates, releasing clay fragments that plug these drainage pores and prevent water movement. This progression results in absolutely useless soils that are simply abandoned for agriculture.

Sodic soils can be toxic or caustic to plants, so the chances of plant survival in such soils are low. Plants that do attempt to grow in sodic soils face the same problems as those growing in saline soils, as well as the additional difficulties of growing in soils with poor structure.

Reclamation of Saline and Sodic Lands

Can saline and sodic lands ever be reclaimed for agricultural use? The answer is a cautious "yes." But it requires a great deal of careful management both during and after reclamation efforts to prevent more serious damage in the future.

- *Reclamation of saline soils:* The goal of this process is to leach the accumulated salts out of the upper soil layers and through the soil. The first step is to improve soil drainage, and then add large volumes of low-salt irrigation water to the soil at regular intervals. This reduces the soil salt content significantly over time. It is very important, however, that the drainage water from this system *not* be returned directly to a stream or river. This drainage water contains high levels of soluble salts, which can damage the river system and hurt other farmers downstream. Additional management tactics that reduce evaporation and reduce the upward movement of salts through the soil can also be used (e.g., mulches), and irrigation systems can be improved.
- *Reclamation of saline-sodic soils:* Soils that are both saline and sodic are quite common. It is important that the land first be treated for sodicity, then for salinity. Attempts to reverse the process will worsen the situation. Recall that sodic soils are, by nature, poorly drained. Adding large quantities of water to them will simply serve to increase Na$^+$ ion levels and increase pH, and probably will have little effect on salinity.
- *Reclamation of sodic soils:* The goal of this process is to remove the Na$^+$ ions from the soil exchange complex. This is most easily done by replacing them with Ca^{2+} or H$^+$ ions, which is often accomplished by adding gypsum (CaSO$_4$ · 2H$_2$O). This replaces Na with Ca, and forms Na$_2$SO$_4$, which can then be easily leached from the soil. The reaction also improves the soil physical structure by stimulating aggregation of particles. The process itself is neither easy nor inexpensive. It usually requires several tons of gypsum/ha, and the soil must be well mixed and must be kept moist.

The management of reclaimed soils is incredibly important—irrigation must be monitored and good drainage must be maintained. If this is not done, the same problem will return fairly quickly. Given the extensive, labor-intensive efforts that must go into the reclaiming of saline or sodic soils, it is obvious that preventing the problem in the first place is preferable to trying to fix it once it has happened.

Waterlogging and the Importance of Drainage Systems

While good drainage is essential for semiarid and arid regions to prevent salinization, it is equally important in areas that receive a great deal of rainfall to prevent waterlogging. Too much water is as bad for agriculture as too little water, except when growing water-tolerant plants like paddy rice. Plant roots need a good supply of oxygen to carry out respiration. If the majority of soil pores are filled with water for an extended period of time, the root cells cannot get needed oxygen, and cannot respire.

Soils with high water tables are likely to be waterlogged for at least some part of the year. When this occurs, the water table rises so high that plants are unable to have much of a root zone (Figure 15-6). The roots have very little soil to exploit for nutrients, so uptake suffers and plant growth declines. This situation can often be prevented by the establishment of a good drainage system in the field. Drainage systems can lower the water table, and therefore speed up the rate of water removal from soils. This, in turn, lowers the moisture content of soils so that oxygen can reach plant roots, and CO_2 can diffuse from the roots to the atmosphere.

There are two major types of drainage systems that can be installed in a field to facilitate the movement of water through or away from the soil: surface drainage systems and subsurface drainage systems. Surface drains are designed to remove water from the field before it penetrates the soil and starts to move into the groundwater. These systems use ditches to collect and remove water from the field. These drainage systems one usually formed with land-forming equipment to provide the correct amount of slope, so initial costs may be moderate to high. Once in place, labor can be done by hand to prevent formation of depressions or ridges that obstruct good water movement away from the field.

Subsurface drains are designed to remove water from the field after it has penetrated the soil through a system of underground channels, which may be made of perforated plastic or permeable clay tiles, or simply by digging tunnels through the soil. Water percolates through the soil and into these channels, which then lead to an outlet ditch to drain the

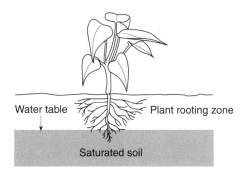

Figure 15-6 When water tables remain high for a significant portion of the growing season, plant roots remain constrained to the upper drier layers of soil. Lack of a deep root system can lead to stunted growth and to easy lodging (tipping) of the plant.

water from the field. Initial costs of these drains are very high and maintenance needs may also be quite high, since each of them tends to get clogged by silt and clay over time. Nevertheless, these drains are very useful because they remove water that has already made its way into the ground, preventing its accumulation in the soil pores.

There are many benefits of good drainage in a field. We have already discussed the use of drainage to prevent salinization in arid areas, and to prevent waterlogging in more humid regions. Drainage also allows plants a deeper root zone for nutrient extraction and improves soil aeration. It provides plants with a good O_2 supply and easy CO_2 removal, and also encourages the activity of beneficial soil organisms, thus increasing nutrient availability. In cooler climates, good drainage also allows soils to warm up earlier in the spring, permitting earlier initial planting dates.

The Politics of Water

While many of the important issues related to irrigation are ecological or chemical in nature, political and socioeconomic issues greatly influence the availability of water and its use in agricultural production worldwide. Availability of potable drinking water varies dramatically from country to neighboring country. For instance, only 12% of the population in Afghanistan has access to potable water, while 81% of the population in neighboring India and 84% of the population in Iran have access to potable water (UNDP 1996).

Currently, no enforceable legislation exists that controls the allocation and use of international waters (Postel 1996), yet nearly 40% of the world's population depends on water from river systems that are shared by two or more countries. At least 214 rivers in the world flow through more than one country, and tensions among nations that rely on the same water resources are growing. The basic question is this: If several nations share a river, who has rights to the water? The nation where the headwaters occur? The nations with the most need? Or should there be equal access along the river for all people? Who makes these decisions? Who enforces them? And what happens when no agreement is reached?

It is obvious that people in upstream nations have a real advantage over those in downstream nations. The water flows past their farms and their cities first, so they can remove as much water as needed. They are often so far removed from the people downstream that water conservation for downstream populations may never enter their minds. Conflicts have arisen all over the globe. In Africa, nine countries share water from the Nile River (see Application 15-3). In the Middle East, the sharing of waters from the Jordan River basin is creating conflict in a land already ripped apart by political differences. Even in the United States, disagreement with Mexico over the use of the Colorado River has caused conflict. In early 1995, Mexico petitioned the United States for the use of more water from the Colorado, but the United States government declined out of concern for growers in Texas who needed the water. As populations around the world continue to increase and water demand increases as well, we will likely see growing unrest related to water availability. Concern about war and instability over water issues so worries the United Nations that the UN Commission on Sustainable Development requested a global freshwater assessment to look into the problems of water availability.

While these conflicts are occurring among countries, other conflicts concerning water are occurring within countries as agricultural and urban areas compete for use of water. Which system has more rights to water use: an agricultural system that produces food for the nation or an urban area where millions of people need water for drinking, cooking, and sewage systems? These conflicts will intensify as urban populations continue to increase.

Application 15-3 *Water Tensions Between Nations: The Nile River*

The Nile River in northeast Africa supports nine different African nations. About 85% of the Nile's water is generated by rainfall in Ethiopia, while the rest originates in Tanzania. The water passes through six other countries, with Egypt as the last country in line. Currently, water demand in Egypt is very near the limits of supply, and demand is expected to increase with population (Egypt's current population of about 60 million people is increasing by about one million every nine months). Egypt itself receives little rainfall, so the nation is very dependent on water from the Nile. About 97% of Egypt's surface water is delivered by the Nile River and over 3.2 million hectares of agricultural land are completely dependent on irrigation by Nile water (Postel 1996). Egypt is therefore quite vulnerable to any reduction in water supplies, and any actions by its closest neighbors to reduce water flow will quickly become a national security issue.

It seems that international conflict in the region is just a matter of time. Ethiopia is developing the headwaters of the Nile, building hydropower stations and water storage facilities, and increasing irrigation to many of its agricultural lands. Approximately 3.7 million hectares of land in Ethiopia could be irrigated, but using Nile River water to irrigate even half of this area could reduce downstream flow of the Nile by 9 billion cubic meters per year, which is 16% of Egypt's water supply! Unfortunately, for downstream nations like Egypt, Ethiopia may feel no obligation to limit its use of the Nile. But Ethiopia is not alone. Uganda is also expanding irrigation, which could potentially use an additional 2 billion cubic meters of water per year (Postel 1996).

There is currently no water-sharing agreement that includes all of the countries sharing the Nile River. Some agreements do exist between neighboring countries, but Sudan's agreement to provide Egypt with 55.5 billion cubic meters of water per year may mean very little if the downstream flow is cut significantly at the headwaters.

What is Egypt likely to do if the Nile River flow is cut? Already, Egypt supplements its river water supply with groundwater extraction, agricultural drainage water, and treated wastewater. Uses of these alternative water sources will likely increase. But the removal of groundwater is not sustainable, so any agricultural productivity that results from this water will disappear quickly when the groundwater is drained. When this occurs, agricultural productivity will decline dramatically, and this, combined with large populations, could result in widespread food shortages or starvation.

While Anwar Sadat was president of Egypt, he stated that the only matter that would take Egypt into war in the Middle East again would be water. We see similar sentiments expressed across the globe: Turkey, Syria, and Iraq argue over water from the Tigris-Euphrates water basin; Israel, Jordan, the West Bank, and Syria face tensions over the Jordan River Basin; Mexico and the United States have not yet resolved disagreements over the use of water from the Colorado River. Conflicts among countries that must share water will only escalate as populations increase, with those nations that are most powerful and affluent controlling the conditions of resolution. Foresight in resolution of these conflicts is needed to avoid future crises and to provide the basic water needs for the entire human population.

It is obvious that water conservation is key to resolving many of these conflicts. If citizens in the United States are truly concerned about the plight of their Mexican neighbors, they will work to conserve as much water as possible so that the Colorado River can carry sufficient water for the needs of both nations. For instance, use of water for lawns can be limited to conserve water for food production. Urban dwellers can cut back on water use to allow more water to go to agricultural fields. Farmers can irrigate their fields with efficient systems, and only when necessary. Pricing water in a rational way to encourage conservation, improving efficiency standards for many of our appliances, and finding innovative ways of reusing wastewater will help to conserve water and prevent some of the inevitable conflicts of the future.

Summary

Worldwide, freshwater supplies are beginning to diminish as agricultural productivity relies more and more on irrigation and as demand for freshwater from urban regions increases. In many areas, deep groundwater supplies are being tapped to meet water needs at rates much faster than rates of renewal. Water use efficiency depends on the type of irrigation system used, with micro-irrigation or drip irrigation systems being the most effective in limiting losses from evapotranspiration. Maintenance of water quality is critical for limiting soil salinity, and, ultimately, reducing the amount of agricultural land lost to salinization and sodicity. Water conservation methods, both nationwide and at the level of a local farm, will become increasingly important in meeting future demand for water.

Topics for Review and Discussion

1. Explain the difference between salinity and sodicity. Which is more of a problem? How would you manage soils that have both saline and sodic depositions?
2. Look through a book of maps. What percentage of major rivers run through more than one country? Focus on areas of current political turmoil. How many countries that are already involved in conflict rely on the same river for fresh water?
3. Drip irrigation was developed in the Middle East, where the climate is arid and water is not readily available. Explain why drip irrigation is a good method for use in that region, and why it is less easily adapted in other areas.
4. What are the differences between subsurface and surface drainage systems? When would you use one over the other? Explain in terms of climate, soil type, economics, and cropping system.
5. Conflict over fresh water is likely to escalate in future years as availability decreases and demand increases. What precautions can individual nations take to decrease the likelihood of conflict?

Literature Cited

Brown, L. R. 1995. *Who Will Feed China?* New York: W. W. Norton and Co.
Brown, L. R., and B. Halweil. 1998. China's water shortage could shake world food security. *WorldWatch* 11: 10–21.
Brown, L. R., H. Kane, and D. M. Roodman. 1994. *Vital Signs, 1994.* New York: Worldwatch Institute.

Delouche, J. C. 1980. Environmental effects on seed development and seed quality. *HortScience* 15: 775–780.

Diez, J. A., R. Roman, R. Caballero, and A. Caballero. 1997. Nitrate leaching from soils under a maize-wheat-maize sequence, two irrigation schedules and three types of fertilisers. *Agriculture, Ecosystems and Environment* 65: 189–199.

Dornbos, D. L., Jr., R. E. Mullen, and R. M. Shibles. 1989. Drought stress effects during seed fill on soybean seed germination and vigor. *Crop Science* 29: 476–480.

El-Hassanin, A. S., T. M. Labib, and E. I. Gaber. 1993. Effect of vegetation cover and land slope on runoff and soil losses from the watersheds of Burundi. *Agriculture, Ecosystems and Environment* 43: 301–308.

Haruvy, N. 1997. Agricultural reuse of wastewater: Nation-wide cost-benefit analysis. *Agriculture, Ecosystems and Environment* 66: 113–119.

Kovar, J. L., S. A. Barber, E. J. Kladivko, and D. R. Griffith. 1992. Characterization of soil temperature, water content, and maize root distribution in two tillage systems. *Soil and Tillage Research* 24: 11–27.

Miller, G. T., Jr. 1990. *Living in the Environment: An Introduction to Environmental Science.* Belmont Calif.: Wadsworth Publishing Co.

Postel, S. L. 1996. Dividing the waters: Food security, ecosystem health, and the new politics of scarcity. *Worldwatch Paper #132.* New York: Worldwatch Institute.

Postel, S. L., G. C. Daily, and P. R. Ehrlich. 1996. Human appropriation of renewable fresh water. *Science* 271: 785–787.

United Nations Development Programme (UNDP). 1996. *Human Development Report.* New York: Oxford University Press.

Bibliography

Altieri, M. A. 1995. *Agroecology: The Science of Sustainable Agriculture.* 2d. ed. Boulder, Colo.: Westview Press.

Brady, N. C., and R. R. Weil. 1996. *The Nature and Property of Soils.* Upper Saddle River, N.J.: Prentice-Hall.

Smith, R. L. 1996. *Ecology and Field Biology.* 5th ed. New York: HarperCollins.

Stewart, B. A., and D. R. Nielson, eds. 1990. *Irrigation of Agricultural Crops.* Madison, Wisc.: American Society of Agronomy.

Tivy, J. 1992. *Agricultural Ecology.* Essex, U.K.: Longman Scientific and Technical.

Chapter 16

Energy Conservation and Animal Agriculture

Key Concepts

- Livestock production systems
- Desertification
- Integrating livestock into agricultural systems
- Grazing management systems

Globally, domestic livestock are almost as numerous as humans, but they consume more than four times as much plant material. The human population around the world obtains about 10% of its total calories from animal products, although this percentage is much greater in industrialized nations like the United States.

Shortening the Food Chain

Recall from Chapter 2 that energy enters a food chain through the primary producers, and as herbivores and carnivores feed on plants and animals below them in the chain, much of that energy is lost from the system as waste products, metabolic energy, and heat energy (see Figure 2-6). For this reason, a carnivore at the top of the food chain relies on a very large number of primary producers to eventually provide it with the energy it needs to sustain itself. If the carnivore "cut out the middleman" and ate the primary producers directly, the energy lost to the intermediate animals in the food chain would be available to the carnivore directly.

In agroecosystems, the primary producers are the crops. Human beings are the top carnivores in the system. The intermediate animals are livestock, including cattle, pigs, chickens, sheep, and goats. When we eat grains and plant products directly, we benefit from the extensive amounts of energy stored in organic compounds in those plants. When humans eat meat, we are increasing the total number of plants needed to feed us, since much of the initial energy available from those plants has been used for the respiration, metabolism, and waste produced by the livestock. For example, it takes approximately 2.8 kg of grain to produce 1 kg of chicken, 6.9 kg of grain to produce 1 kg of pork, and about 4.8 kg of grain to produce 1 kg of beef (Durning and Brough 1991) (Figure 16-1). It is clear that a diet rich in meat will use far more grain than will a diet of primarily vegetables, legumes, and grains. By shortening the food chain, then, we are reducing energy losses from primary producers to top consumers, and decreasing the amount of agricultural land needed to sustain each person on earth. In fact, some societies with heavy population pressures are beginning to eliminate livestock from their agricultural systems, using land for production of useful plant products rather than for grazing animals.

Despite a trend in some industrialized nations toward increased vegetarianism, we are unlikely to see any major declines in the livestock industry in the near future. In fact, we may actually see increases in demand as incomes increase in many developing nations around the world. As personal income increases, people diversify their diets and begin to eat meat. For example, in 1978, only 7% of the grain produced in China went for animal feed. By 1990, 30% of grain in China went to feed animals, which were becoming more popular in the diets of the affluent. If this amount continues to increase and grain supply cannot keep up with demand, food prices will rise accordingly (see Application 16-1).

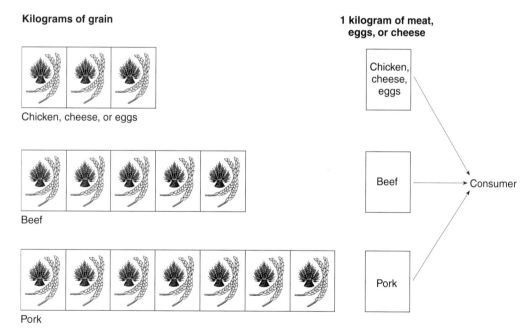

Figure 16-1 Grain use comparison for various animal production systems. One kg of grain can go directly for human consumption, or can be used to produce 0.36 kg of chicken, 0.15 kg of pork, or 0.21 kg of beef.
Based on data from Durning and Brough 1991

Application 16-1 *Animal Production and Food Security*

As nations grow more affluent and diets begin to diversify, animal products will be incorporated into daily menus for a growing population of people. To meet the growing demand for meat, more grain will be used for livestock feed, and less will be used for direct human consumption. For many nations, meeting this feed demand is possible only by importing feed from other countries or by converting agricultural land to pastures.

Importing feed from other nations is the primary solution for nations that are demanding more grain to feed livestock. As feed demand increases, many nations must turn to their neighbors for help, reducing self-sufficiency and turning some exporters into importers. This need to import grain can result in serious economic consequences for some nations by increasing their burdens of debt to other nations. Animal agriculture also relies on fewer employees per unit of land than does traditional cultivation; thus, employment opportunities decrease, contributing to inequity among classes in a nation.

This inequity can also be reflected in the food availability within a nation. Unfortunately, as the more affluent individuals in a nation demand more meat, food availability for the poor is reduced. Farmers, anxious to maximize profitability, may convert their land to grain production systems for animals rather than using it to produce staple foods used by the poor. During and Brough (1991) report that in many developing nations, the percentage of land used for production of animal feed has steadily increased, with many farmers converting their systems from food crops to feed crops. This can, of course, lead to shortages of staple food crops for poorer citizens, and it may also lead to further environmental degradation if pastures are poorly managed.

Having said all of this, it is important to realize that livestock production, both in developing and industrialized nations, has clear advantages. Before we look at the positive aspects of animal agriculture, we must first examine the differences between livestock production in these two groups of nations.

Livestock Production in Industrialized and Developing Nations

Livestock production can differ substantially between industrialized and developing nations. Although grazing systems are very important in industrialized nations, many producers in these countries use animal feedlots to raise huge numbers of livestock on a relatively small area of land. They usually produce only one kind of animal or animal product, and may rely on growth hormones to encourage rapid increases in growth and antibiotics to keep diseases associated with crowded conditions to a minimum. Many of these feedlots require large inputs of fossil fuels to produce animal feed, to heat and cool facilities, to transport livestock and feed, and to pump water to a central facility. Wastes produced in these systems can be spread on neighboring fields, but may wash into nearby waterways or groundwater, polluting the water with bacteria and excess nutrients from a concentrated source.

Many farmers in developing nations produce only enough meat and animal products to feed their families and to provide labor and transportation. Rather than being confined to a small location and fed imported feed, animals in these systems may use nearby natural forested areas or grasslands as food and water sources, and the flocks and herds are moved often to avoid decimating one region. Since animal labor is often used on farms, no fossil fuel inputs are required. Waste is either returned to the soil directly where the animals graze, or it is collected and used as a fertilizer or as a fuel.

Both systems have advantages and disadvantages. In both industrialized and developing nations, animal agriculture can occur on land that cannot be used for much else. Land on rocky terrain, at high elevations, or in marginal areas can often support enough grasses and forbs to sustain a small herd of livestock. For example, in South America, agropastoral systems have replaced hunting and gathering as a way of life in the high elevations of the Andes. The clear differences in climate and vegetation with elevation have resulted in the adaptation of a vertical agricultural system, with different production systems spread out along an elevational gradient. At the lowest elevations, farmers produce many of the typical tropical crops that are common to the region, including cocoa, cotton, corn, sugarcane, and numerous tropical fruits. As elevation increases, farmers grow cereal crops that are well adapted to slightly cooler climates, including most of the important grains, followed by tuber crops farther up the slopes. At the highest elevations, very little will grow, apart from grasses and forbs. The climate is cold, the terrain is rocky, and the land is marginal. It is here that animal agriculture thrives, and animals that eat the native grasses are spread across the countryside. Animals that graze in this area are responsible for wool, an important export product of the region. Clearly, this high-elevation land would remain unproductive (agriculturally) if animals were not used in this system.

This last point emphasizes what may be the most important advantage of animal production in regions with harsh climates or low rainfall where vegetation is limited to grasses or native rangeland plants. Without importing food, human populations cannot be sustained in these ecosystems because traditional agriculture may not be possible. However, animals can use marginal food sources that humans cannot eat. Humans cannot digest cellulose, a common component of grasses and other plant products, but ruminants like cows and sheep can. Therefore, shortening the food chain is *not* an option where native vegetation consists of rangeland or permanent pasture.

Another clear advantage of animal agriculture in industrialized nations is the scale of production. Livestock production in such nations provides huge amounts of meat and dairy products for an affluent population at very low prices. The availability of such an inexpensive variety of protein-and nutrient-rich foods certainly contributes to the overall good health and longevity of people in industrialized nations.

Nevertheless, there remain a number of disadvantages of animal agriculture in industrialized nations, at least with some of the more intensive production methods. Perhaps the most serious disadvantage has already been discussed: Raising animals for food requires a great deal more grain and fossil fuel energy than the animals return to us as food energy. The conversion rate is low and the process itself is not sustainable. If the entire world's population ate the same diet (produced the same way) that a typical citizen of the United States eats, all of the earth's known energy reserves would disappear in only thirteen years (Miller 1990)!

Other disadvantages of animal production include overgrazing of lands, the use of limited water supplies to provide water for livestock, the loss of forested land to provide grazing

Application 16-2 *The Problem of Manure*

Animal manure can be an excellent addition to agroecosystems. It supplies both organic matter and plant nutrients to the soil, thereby enhancing crop productivity. In many small systems in developing nations, animals are fully integrated into the farming system, and any manure produced by animals goes directly back to the fields. In many industrialized nations, however, animal production systems have become divorced from crop production systems, with alarming consequences.

Large-scale animal production means large-scale manure production, and in large quantities, manure can become a serious pollutant. In many cases, the manure is kept in large outdoor piles, with no protection from the elements. When it rains, the manure can run off the pile and nutrients can leach out of it, accumulating in and on the soil and eventually moving into groundwater or open water systems. This can lead to such problems as eutrophication or nitrate accumulation in soil and water systems (see Chapter 12). When manure piles are left exposed to the air, nitrogen in the pile will volatilize, escaping into the air as ammonia, which can contribute to acid rain and acid deposition. The manure pile provides an unusually concentrated supply of potential pollutants at one point source, from which excessive levels of nitrates and other materials can radiate.

In large-scale production systems, manure disposal becomes a problem. This is quite ironic, given that farmers spend billions of dollars each year on inorganic fertilizers that are made of finite resources, while animal producers spend large amounts of money to have their nutrient-rich manure carted away and disposed of. Durning and Brough (1991) estimate that the nutrients lost in wasted manure in the United States could supply 12% of the nitrogen, 32% of the phosphorus, and 30% of the potassium that farmers apply to their fields using inorganic fertilizer products. It is clear that integrating plant and animal systems, a common practice in developing nations, could contribute greatly to sustainability and productivity in industrialized nations, while reducing the risks of pollution and health hazards.

---

lands or fields for feed production, and even the release of methane from cattle, which has been implicated as a contributor to global warming (although to a much lesser extent than CO_2). In addition, the question of what to do with the huge amounts of manure produced on many feedlots still remains unanswered (Application 16-2).

The use of animals in traditional systems has many more advantages. In most traditional systems, the number of animals produced is small, and the animals themselves are well-integrated into the overall farming system, combining animal and plant production. For example, the manure produced by the animals in these systems is returned directly to the soil, allowing for efficient nutrient recycling. Much of the feed provided to the animals is worthless for human consumption, so resource use is actually more efficient with animals present! Chickens in traditional systems are fed kitchen scraps while cattle graze in fallow fields or consume crop residues from fields that have already been harvested. In essence, the livestock in these systems turn things that people cannot eat into things that they can eat—milk and meat produced on these farms can be eaten or sold. In addition, animals in some of these systems are used for labor (e.g., cultivation of fields) and transportation. Manure can be used as a convenient fuel source in some places. Animals may also serve as investments if the farmer buys livestock during a good year and sells them during a bad year.

Nevertheless, animal agriculture in traditional systems is not without its disadvantages. A return to these traditional systems is not possible in the United States, for example, because they could never support or sustain the demand for meat and animal products that currently exists in this country. While traditional systems work well for small farms and provide for individual families, they are not practical for large-scale production.

Perhaps a more serious disadvantage of animal production systems in developing nations is the possibility of desertification occurring when marginal lands are used for animal grazing. As population continues to grow in many developing nations, marginal lands are used more frequently for animal grazing, leading to serious land degradation.

Desertification

Desertification refers to the conversion of arid or semiarid lands to deserts as a result of decreasing rainfall coupled with the progressive loss of vegetation in a region. Desertification can be the result of natural changes in climate, or can be anthropologically induced as a result of poor land management decisions. Some of these poor management decisions are overgrazing, overcultivation of marginal land in semiarid regions, and harvest of trees for firewood.

How does animal agriculture relate to desertification? In most cases, the conversion of agricultural land to desert land is caused, in part, by overgrazing of land by domestic animals including cattle, sheep, and goats. This overgrazing, combined with natural drought cycles, results in the inability of plant species to recover from stress, eventually leading to a denuded landscape.

When too many animals graze on a system, plants are continually eaten by large herbivores, and cannot recover sufficiently before being grazed on again. If this continues, the plant eventually loses all of its stored resources, is unable to recover at all, and dies. During periods of drought, the problem is exacerbated because plants are under water stress as well as stress from large-scale herbivory.

As the ground cover begins to die, the bare ground left behind becomes more susceptible to both wind and water erosion and to temperature changes. Surface temperatures begin to rise, and some heat is reflected from the ground to the atmosphere, creating warm, thermal updrafts. As this occurs, humidity decreases, cloud formation becomes less likely, and rainfall declines, thereby interrupting the natural hydrological cycle of the region.

Desertification is occurring in many areas of the world at a rate of approximately 6 million hectares per year. It is most common along the margins of established deserts (e.g., the Sahel), but it is also taking place in many semiarid savannas and grasslands. As population increases in many semiarid regions of the world, there will be more pressure for grazing sites by a higher population of livestock. The United Nations estimates that areas at risk for extreme desertification in the future currently support over 900 million people (Southwick 1996).

Desertification can be caused by a variety of poor land management practices, not just overgrazing. As such, desertification was an important part of agricultural history in the United States. The Dust Bowl of the 1930s was a prime example of desertification that occurred as a result of poor land management and excessive tillage that left land with no cover. The vast areas of degraded soil left behind after the Dust Bowl sparked the development of extensive soil conservation programs and new farming practices that have prevented such serious destruction from recurring in that area. Even today, some rangeland in the western United States is at risk if populations of grazing animals are not managed carefully.

Once desertification has taken place, restoration of the land is difficult, but not always impossible. Most scientists agree that the key to restoring the land is reestablishment of plant cover. Unfortunately, in areas where drought is a frequent problem, reestablishment is not easy. Planting native grass and forb species, rather than plants not adapted to the region, is an important first step in reclaiming the land. If native grasses can establish during a period of adequate rainfall, and steps are then taken to reduce grazing pressure in the area, slow reclamation of the land may occur. Obviously, this problem is more easily prevented than cured, so taking initial action to keep grazing pressure low, to reduce soil erosion, and to keep native grasses covering the ground is a wise management practice in vulnerable areas.

Climate change can also affect desertification in many areas of the world. If climate change is a reality and temperatures continue to increase globally, we are likely to see changes in rainfall patterns as well. It is projected that many of the dry regions of the world will become drier, and many of these areas are in developing nations. As population in these areas continues to increase, there will be more pressure for food sources, including animal products. Grazing pressure and overcultivation of marginal land is likely to increase, and this, combined with more frequent and more severe drought, could rapidly increase rates of desertification.

Integrating Animals Into Agroecosystems

It has long been recognized that many farming systems that are among the most ecologically sustainable are those that integrate animals into the system. This has been accomplished in many traditional agroecosystems, but is done to a much lesser degree in systems found in industrialized nations.

Even those agricultural systems that are most dependent on technology and fossil fuels can integrate animals in such a way as to increase productivity and sustainability. For example, adding nitrogen-fixing leguminous forages in crop rotations will not only increase soil nitrogen, which will result in higher productivity for the next crop in rotation, but will also provide a highly nutritious forage for livestock. High-quality forages are directly related to animal performance, and legumes are generally considered to hold greater nutrient value than grasses. Animals may also provide an interesting means of biological weed control since cattle may graze on plants that are in competition with the main cash crops in a field. In agroforestry systems, herbicide use in young tree plantations can be reduced by introducing animals into the system and allowing them to control competing vegetation. Of course, proper grazing management is essential to prevent grazing on the trees themselves.

When it comes to increasing sustainability of an animal-based agroecosystem, the importance of recycling animal manure cannot be understated. Animal manure is an excellent source of nutrients, particularly in areas where availability of chemical fertilizers is limited or prices are prohibitive. Keeping the nutrient cycle closed is a good way to maintain productivity of land without relying too much on external sources of fertilizer. For example, when manure that is produced in a system is reapplied to the land, a large proportion of the plant nutrients ingested by the cattle are returned to the soil in the waste products of the animals. Soil fertility is relatively well-maintained in these systems, even though manuring alone cannot satisfy all of the nitrogen needs of an agroecosystem (there is always some loss of N from volatilization or other means). Because of this, animals may occasionally play a role in renovating marginal lands under very careful land management regimes (see Larney and Janzen 1997). However, in many systems, waste is removed from the farm rather than applied to fields. This not only results in great expense and a serious loss of nutrients, but it also causes soil nutrient status to decline steadily over time. This often requires the addition of synthetic fertilizers to the field, which are produced off-site and then transported, using enormous quantities of fossil fuels.

Another great advantage of integrating animals into agroecosystems is the role that they often play in "cleaning up" after humans. Animals can use crop residues and low-quality cereal grains that humans cannot use; in fact, crop residues are a major portion of the diet of cattle in many nations of the world. Chickens and pigs can eat kitchen and table scraps rather than fossil fuel-produced feeds, thus providing nutritious, protein-rich foods from human garbage. This, of course, can only be done on a small scale—most households do not produce enough scraps to sustain more than a few chickens and a few pigs at one time.

Integrating animals into agricultural systems can provide other interesting side benefits. In many nations of the world, very few rural people have access to reliable power and renewable cooking fuel sources. In some of these nations, cow dung is converted into biogas, a renewable source of energy and nutrients for agriculture. In India, for example, over 10 million rural people rely on biogas for their energy needs. The major advantage of biogas is that it provides a locally controlled fuel supply from a material that is readily available in large quantities.

Sealed containers known as biogas digesters convert cow dung into a flammable gas, which can then be used to provide electric light, pumped water, and clean cooking fuel. The process destroys most of the pathogens that are commonly found in dung and waste. The by-product of the digestion process creates a slurry that is a useful fertilizer because it contains many more nutrients than does composted dung. For example, the slurry contains twice as much nitrogen as that contained in composted dung because the composting process takes place in open air, allowing much of the nitrogen in the compost to escape to the atmosphere. In addition, the slurry releases nutrients more readily than does composted dung.

There are many environmental benefits to using biogas. In those areas that traditionally rely on fuelwood for cooking fuel, biogas provides an ecologically sustainable alternative, and thereby helps to decrease the need to cut wood, preventing deforestation. In addition, biogas burns much more efficiently than do solid fuels, resulting in the release of fewer contaminants.

The integration of animals into an agricultural system does not need to be limited to the four-legged or two-legged kind. In some cases, integrating water resources into farming systems (aquaculture) can result in a highly productive and very diversified agroecosystem. Canals cut through a field can circulate water and nutrients around from platform to platform, and also provide a good habitat for fish production. Occasional dredging of the accumulated organic matter in the canals results in a high-quality fertilizer that can be applied as a top-dressing to the soil, thus keeping nutrients within the system. Platforms vary in size, and usually contain a wide variety of intercropped plants, including corn, squash, alfalfa, and other annuals. Other animals that feed on crop residues or alfalfa produced on the platforms can also be integrated into the system, and provide manure to help maintain soil fertility.

Livestock Production Systems

The characteristics of the livestock production system determine both the sustainability and the potential ecological consequences of the system:

- *Grazing systems:* In these systems, the main source of animal feed is the forage that is produced within the system. The amount of forage produced varies with land quality, climate, and management, and determines the **stocking rate** of the land (carrying capacity). Stocking rate refers to the number of animals that a system can support, and is usually reported per area of land. Stocking rate can be either fixed (remains constant over time) or variable (changes according to level of available forage). Sustainability depends on how closely the stocking rate is adhered to by the grower. Manure produced by livestock in a grazing system will be dispersed according to cattle dispersal, so many of the nutrients that cattle consume will eventually be recycled back into the system.
- *Outdoor confined systems:* In these systems, livestock (e.g., cattle, sheep, pigs) are contained in outdoor pens at relatively high densities. There is little or no vegetation growth within the pens, so all feed must be provided from external sources. Because all feed and water is brought into the system, stocking rates are not limited by the carrying capacity of the land area that the animals live on, so the number of animals that can be crowded into a given land area is much higher than the number in grazing systems. Manure produced by the livestock accumulates on the ground of the pens and is occasionally pushed into large mounds. It can then be applied to nearby fields or disposed of. Conditions are often quite crowded. Diseases and parasites spread easily, so antibiotics and antiparasitic drugs are commonly used.
- *Indoor confined systems:* The difference between outdoor and indoor confined systems is rather intuitive. In indoor confined systems, the animals are contained throughout the year in environmentally controlled buildings. Again, there is no vegetation grown within the pens, so all feed and water must be provided from external sources. The animals are usually supported by steel floors, and the manure, which accumulates in pits beneath them, is collected and disposed of or used on fields as a fertilizer. As in the outdoor system, space requirements for animals are minimal, so densities can be very high. This can lead to pest problems similar to those described for outdoor confined systems.

Of these systems, grazing systems are most likely to be sustainable over the long term because the nutrient loop is closed, and population densities are limited ecologically by the system itself. Improper management of a grazing system, however, can result in serious ecological consequences, as just described.

Grazing Management Systems

Pasture plants—commonly grasses, forbs, or legumes—intercept and store enormous amounts of solar energy each year. If they are managed correctly, they can support high levels of livestock production at fairly low cost, both ecologically and economically. The basic goal of grazing management is to produce a uniform, high-quality forage at a reasonable price, and to use that forage to maximize animal production. Intuitively, then, it follows that the best management method for sustainable production is to select animals based on how they do on the forage best adapted to the soils they are grown on, rather than a method based on choosing a particular breed and trying to grow a certain forage to feed it, regardless of cost and adaptation to the region. In the United States, most grazing systems conform to this management rule and are based on nitrogen-fixing legumes, rather than fertilization with commercial fertilizers.

There are two basic grazing management systems in place in industrialized nations: continuous grazing (i.e., set stocking) and rotational grazing. Continuous grazing attempts to determine the number of animals to be placed into a field based on the amount of pasture growth in that area (Murphy 1990). Plants are continuously exposed to grazing, and animals are always present in the pasture. Rotational grazing takes both the needs of the pasture plant and the needs of the animals into account. Plants are given a rest from grazing according to a specific rotation schedule. In general, more frequent and more intensive grazing results in lower productivity of the pasture plants over time.

Continuous Grazing Systems (Set Stocking)

In the United States, many continuously grazed systems are not managed for their most efficient production. In these systems, no adjustments are made for the variation in plant growth rates throughout the year, resulting in alternate overgrazing and undergrazing depending on season. In continuous grazing systems, the number of livestock within a pasture is set and does not change over time. Therefore, when the plant growth rate is low in the early spring and mid-to late summer, the pasture tends to be overgrazed. The same number of animals present in the field when the plant growth rate is high (late spring, early summer, fall) will undergraze the pasture.

Overgrazing is a problem because plants do not have sufficient time to recover between foraging incidents. But why is undergrazing unacceptable? After all, is it not better to have too much forage than too little? Unfortunately, it is not that simple, and the ecological consequences of undergrazing can result in lowered overall productivity of the system. Animals are selective grazers and, if given the choice, will selectively choose to graze on the more palatable plants in a field. Those plants that are left behind will be allowed to mature, flower, and set seed, perpetuating the plants that are not preferred by the animals. These plants will compete with the preferred forage plants for sunlight, water, nutrients, and space. As animals continue to graze on the preferred plants, the uneaten forage plants will gain a competitive advantage. Over time, then, the undergrazed patches of land will become less palatable and total production will fall. This selective grazing will occur when there are too few animals present in a field to use all of the forage produced, including the less palatable plants. The solution is to increase stocking density to use all of the available forage, but that creates a problem during times of the year when the pasture is less productive. However, if stocking density is continuously adjusted throughout the year, the system can be well-managed, productive, and sustainable.

Rotational Grazing

Rotational grazing can mean many things in the United States, and the principles behind it are often not well understood. It is unfortunate that some managers with the best intentions simply move livestock from field to field without a good understanding of the ecology of rotational grazing and how their actions will affect the overall sustainability and productivity of the system. Rotational grazing refers to a regular sequence of grazing and rest for a group of fields within a pasture.

Perhaps the best method of grazing management in a rotation system is *Voisin's grazing management,* which takes into account both the requirements of the plants in the pasture, and the needs of the grazing animals. This system has been referred to as rational grazing because forage in a pasture is "rationed out" to meet animal's nutritional needs while simultaneously protecting the pasture itself from both over-and undergrazing. This management technique takes both plant growth rate and forage availability into account (Murphy 1990).

Mob stocking is a modified version of Voisin management (Voisin 1959, 1960). This method is used to improve pastures by cleaning up much of the low-quality forage left behind by undergrazing. Large numbers of animals are moved into small pastures for short periods of time; by the time they are moved into the next pasture, all of the coarse and fibrous forage left behind in the pasture is cleaned up, thus transforming low-producing areas into more productive areas. Of course, because the animals themselves are eating low-quality plant material, this is not a good method to use to increase animal production, but it works well for cleaning overgrown pastures.

Voisin Grazing Management: The Ecology of Sustainable Animal Agriculture

When plants are grazed by an animal, much of the leaf area of the plant is removed. In order for the plant to be able to begin photosynthesizing again, it must replace the leaf area that has been removed. It does this by drawing on the food reserves left in the remaining plant. These food reserves supply energy for regrowth of the plant, and for the storage of more food reserves for the future. If the plant is regrazed before these reserves are replenished, the plant cannot regrow very well, if at all (Figure 16-2). The premise behind rational grazing management is to allow pasture plants to replenish their reserves and to photosynthesize and regrow before they are regrazed. This time of regrowth is referred to as **rest.** If rest time is significantly shortened (move left along the *x*-axis in Figure 16-2), the amount of regrowth that the grass will be able to accumulate during the shorter time is significantly reduced as well. While the plants are resting, livestock forage on another section of pasture, which will then be allowed to rest as the livestock move on to yet another section (Figure 16-3). The entire pasture is thus divided into small areas known as paddocks (often with

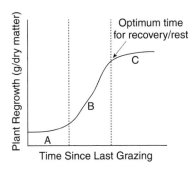

Figure 16-2 Plant regrowth curve. Early in the year or immediately following grazing, plants have very little leaf surface, so photosynthetic production is low and growth is slow (A). As leaf area develops and photosynthetic rates begin to increase, a period of rapid growth (B) follows the initial period of slow growth. Growth rates eventually slow after food reserves have been stored and shading of lower leaves has occurred (C). If a plant is regrazed before the period of rapid growth has occurred, the plant will not recover well, if at all, and overall productivity of the system will decrease.

easily moved electric fencing), and the animals are rotated from paddock to paddock, according to both the needs of the animals and the needs of the pasture plants. This method not only allows plants to recover from grazing without further stress, but also minimizes forage waste because each paddock is essentially "cleared" before the animals are moved to another paddock. The key elements to Voisin management are rest periods for the plants between grazing activities, and the length of time the animals remain in any given paddock.

Rest Periods. Because plant growth rate varies with season, rest periods between grazings should change according to changes in plant growth rates. For this reason, rest periods should not be set in stone, but instead should be adjusted over time. For example, if the growth rate of pasture plants is twice as fast in May as it is in July, rest periods in July must be twice as long in July as they are in May. Of course, plant growth rates vary according to climate and must be determined for each individual situation.

Why are rest periods so important, and what will happen to the productivity of the system if they are shortened? First, it is important to realize that the productivity of the plants in a pasture will determine the overall amount of forage available to animals entering a paddock. If very little time is allowed between grazings, regrowth will be minimal, and the total amount of available forage that has accumulated per hectare will be very low. Animals will move through the system faster since there will not be enough forage available, and eventually plants will simply stop growing (Murphy 1990).

In the northeastern United States, ideal rest periods for pastures are about eighteen days in the spring (May and June) and about thirty-six days in late summer (August and September) (see Murphy 1990). When this amount of time is allowed between grazings, total accumulated regrowth of forage is about 4700 kg/ha. Suppose that you do not adjust rest

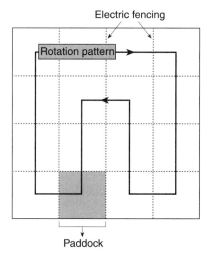

Figure 16-3 Rotational grazing allows plants in one section of a pasture to rest while the livestock forage on a different section of pasture. This allows the plants sufficient regrowth time to maximize overall productivity.

periods with season, and still only allow eighteen days of rest in August (half of the optimum rest period for that month). Dropping the rest period by half actually cuts productivity to only one-third of that which is normally produced (1560 kg). If you simply allow pastures to be grazed as soon as the plants are tall enough for the animals to grip them (six to twelve days, depending on season), this would shorten the rest period even further, and could drop productivity to 10% or less (less than 470 kg/ha). With such small amounts of forage available to animals, weight gains will be minimal and growth rates will be very low.

What will happen to the productivity of the system if longer rest periods than necessary are allowed? Generally, if rest periods are much longer than the optimum, forage continues to accumulate in the pasture, but the increase is mostly due to fiber, which decreases the nutritional value of the plants (Murphy 1990).

Adjusting rest periods is a critical step in managing forage growth and animal productivity, but in setting up a schedule, you will see there are some problems that must be addressed. For instance, suppose that you have six paddocks to rotate cattle through. In May and June, you will need a shorter rest period than in August and September, which means that the pasture grass is growing faster. But you still have the same number of cattle, and they can only eat so much per day. By the time they are ready to move from one paddock to another, you have exceeded the rest period of your next paddock. On the other hand, when the rest period is longer in August, your cattle may be ready to move onto the next paddock before it has had an adequate rest. Local conditions can also affect rest periods dramatically. For example, if one month is very hot and dry, a longer rest period may be necessary. Animals may be able to reenter a paddock sooner if conditions for plant growth are favorable. How do you deal with these differences in rest times that occur both seasonally and even within a season?

The best indicator of adequate rest is, of course, plant growth. Paying careful attention to plant growth rates allows a grower to make adjustments to rest periods based on local conditions. If the plants in the paddock that animals are about to enter have not grown enough (generally about 10–15 cm, depending on forage type), rest periods could be increased by one of several methods:

1. Increase the amount of pasture available for grazing animals, thus increasing the number of paddocks.
2. Graze only half of the total pasture area during periods of rapid growth; hay or ensile forage produced on the rest of the land. Return the land to rotation after an adequate rest period.
3. Remove all animals from the pasture and either sell them or feed them elsewhere while the pasture rests.
4. Add stored hay to the diet of the animals in a paddock as a supplement until the pasture has had an adequate rest period.

This system is very sustainable provided that the rest periods are adequate and the number of cattle grazing the system remains constant. If too many animals are present, they will graze each paddock faster and force movement from paddock to paddock. The only way to increase animal numbers and maintain sustainability is to increase pasture size and number of paddocks. Of course, dramatic changes in climate, water availability, or soil health can also adversely affect sustainability. Still, the length of time that a given number of animals remains in each paddock ultimately determines how much land will be necessary to provide adequate nutrition for maintenance and growth of the livestock.

Periods of Stay. The amount of time that each group of animals is in a paddock before it moves on to the next is known as the **period of stay.** For any given group of animals, periods of stay should generally not exceed three days (Murphy 1990). If animals are left in a paddock for longer than three days, they spend a great deal of time and energy looking for good feed because they have already eaten the best forage plants and the remaining plants are usually less palatable. For this reason, shorter periods of stay are better than longer periods of stay. In general, the shorter the period of stay, the more productive the system will be, both for plants and animals.

Shorter periods of stay (two days or less) are most important for animals that are milking, growing, or being fattened for market. Milking animals are more productive (and let down their milk better) if they are given a fresh paddock after every milking. Growing animals also gain weight most rapidly if they are moved to a fresh paddock every twelve to twenty-four hours (Murphy 1990). These shorter periods of stay ensure that forage provided to animals is of better quality, is consistently nutritious, and is grazed more uniformly than forage that is grazed for a much longer period of time.

Size and Number of Paddocks. Given that short periods of stay are most effective for the productivity of livestock, and that undergrazing of a pasture can adversely affect forage quality, it follows that the size of the paddock must be adjusted to ensure that all forage in each paddock is grazed uniformly and completely during each period of stay. If a longer period of stay is required for grazing to be uniform and complete, this indicates that too much forage is available in the paddock, so the area may need to be subdivided or removed from the rotation and harvested as hay. If the period of stay becomes too short, this indicates that larger (or more) paddocks are needed. The sizes of paddocks may change over time as plant growing conditions change.

The number of paddocks to include in a rotational grazing program depends on both the length of the rest period and the periods of stay. In general, shorter periods of stay lead to higher forage yield and animal productivity, so productivity of the system increases as the number of paddocks increases. Other factors that affect a grower's decision about number of paddocks are:

- Topography of the land;
- Rest periods needed;
- Plants present in the pasture, and potential yield; and
- Financial constraints, including cost of livestock, fencing costs, and labor costs to move livestock (Murphy 1990).

Calculations. The best way to calculate the size and number of paddocks needed in your system is to start by estimating the rest period that your pasture will likely need during the *slowest* time of growth for your region (Murphy 1990). In some areas of the world, this rest period will be short (<18 days), but in others it may be much longer (>120 days). Remember that your pasture plants will need the most rest during times of slow growth. During times of rapid growth and shorter rest periods, you can use fewer paddocks and harvest the forage in the rest for hay or silage. Next, estimate the period of stay for your animals, remembering to keep this fairly short to maximize productivity. With these two estimates, the total number of paddocks needed will simply be the rest period/occupation period. For example, suppose that you estimate that your pasture needs forty-two days to recover during periods of slow growth. You plan to move your animals from one paddock to another every two days. The number of paddocks you will need is 42/2 = 21. You may, of course, use more than twenty-one paddocks and, in fact, you will not want to use Voisin's method with fewer than ten paddocks (Murphy 1990).

Now that you have estimated the number of paddocks that you need, the next step is to determine how large those paddocks should be. You can do this easily by simply dividing your total pasture land area by the number of paddocks needed to obtain an average area for each paddock (Murphy 1990). However, paddocks do not need to be equal in area, but should produce a fairly similar amount of forage within each area. If the paddock areas are too big for the number of animals, you can simply decrease the size of the paddocks to whatever size is necessary to reduce the amount of forage, or increase the stocking density. As stocking density increases, there will be more competition for feed, and less selective and more uniform grazing. However, if you let the density get too high, you will find that your periods of stay will shorten, and you may need to add more paddocks in order to provide the forage plants with adequate rest periods. Indeed, paddock size is not as important as providing an adequate rest period for plants within these paddocks (Murphy 1990).

Other Types of Rotational Grazing

In some cases, farmers may wish to allow some of their animals access to high-quality forage to meet high nutritional requirements, while limiting access to other animals. This can be accomplished by one of two methods. In some cases, farmers may allow animals with high nutritional requirements to have first access to a new paddock. After they have consumed a great deal of the high-quality forage, they are moved to a second field, and the animals with lower requirements are moved into the first paddock to finish grazing.

Creep grazing accomplishes the same general goal, but in a slightly different way. Farmers may choose to plant a higher quality forage (the creep pasture) immediately adjacent to the base pasture used by most animals. These pastures are separated by a gate that limits access to the creep pasture to those animals that can fit through a specially constructed fence. Usually, smaller animals that need a high amount of nutrients for growth have free access to the creep pasture, while larger animals that no longer need such high-quality forage are kept out of the creep pasture by a gate.

Summary

Animal production systems can be an important component of agricultural programs, provided that they are ecologically sustainable, and that the by-products of the production system do not produce an environmental hazard. Integrating animals into existing agroecosystems is an excellent way to cycle nutrients within a system, reducing fertilizer needs while preventing pollution from animal manure. If integration into cropping systems is not possible, rotational grazing of animals can reduce losses from the system and can keep productivity high. Rotational grazing ensures an adequate supply of a high-quality food for cattle, while keeping the pasture grasses healthy and in a continual state of growth.

Topics for Review and Discussion

1. As countries become more affluent, they begin to diversify their diets by adding meat and animal products. Explain why this may eventually lead to grain shortages in

many countries, and how retaining a largely vegetarian diet among their population may help many nations to retain some food security.
2. What problems are associated with livestock production in developing nations? In industrialized nations? Which do you think is more likely to cause potentially serious environmental destruction in the future? What might be done to alleviate these problems?
3. What are the differences among outdoor confined, indoor confined, and grazing livestock production systems? What do you see as the relative advantages and disadvantages of each one? Which system is most common in particular parts of the United States and why?
4. Explain why rest periods are so important in any kind of outdoor grazing system. What happens if the rest period is too short? Too long? How can you adjust rest period without continually increasing and decreasing the number of livestock animals you own if forage growth changes throughout the season?
5. What is desertification and why does it occur? What can be done to reclaim land that has been subjected to conditions leading to desertification?

Literature Cited

Durning, A. B., and H. B. Brough. 1991. Taking stock: Animal farming and the environment. *Worldwatch Paper 103*. Washington D.C.: Worldwatch Institute.
Larney, F. J., and H. H. Janzen. 1997. A simulated erosion approach to assess rates of cattle manure and phosphorus fertilizer for restoring productivity to eroded soils. *Agriculture, Ecosystems and Environment* 65: 113–126.
Miller, G. T., Jr. 1990. *Living in the Environment: An Introduction to Environmental Science*. Belmont Calif.: Wadsworth Publishing Co.
Murphy, B. 1990. Pasture Management. In *Sustainable Agricultural Systems,* eds. C. A. Edwards, R. Lal, P. Madden, R. H. Miller, and G. House, 226–237. Delray Beach, Fla.: St. Lucie Press.
Southwick, C. H. 1996. *Global Ecology in Human Perspective*. New York: Oxford University Press.
Voisin, A. 1959. *Grass Productivity*. New York: Philosophical Library. Reprinted 1988. Washington, D.C.: Island Press.
———1960. *Better Grassland Sward*. London: Crosby Lockwood and Son.

Bibliography

Altieri, M. A. 1995. *Agroecology: The Science of Sustainable Agriculture*. Boulder, Colo.: Westview Press.
Durning, A. B., and H. B. Brough. 1991. Taking stock: Animal farming and the environment. *Worldwatch Paper 103*. Washington D.C.: Worldwatch Institute.
Parker, C. F. 1990. Role of animals in sustainable agriculture. In *Sustainable Agricultural Systems,* eds. C. A Edwards, R. Lal, P. Madden, R. H. Miller, and G. House, 238–245. Delray Beach, Fla.: St. Lucie Press.
Soule, J. D., and J. K. Piper. 1992. *Farming in Nature's Image: An Ecological Approach to Agriculture*. Washington, D.C.: Island Press.
Tivy, J. 1992. *Agricultural Ecology*. Essex, U.K.: Longman Scientific and Technical.

Chapter 17

Political and Socioeconomic Issues of Agroecology

Key Concepts

- National agricultural support policies
- International agricultural policies
- On-farm agricultural/economic issues
- Use of natural resources and effects on the environment
- International conflict and social justice

Applying ecological principles to agricultural systems generally means making some changes to existing systems in an attempt to make them more like natural systems, and, thus, more sustainable over the long term. The ability of farmers to make these kinds of changes, however, depends greatly on the politics, social structure, and economic policies of the nation in which they live. In some cases, their ability to make changes may even depend on the economic policies of other nations that are perhaps thousands of miles away.

We live on a planet with a finite amount of many natural resources. While some resources are renewable (e.g., sunlight), other resources take so long to be replenished naturally that they are considered nonrenewable. Distribution of these resources across the globe is uneven, yet an equitable distribution of global resources would suggest that no individual, corporation, or nation should have the right to a continuously increasing share of the earth's finite resources. However, the rates of resource consumption vary greatly among different nations of the world. For example, a person born in the United States or Canada today will use far more natural resources over his or her lifetime than someone born in

Table 17-1 National commercial energy use (oil equivalent) in kilograms per person per year (1971 and 1993 data).

	Kilograms per Capita	
Nation	**1971**	**1993**
Albania	604	455
Australia	4079	5316
Bangladesh	18	59
Brazil	361	366
Canada	6233	7821
China	278	623
Egypt	200	539
Ethiopia	19	23
France	3025	4031
Iran	714	1235
Japan	2553	3642
Nepal	6	22
Nicaragua	248	241
Nigeria	39	141
Norway	3565	5096
Rwanda	11	27
South Africa	1993	2399
United States	**7633**	**7918**

Data from United Nations Development Programme, *Human Development Report* (1996)

Bangladesh. In fact, in the mid-1990s, 22% of the world's population was consuming 60% of its food and 70% of its energy (Table 17-1). Resource consumption rates continue to increase in most countries. Note that the *increase* in U.S. energy consumption from 1971 to 1993 (Table 17-1) of 285 kg per capita, while representing only a 3.7% increase in U.S. consumption, was still larger than the per capita consumption rate in Nicaragua in 1993. Consumption rate and population size create pressures on the supplies of resources required for agricultural production.

Both national and international politics are important to consider when designing agricultural systems. Unfortunately, political stability is an issue in some areas of the world. Political instability can lead to difficulties in obtaining necessary inputs, as well as in the ability to work the land properly and to harvest and market produce. Trade relations between nations can also influence marketability of products and volatility of the market.

Economic considerations are perhaps the most important factors to examine when making agricultural management decisions. Many nations protect their own farmers by various policies, taxes, and subsidies. In this way, governments themselves can encourage the internal production of various farming products, and can manipulate market prices by limiting imports of such products from other nations. Individual farming decisions are, of course, influenced directly by economics. Farmers will not choose to grow crops that are unprofitable, and may not be willing to make changes in production systems that will require changes in technology or in expertise because such changes can be costly.

As we consider the design and management of agroecosystems, socioeconomic and political influences on these systems cannot be ignored. If we lived in a perfect world, we could plant each field with crops that are well-adapted to each region, using technology that would be sustainable for centuries. Unfortunately, such a world does not exist, and we must often plant what is available, what is most profitable, and what will feed our population. Still, if we are to continue to feed our growing population over the long term, sustainability of production and resources will be of utmost importance (see Chapter 19), yet we must meet the challenge to design sustainable systems that work within our own political, social, and economic structures.

Agricultural Support Policies

The governments of almost all industrialized nations currently have some sort of agricultural support policies, or **subsidies,** in effect. Such subsidies are usually offered with the objectives of stimulating national economic development, helping the poor or rural farmer, or advancing national food and energy independence. Probably one of the most compelling reasons for offering subsidies is to guarantee domestic productivity in times of conflict or instability. The different policies in effect in different countries are all quite different, but all have the goal of manipulating the market to aid national and individual interests.

In some cases, farmers are guaranteed a minimum price on their farm products. If they are forced to sell their products in the market at a lower price, perhaps because a bumper crop yield increases the supply, the government pledges to make up the difference between the market price and a predetermined *support price* in cash (Table 17-2). Farmers are therefore always guaranteed a minimum amount on their yield, regardless of market demand. Domestic prices of products may also be fixed at an arbitrary amount, regardless of the price on the international market, and this difference between international and domestic price is picked up by the consumer. This is known as the **consumer subsidy equivalent,** and it has a much greater effect on poor citizens of a nation than it has on the rich. Since a larger

Table 17-2 Price supports granted on selected crops in the United States (1995 data).

Commodity	Price Support Granted on 1995 Crop (in millions of $US)
Barley	27
Beans, dry edible	0
Corn	1233
Cotton, upland	876
Peanuts	256
Rice (rough)	649
Sorghum, grain	13
Soybeans and other oil crops	900
Sugar	
Beets	795
Cane	170
Wheat	295

Data from USDA, *Agricultural Statistics 1998*

percentage of a poorer citizen's income is spent on food (richer citizens spend excess income on entertainment, conveniences, etc.), any increase in food costs will impact the budgets of the poor more than the budgets of the rich.

Another common support program in the United States is the *subsidizing of natural resources*. With this program, farmers are able to obtain resources produced and distributed by the government for prices well below market value. In California, for example, farmers purchase water for irrigation at less than 10% of what it costs for the government to provide it, and less than 1% of its actual market worth (Roodman 1996)!

Some countries, including the United States, take land out of production by paying farmers *not* to farm their land. While this often gives the agricultural land a much-needed break, it also decreases supply, which raises prices that are then passed on to the consumer. Many nations also have *import quotas,* which effectively block imported farm products from entering the country by heavily taxing all incoming products above a certain quota. For example, the United States has trade restrictions to prevent cane sugar grown in other countries from entering the national market, which artificially raises domestic sugar prices above international market prices (see Application 17-1). Such import quotas are quite controversial among economists in industrialized nations, with some advocating trade restrictions on farm products on the international market, and some opposed to such restrictions. Those opposed to government trade restrictions believe that if consumers can buy food more cheaply from foreign sources, they should be able to do so without penalty. Those in favor of such restrictions are concerned that opening the market will hurt the small farmers in industrialized nations, leading to the bankruptcy of many family farmers. While they acknowledge that many subsidy programs actually help large-scale farming operations more than small farmers, they do not see any other feasible option for aiding small farmers.

It is true that while agricultural subsidies in the United States are often argued to be for the direct benefit of small farmers, they favor large farmers more. Since most U.S. subsidies are based on quantity produced (on a per unit basis) rather than farm size, those farms that produce more (i.e., large-scale operations) actually benefit more from these subsidies than do small farmers. This is not to say that small farmers do not benefit, but the benefits go far beyond the group that the subsidies are aimed at. Some economists argue that if the goal is to keep small farms in business, it would make more sense to send payments directly to qualifying farmers, rather than base them on farm output.

Another commonly cited advantage of these agricultural subsidies is the agricultural self-sufficiency that results from production of large quantities of food products. This is particularly important in times of conflict or war, when nondomestic supplies of certain food products might not be available. However, some of these production subsidies have led to surpluses that may then go to feed livestock, or may be bought by the government and destroyed or dumped onto the international market. The addition of these products to the international market, of course, will lower global prices, which will do two things. First, it will make the difference between guaranteed prices and international market prices greater, actually costing the government more. Second, it will decrease profits for farmers in other nations. This is particularly true when a surplus of food products is combined with an *export subsidy* program. These programs subsidize many basic foodstuffs produced by industrialized nations, making the prices of these products even cheaper. This, in turn, increases supply and lowers prices in developing nations (see Application 17-2), which may not be bad for urban dwellers in these nations, but greatly harms the farmers and rural poor who must sell their goods for a great deal less than they normally would in order to compete in the local market.

Application 17-1 *Subsidization of Sugar Production in the United States*

Domestic sugar production (Figure 17-1) began in the 1700s when the United States government imposed strict tariffs on many imports, including sugar, to generate revenue for the young nation. A small sugar industry began, producing enough sugar to meet most domestic needs at a price that was substantially lower than the heavily taxed imported sugar. However, when world sugar production increased substantially in the 1930s, prices on the world

Figure 17-1 A collection of sugarcane varieties and germplasm in Florida.

(continued)

market were so low that even the strict U.S. tariffs were too small to protect domestic production. Sugar producers and refiners pressed Congress for protective measures to prevent their farms and refineries from going bankrupt.

The 1934 Sugar Act, which eventually became incorporated into the Farm Bill in the 1980s, placed a heavy tariff on sugar imported into the United States from other nations. The Agriculture and Food Act of 1981 provided for a domestic loan program that set a support price on sugar, and allowed growers to use their sugar crop as collateral in the event of a loan default. It also provided an import quota system to restrict competition from developing nations, and to ensure a high domestic price for sugar.

The United States is not alone in its sugar subsidies. Japan and most European countries also subsidize domestic sugar production, and these international subsidies are cited by U.S. sugar corporations as the main reason that U.S. subsidies should remain in place. But there are strong arguments both for and against subsidizing U.S. sugar production.

Response of Sugar Corporations and Their Supporters

The staunchest defenders of the sugar subsidies are the sugar companies themselves. They state that U.S. sugar prices cannot be compared to prices in other nations because many other nations also subsidize. They argue that they could compete fairly against farmers in other nations if other nations did not subsidize their farmers to grow more sugar than needed, and then sell their surplus on the world market at very low prices. If all nations eliminated their subsidies simultaneously, the United States could also eliminate its subsidies without fear of destroying domestic production.

The sugar corporations also point out that:

- Current U.S. sugar policies ensure a reliable supply and stable prices for North American consumers. To support this, they cite the large jump in sugar prices that occurred when the Sugar Act of 1934 expired in 1974.
- Consumers in the United States pay less for sugar than the amount paid in almost all other industrialized nations, including the European Union.
- Commercial sugar users in the United States can be assured of reliable deliveries and supplies as needed.
- Domestic sugar production provides over 400,000 agricultural jobs across the nation, and provides about $30,000,000 per year to the U.S. government to operate the sugar subsidy program.

Response of the Opponents of Sugar Subsidization

Why, then, are many people opposed to sugar subsidies in the United States? If the government price support ensures a reliable supply; steady, low prices; and provides so many jobs, why is there any argument at all about the current policies?

For one thing, most people in the United States view agricultural subsidies as a means of protecting small farmers from bankruptcy or financial problems. In practice, however, the government price support of sugar tends to support the large corporations of sugarcane and

(continued)

sugar beet producers and refiners (more than half of the benefits of the program are given to only a small number of growers).

Opponents of sugar price support also argue that:

- In most years, U.S. sugar prices are substantially higher than world sugar prices. A small quota is allowed to be imported duty-free, but anything beyond that is taxed heavily. Because of this, growing sugarcane or sugar beets is much more profitable than it would normally be.
- Most production in the United States is currently limited to only a few large corporations located primarily in Florida and Hawaii. Limited production means limited supply, which also increases domestic prices. The lack of foreign competition also increases prices because market forces are not used to determine cost.
- The Florida land on which sugar is grown was initially part of the Everglades wetland system, which was drained and sold to two large sugar corporations. The land is very productive, with good soils and high organic matter content, but any harmful runoff from fertilizers or pesticides goes into the Everglades natural system. Keeping price supports in place encourages sugar production in this environmentally sensitive area.
- Taxpayers bear the cost of higher prices for sugar and sugar-containing products used in the government's school lunch program, military galleys, and food stamp programs.

What Would Happen If Sugar Subsidies Ended?

Proponents of the sugar subsidies state that if sugar subsidies ended, one of two conditions would result. First, world prices for sugar could increase dramatically, and those prices would then be passed on to the American consumer. Second, if prices decreased, many American sugar producers and refiners would be forced out of business, causing the loss of hundreds of thousands of jobs. In addition, since a special tax is added to every pound of sugar marketed in the United States, the sugar industry pays more than $40 million per year to the U.S. government, a source of revenue that would be lost if U.S. producers and refiners were forced out of business.

Opponents of the subsidies argue that a 1995 report by the Food and Agricultural Policy Research Institute stated, after extensive studies of the sugar program, that the U.S. sugar industry would still thrive after the subsidies were eliminated. Only those companies that are most inefficient would be forced from the market by competition. If only the loan program was removed, and the import quotas were kept in place, sugar prices would decline only slightly. If the import program was removed as well, increased supply would decrease prices, and acreage planted to sugarcane and sugar beets would decrease somewhat. Sugar production would expand in many nations and world sugar prices would rise as foreign markets began to respond to the needs of the United States. This would aid the economies of sugar-producing nations that would benefit from a higher demand and price.

Application 17-2 *The Economics of Supply and Demand*

Consumer demand for any particular product is based on the price of the commodity. If the product is expensive, only a small proportion of the population will be able to afford it, so demand will be fairly low. If the product is relatively inexpensive, more of the population will be able to afford it, and demand will increase (Figure 17-2). When supply exactly meets demand (S_0) it fixes the price at a certain acceptable level (the highest price the producer can get and still sell all of his or her product) (P_0) and the quantity sold is fixed at Q_0 (the number sold at the price that consumers are willing to pay for the product). Consumer demand is based solely on price of the commodity, which, in turn, determines the quantity sold.

If the supply of the commodity increases for some external reason (e.g., good weather produces a bumper crop of corn in the midwest), the supply line of the graph will shift to the right (S_I). When this happens, the point where supply and demand intersect will be shifted as well. Since the price of the commodity is shifted lower, a larger proportion of the population will be able to afford the product, so demand will increase, and a larger quantity of that item will be sold. On the other hand, if the supply decreases (e.g., a freeze destroys a citrus crop in Florida), the supply line of the graph will shift to the right (S_D). This causes the price to increase, and decreases demand because fewer people will be able to afford the commodity.

On the international market, then, if an overproducing nation dumps its surplus on the market of another nation, supply in the receiving nation will shift to the right, which allows more people there to buy that product at a lower price. But it also decreases the value of the crop so local farmers will earn less from their crop. On the national market, even when supply of a commodity is very high (which would normally push prices lower), government subsidies can hold the price of that commodity at an artificially high level. As long as enough consumers can afford to pay that higher cost, producers will benefit from the subsidies because they will consistently receive a high price for their product.

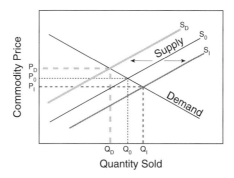

Figure 17-2 Economic supply and demand. Increased supply (S_I) will decrease price (P_I), which will increase consumer demand and therefore increase the quantity sold (Q_I). Decreased supply (S_D) will increase price (P_D), which will decrease consumer demand and therefore decrease the quantity sold (Q_D).

The primary purpose of price supports is to keep domestic prices high, even when international prices are low. This serves to both increase profits for domestic farmers, and to decrease domestic market volatility by protecting the national market from international fluctuations. Unfortunately, while national farming interests are being protected in the United States and other industrialized nations, farmers in developing countries face serious market volatility. Their interests may not be protected by their governments, which strive to keep prices low for their urban populations. Because of this, farmers in developing nations are quite susceptible to changes in market prices and productivity, brought on by such uncontrollable factors as weather changes or political stability. These farmers are further hurt when their products are excluded from the markets of industrialized nations by heavy tariffs and import quotas.

What would happen, then, if price supports were simply ended? Since all agricultural products would then be available on a global market, world agricultural prices would experience less fluctuation than they do now. In addition, world agricultural production would likely shift toward developing nations where climate is conducive to year-round crop growth and where labor costs are often less expensive. This would actually increase the economic base of developing nations because agriculture has a much greater share of national production in poorer, developing nations than it does in wealthy, industrialized ones. Still, such a shift would put many farmers in industrialized nations out of business, so the elimination of price supports in these nations is unlikely to occur in the near future. Still, even without ending price supports, industrialized nations could help developing nations to become more self-sufficient by simply eliminating export subsidies. This would encourage farmers in developing nations to produce and sell more, even if only in their own domestic markets.

One of the arguments against eliminating import quotas, and therefore increasing international trade in agricultural products, is that this would inevitably lead to a push for higher productivity in many developing nations. This could lead to even more monoculture of cash crops for export, with many of the disadvantages discussed previously (see Chapter 13). In turn, industrialized monoculture could lead to a heavier use of pesticides and herbicides, which may be potentially harmful to local environments. On the other hand, if the increased production was integrated into existing local farming systems, then as agricultural production shifted toward developing nations, worldwide chemical use may actually drop because industrialized nations tend to be the heaviest users of agricultural chemicals. Which of the scenarios would be more likely to occur? The answer depends on economics, since economics will always drive the market, and increased demand will encourage farmers to use whatever means necessary to increase their productivity.

In some cases, subsidies may encourage environmentally and ecologically destructive methods of agriculture. For example, subsidizing the use of agricultural inputs such as irrigation water, pesticides, and fertilizers can encourage waste and higher rates of use, leading to salinization, soil degradation, and pollution. Subsidizing livestock and crop production through such programs as low fees for grazing public lands or tax breaks for clearing land, encourages environmentally destructive methods of farming and overgrazing. When environmentally unsound methods are subsidized, it can be economically disastrous for farmers to adopt alternative farming methodologies based on ecological principles. It is clear that sustainable agricultural practices will be adopted only if the policies and economic structure are in place to promote them, and if farmers are not placed at a competitive disadvantage when they use these practices.

Economic Issues of Individual Farmers

While many agricultural management decisions are greatly influenced by government policies and national economic incentives, most decisions are made at an individual level, based on individual economic situations and on a farmer's willingness to take risks. Every farmer in every nation has to choose a particular mix of crops and the technologies used to grow them. This choice always involves a risk to the financial well-being of the farm. In the case of farmers in industrialized nations, entire life savings can be at risk. For farmers in developing nations, these choices may make the difference between feeding their families or allowing them to go hungry.

Many of the methodologies that have been discussed in this text will serve to make agricultural production more sustainable over the long term, but may involve a loss of yield or may necessitate a shift in technologies used or crops produced. Over the short term, this can provide serious disincentives for their adaptation. Shifting from one crop to another or from one farming system to another can be very expensive. Farmers must bear the cost of new machinery, related technology, and often the expertise that goes with these changes. In some cases, current economic incentives may favor continued large-scale production of a single commodity, so that farmers who choose to make the move toward sustainability and diversified cropping systems will find themselves at a serious economic disadvantage.

Small farmers in industrialized nations like the United States also face a number of external influences that encourage the growth of large-scale operations and subtly discourage new, sustainable production methods of agriculture on family farms. For instance, national policies and supportive funding are often geared toward developing technologies for large-scale farms. Many research dollars are aimed at increasing national productivity of various commodities, and many of the methods cited in published articles and extension bulletins emphasize practices that are more compatible with large-scale production rather than with small farming systems. Specialized knowledge is needed for the development and use of many alternative, sustainable farming practices, yet as pointed out in many places in this book, this knowledge base is deficient or lacking in many situations.

Can subsidies and national price supports help these farmers? After all, the entire rationale behind them is to reduce risk to small farmers. The answer is both yes and no. While the subsidies just discussed do help small farmers to stay in business and prevent them from bankruptcy or loan default, they may also subtly discourage the development of sustainable, small farming systems. Since the subsidies are based largely on productivity, farmers recognize that as their productivity drops, their subsidies will drop too. Such subsidies encourage small farmers to adapt the same techniques and technologies of large-scale farmers, but many small farmers simply cannot compete with large-scale producers. Their per acre costs of equipment and infrastructure may be much greater than those of the large producer who farms vast acreages of land.

Still, many of the small farmers could potentially find alternative markets for their produce. One example is the local distribution and sale of organic produce, which is a rapidly growing segment of U.S. agriculture today. Unfortunately, many local markets are not prepared to buy local produce, and actually may prefer to deal with wholesalers who buy from a number of farms and then distribute farm products from one centralized location. In some communities around the nation, however, citizens are forming small community-supported agricultural organizations. These vary in structure and size, but basically consist of community members buying or renting "shares" in a local farm, and then reaping the benefit of fresh produce as it becomes available. In this way, farmers can be assured of a local market, and may even have the freedom to experiment with organic production and other alternative cropping systems.

Agriculture and the Environment

According to many economists, one of the most serious problems of market-based economies is that the prices attached to various goods and services seldom reflect the costs of environmental damage associated with the production and marketing of the products. The cost of Chilean peaches that are sold in New York in February may be a bit higher than the cost of fresh Georgia peaches in July, but those prices may not reflect the true cost of the nonrenewable fossil fuels used to transport the peaches from South America to North American markets in the middle of winter. Currently, the prices of many nonrenewable resources such as fossil fuels are relatively low, but could be much greater in the future as demand depletes their limited supplies (see next section). For this reason, consumers make buying decisions that are based on prices that do not adequately reflect the long-term costs. If the true environmental and long-term costs were passed on to consumers, this may encourage environmentally sound choices and may serve to increase markets for local farmers when consumers decide that locally produced fruit is more economical than out-of-season produce shipped from other places. People who still decide to purchase this produce would be paying the full cost of resource usage that results from the transportation of the peaches from Chile to New York. The problem is that such long-term costs are very difficult to estimate, so it is difficult to implement economic schemes that do not involve current costs only.

The estimation of costs and environmental damage is a particularly complex and controversial area. If societies insist that people and businesses be held responsible for their own pollution and environmental degradation, and that they must pay the full cost of environmental damage, what would be the result? First, this would provide a financial incentive for producers to operate in ways that are far more environmentally sound. Second, the increased cost to consumers for products that hurt the environment would decrease demand for these products, and would encourage substitution, recycling, reuse, or conservation. In addition, taxes on activities would actually discourage those activities, and society could therefore choose to discourage certain activities by taxing them (e.g., taxes on cigarettes to discourage smoking). If taxes were increased on practices that are harmful to the environment, ecological destruction would be discouraged. However, the estimation of the true costs of environmental degradation to the present *and future* generations is difficult.

However, sometimes governments unwittingly subsidize environmental destruction, even in agriculture. Subsidies or payment programs that are based on productivity may encourage the overuse of fertilizers, pesticides, and irrigation water as growers strive to maximize short-term yields. Governments may also subsidize the use of natural resources like water, which encourages waste. Movement away from these practices toward programs that reward producers who reduce resource waste, pollution, and environmental degradation could encourage conservation of resources and eventually lead to more sustainable systems of production. National and global economics are important in making agricultural and environmental decisions, but sustaining agricultural production for generations also requires the maintenance of a healthy environment and the conservation of resources.

Using Nonrenewable Resources

If the market prices of all goods and services were to include the present and future costs of any pollution, environmental degradation, or other harmful effects passed on to society and the environment by the production, distribution, and marketing of the product, prices on many common goods would rise dramatically. If these rising prices were passed on to consumers, there would be serious incentive for individuals to make changes in their lifestyle, perhaps by changing various aspects of their diet, having fewer children, refusing to buy articles that are environmentally destructive, recycling, reusing, etc. These short-term changes in lifestyle would potentially have dramatic long-term effects on the economy and environment.

We can compare nonrenewable resources, such as fossil fuels, aquifer waters, and some sources of nutrients, to capital in a bank. Renewable resources, such as solar or wind power, rain water, and atmospheric sources of nutrients, are comparable to the interest that results from an investment of capital. As long as we are living off of the interest, and perhaps only occasionally dipping into the capital, the money we have in the bank will support us indefinitely. However, if we start using the capital regularly, making large withdrawals and not replacing them, we will eventually deplete the bank account, leaving us little to live on. The same is true for our natural resources. Using fossil fuels, waters in aquifers, and phosphate from deeply buried rocks is analogous to depleting the capital. These resources took thousands to millions of years to form, and once they are gone, they cannot be replaced quickly (see Application 17-3).

Currently, many people do not consider the long-term results of short-term actions. But will the decisions that we make now be good decisions for future generations? Or will unsound, short-term decisions lead to serious economic and environmental problems over the long term? Both the short-term and long-term economic consequences are important considerations in agricultural decision making and planning, especially if resources and agricultural systems are to be sustained across generations.

Social Issues and International Justice

Most people are well aware that, globally, we produce enough food for every individual on the planet to receive an adequate diet for growth and well-being. The problem is not so much with production as with distribution. Distribution is not equitable; some nations destroy food surpluses while citizens of other nations go hungry. When distribution of farmland is not equitable, increased population can eventually produce a multitude of people with no land to call their own (Durning 1989). As families divide their land among their children, plot size decreases so much that even subsistence production is no longer possible. Nevertheless, while inequitable distribution is a major reason for hunger to exist in the world, it is not the only reason.

Over the past several decades, agricultural productivity has increased tremendously in many of the industrialized nations of the world, but not in many developing nations. What is the reason for this difference? Most nations cannot afford to put a great deal of money into agricultural research or plant breeding programs that will improve national agricultural productivity. Those that can, such as the United States and Canada, make certain assumptions about how crops are grown, what technologies are used, what level of farmer expertise is available, and what crop varieties are adapted to the regions studied. They assume that a certain level of inputs will be used to produce a good yield, including irrigation water,

Application 17-3 *Phosphorus as a Nonrenewable Resource*

When most people think of nonrenewable resources and agriculture, they think of the fossil fuels that are used to produce pesticides, herbicides, and fertilizers that are applied to crops, or perhaps the fossil fuels used to power their tractors, plows, and other farm equipment. Some even realize that the water that is pumped from underwater aquifers is finite in its availability, and must therefore be conserved. But few people realize that one of the resources that is most threatened in terms of its availability is phosphorus.

The sustainability of an agricultural system relies on a delicate balance of inputs and outputs to the system. Some macronutrients can be added to the system naturally, and others, once present in the system, can be recycled and made available to plants under the appropriate conditions. Although required in large quantities by plants, phosphorus is notoriously unavailable in most agroecosystems (see Chapter 4). Any natural input of phosphorus to a system tends to be very slow, and occurs in very low amounts, so it is usually added to agroecosystems as a fertilizer.

Except in closed sustainable systems where organic fertilizers are used, almost all systems rely on inorganic sources of phosphorus made from rock phosphate. Rock phosphate is a nonrenewable resource that is slowly being depleted from the earth. Most of the farming systems in production today cannot continue generating high yields without the levels of phosphorus currently being applied. Continued reliance on the nonrenewable supplies of rock phosphate to keep systems productive may threaten the long-term sustainability of our systems.

---

inorganic fertilizers, pesticides, and herbicides. Sustainability may not be an issue, and many of the systems developed will be based on productivity and profitability only.

Often, this research cannot be used to improve agricultural programs in developing nations without considerable modification and adaptation. Many farmers in developing nations do not have a ready source of water, cannot afford or cannot obtain pesticides and fertilizers, and rely on organic fertilizers only. Methods and crop varieties developed specifically to improve profitability for farmers in industrialized nations may be very poorly adapted for other regions, and may actually decrease productivity. We cannot assume that what works in industrialized nations must be the solution to reducing hunger and increasing productivity in developing nations.

How, then, can we reduce this inequity in agricultural productivity among nations? Research must be designed specifically to address the problems of agriculture in the environments and agricultural systems that are present in developing nations. Research in agroecology, focusing on both sustainability and profitability under local conditions, can help the poorer people in rural areas of many nations. The design of ecologically sustainable systems that are profitable, and that work within the existing socioeconomic and political frameworks of developing nations, can help address local needs and lessen global inequities in food distribution. In time, small farmers in developing nations may eventually export their knowledge to small family farms and even large-scale operations in industrialized nations, reversing the usual flow of knowledge!

Research must also be aimed at the staple crops that are grown in many developing nations. While rice, wheat, and corn constitute major parts of many diets in the world, there are nations that rely more heavily on relatively obscure crops that have received very little research attention from scientists. However, development of a research program is expensive, and therefore may not be worth doing on crops with a very limited range of acceptability, adaptation, or economic value. Local governments may not be able to fund such research projects. The main issues are whether and how agricultural research in industrialized nations could benefit the people and countries that cannot afford to fund local research programs.

It is generally accepted that in many countries of the world, traditional methods of cultivation alone cannot support present population levels. In the past, industrialized nations have exported their technology and methodology to most countries in order to increase productivity. However, while scientific knowledge gained in industrialized nations may benefit developing nations, the use of that knowledge can (and should!) vary from country to country. The nonsustainable methods of industrialized nations usually cannot be moved intact to developing nations, but must be carefully selected and adapted to local conditions and agroecosystems.

Human Aggression: Competition, Conflict, and War

The need for food and water is the most pressing need that human beings face. When food supplies are threatened, or access to fresh water is denied, aggression is a very natural consequence. Many nations in the water-starved areas of the Middle East and East Africa have acknowledged that they would go to war if necessary to protect their water rights from being usurped by neighboring nations. As population in these areas continues to increase and as more water is needed to irrigate land to produce food to feed more people, the potential for conflict can only increase.

War and conflict, both national and international, can have devastating effects on the stability of food supplies. When farmers are uncertain whether they will be able to harvest what they plant, or when the farmers themselves are drawn into conflict, the food supplies of the nation become less stable. Access to local and international markets may also be denied because international sanctions can prevent all trade—both import and export trade—from occurring on the international market.

Even within a nation, conflict surrounding agricultural production can create unrest. For example, as cities continue to grow and more people move from rural areas to urban areas, the cities themselves will begin to compete with agricultural land for resources. In China, for example, water is being diverted from agricultural land to cities to meet the needs of millions of urban residents. Water from the Colorado River is used by residents of Los Angeles and San Francisco, often leaving only a trickle for farmers in northern Mexico (Postel 1996). In some cases, food prices are held down by government restrictions to allow urban dwellers to buy food at low cost, which hurts the rural farmer. These kinds of conflicts are likely to grow as population increases.

Summary

Most agricultural decisions are made based on the economic situation of individual farmers, which may be influenced by national policies of price supports or subsidies. Shifting from current resource-intensive agricultural practices to more sustainable production methods under the current economic climate in industrialized nations may actually result in economic hardship for the farmer or grower trying to make the shift. Long-term economic consequences and future value of scarce resources are often not considered in agricultural decision making. If the future value of natural resources is considered, it may be economically advantageous for farmers to shift to sustainable production methods, and to conserve valuable resources for future use. To sustain our current population as well as future generations, it is important to guard against overuse of resources, and to seek national and international economic and research policies that do not discriminate against small farmers in developing nations.

Topics for Review and Discussion

1. What are agricultural subsidies, and why do you think they were initially implemented in industrialized nations?
2. Choose an agricultural commodity that is produced in the United States and research its economic history. Are there national price support systems in place to protect farmers? How do national prices of that commodity compare with prices on the international market?
3. If an industrialized nation produces a surplus of a grain crop, and is unable to use all of it within its own borders, it may dump that surplus on the international market. From what you know about the concept of supply and demand, explain how this could directly affect small farmers in developing nations. Why do you think that international governments allow this to occur without protecting their own farmers with an import quota or price support?
4. Examine Table 17-1. Given that the average life expectancy for someone in the United States is about 77 years, and the average life expectancy for someone in Nepal is about 54 years, how much more energy will someone in the United States consume over his or her lifetime than someone in Nepal, assuming consumption rates similar to 1993 rates?
5. You are the agricultural advisor to the president of a developing nation located in tropical Africa. Your nation has very few economic resources to put into scientific experiments that will help to improve crop production, and the main crop in your country is not commonly grown in industrialized nations. What suggestions can you make to the president to increase long-term food security for your nation, and how will you recommend that these suggestions be implemented, given the economic condition of your country?

Literature Cited

Durning, A. B. 1989. Poverty and the environment: Reversing the downward spiral. *Worldwatch Paper 92.* Washington D.C.: Worldwatch Institute.

Postel, S. 1996. Dividing the waters: Food Security, ecosystem health, and the new politics of scarcity. *Worldwatch Paper 132.* Washington D.C.: Worldwatch Institute.

Roodman, D. M. 1996. Paying the piper: Subsidies, politics, and the environment. *Worldwatch Paper 133.* Washington D.C.: Worldwatch Institute.

United Nations Development Programme (UNDP). 1996. *Human Development Report.* New York: Oxford University Press.

United States Department of Agriculture. 1998. *Agricultural Statistics 1998.* Washington, D.C.: U.S. Government Printing Office.

Bibliography

Durning, A. B. 1989. Poverty and the environment: Reversing the downward spiral. *Worldwatch Paper 92*. Washington D.C.: Worldwatch Institute.

Finn, D. R. 1996. *Just Trading: On the Ethics and Economics of International Trade*. Nashville, Tenn.: Abingdon Press.

Hayami, Y., and V. W. Ruttan. 1985. *Agricultural Development: An International Perspective*. Baltimore: Johns Hopkins University Press.

Horwich, G., and G. J. Lynch, eds. 1989. *Food, Policy, and Politics: A Perspective on Agriculture and Development*. Boulder, Colo.: Westview Press

Madden, J. P., and T. L. Dobbs. 1990. The role of economics in achieving low-input farming systems. In *Sustainable Agricultural Systems,* eds. C. A. Edwards, R. Lal, P. Madden, R. H. Miller, and G. House, 459–477. Delray Beach, Fl: St. Lucie Press.

Penson, J. B., Jr., and D. A. Lins. 1980. *Agricultural Finance: An Introduction to Micro and Macro Concepts*. Englewood Cliffs, N.J.: Prentice-Hall.

Schusky, E. L. 1989. *Culture and Agriculture: An Ecological Introduction to Traditional and Modern Farming Systems*. New York: Bergin & Garvey.

Turner, R. K., D. Pearce, and I. Bateman. 1993. *Environmental Economics: An Elementary Introduction*. Baltimore: Johns Hopkins University Press.

Vickers, D. 1997. *Economics and Ethics: An Introduction to Theory, Institutions, and Policy*. Westport, Conn.: Praeger.

Chapter 18

Issues in Tropical Agriculture

Key Concepts

- Definition of the tropics
- Managing tropical agroecosystems
- Common tropical crops

Introduction

Tropical agroecosystems can be some of the most rewarding systems to work with, but also some of the most frustrating. While proper management can increase yields, improve local diets, and establish sustainability of agriculture in a region, improper choices can lead to economic disaster. While this is true in temperate systems as well, the tropics have their own set of problems, issues, and concerns. Monthly variations in rainfall mean that soils may be very dry for a good portion of the year. Soil acidity, aluminum saturation, and erosion can all be worsened with improper management. Warm temperatures throughout the year allow year-round crop growth, but also allow year-round increases in pest and disease populations. In order to develop a good tropical agroecosystem, it is important to take all of these factors into account. It is also critical to remember the importance of agricultural systems in the economics of tropical nations, where high-yielding export crops can be a major source of income.

While many countries of the tropics have large cities with some industry, economies are often largely based on agriculture. In fact, many nations rely on the income they receive from agricultural exports to finance any industrial advancements that occur. For this reason, the transformation from developing to industrialized nation can depend largely on agricultural productivity. Of course, any nation's agricultural system must first meet the needs of its own people before any products are made available for export. Unfortunately, population is growing very rapidly in the tropics, which means that more of the food that is produced locally needs to be used for local consumption, rather than contributing to a country's gross national product (GNP) as export commodities.

Definition of the Tropics

What are the tropics? The tropics can be defined in many different ways, according to such factors as geographical location, biological diversity, and climate. Geographically, the tropics are located between 23.5°N and 23.5°S latitude, and include approximately 25% of earth's land surface, and 40% of the total earth surface (including water). This region receives more than 50% of all rainfall that falls on the earth, and houses more than 75% of its people. It is estimated that more than half of all species of organisms on earth live in the tropics, in part because of the wide variety of niches and habitats available to both plants and animals.

Temperature is probably the most important factor that distinguishes lowland tropical regions from other ecological regions of the earth. The mean annual temperature in the tropics ranges from 20–23°C. The range in monthly temperature means may be as low as 2°C in humid areas, or 5–10°C in arid regions. Because of these relatively stable temperatures, plant growth can occur year-round. Of course, this means that insects and disease organisms can survive from year to year as well.

The amount of rainfall received and its seasonal distribution can also define tropical regions. The humid tropics are those areas where rainfall in the driest month is greater than 60 mm; arid tropics receive less than 60 mm rainfall in the driest month. Agricultural management choices will depend on the seasonal distribution of rainfall (are there one or two rainy seasons?), and on the intensity of the rainfall, since intensity will determine how effective the rain will be for plant growth and development. Very light, gentle rains will penetrate the soil and become available for plant uptake. Very heavy, torrential rains are more likely to run off the soil surface, often carrying soil particles and nutrients.

Most of the rain that falls in the tropics during rainy seasons is convectional precipitation. With convectional precipitation, warm, moisture-laden air rises and cools as it moves upward. Because warm air holds more moisture than cool air, moisture condenses out of the air as it cools, causing regional precipitation.

While temperature and moisture are important determinants of plant growth in tropical regions, solar radiation also plays an important role. The total energy received at a given latitude decreases with increasing distance from the equator (Table 18-1). Both altitude and regional cloudiness can also affect the quantity of radiation received at any given latitude. In the tropics, cloudiness and humidity play the most important roles in determining how much radiation reaches the earth's surface at any given location.

Table 18-1 Energy received at various latitudes compared to that received at the equator (0°).

Latitude	0°	10°	23.5°	40°	60°	80°
Energy Received (% of that received at the equator)	100	98	92	79	57	43

Day length also has a major effect on plant reproduction. Because seasonal variation in day length at the equator is very small throughout the year, plants that require very long or very short days to trigger reproductive cycles often do not produce well at these latitudes.

The Difficulties of Managing Tropical Agroecosystems

Tropical soils have a notoriously bad reputation for being difficult to manage. In truth, soils of the tropics vary as widely in their quality as do soils anywhere else in the world. In many areas, however, management is much more difficult due to poor nutrient status, low water-holding capacity, and high erosion.

A great deal of the parent material found in tropical soils contains large quantities of aluminum and iron. In areas of high rainfall, many other nutrients are leached from the soil, leaving behind aluminum and iron ions and decreasing pH. As the pH of the soil declines, H^+ reacts with the soil silicates, releasing Al^{3+} ions and increasing both exchangeable Al^{3+} and percent Al^{3+} saturation (percent of exchange sites within soil colloids that are occupied by Al^{3+} ions). As Al^{3+} saturation increases, the base saturation of the soils decreases, and nutrient levels decline.

In addition to the problems caused by increased levels of aluminum and iron, oxides of these elements can become concentrated in a layer just below the soil surface as a result of water movement in the soil. When vegetation cover is removed and subsequent erosion brings these layers to the surface, the sun can literally bake these soils to form a very hard substance known as laterite. Over time, laterite will eventually weather and form a very thin depth of soil that can support minimal vegetation, but the process essentially removes the soil from agricultural production for a very long time.

Liming soils to a pH of about 5.5 will prevent the buildup of aluminum in these soils, but the availability of lime or prohibitively high prices may keep this from being a viable option in many areas. Nevertheless, Al^{3+} saturation is a serious matter. Al^{3+} ions inhibit root growth of many plants, resulting in decreased ability to take up water and nutrients from soils that are already deficient in nutrients. Different plants vary widely in their tolerance for Al^{3+} saturation. Plants originating in dry climates tend to be more sensitive than those originating in more humid areas (rice is actually quite aluminum tolerant). In some of these high Al^{3+} areas, farmers may be tempted to plant legumes to add nitrogen to the soil and counter some of the problems of low nutrient status, but unfortunately, legumes are not very aluminum tolerant.

In the arid tropics, the effects of leaching are not nearly as pronounced. In these areas, salinity may be more of a problem than aluminum toxicity, and pH tends to be high (alkaline) rather than too low.

Nutrient Availability

Availability of nutrients may be directly related to seasonality of rains in the tropics. For example, the availability of nitrogen to plants increases immediately after rains begin, due to an increase in the activity of numerous soil organisms. During the dry season, however, less microbial activity occurs, so any nitrogen present in the soil may easily volatilize and become depleted in the soil. Planting dates can be adjusted to take advantage of these shifts in nutrient availability. The best planting dates for cereal crops are near the very beginning of the rainy season so that the plants can take full advantage of the increased nitrogen. However, since legumes are nitrogen-fixers (have a symbiotic association with nitrogen-fixing bacteria), their planting dates are much more flexible. Still, it is important that some nitrogen be present in the soil when the legumes are first planted for their uptake and use until they establish their symbiotic associations with nitrogen-fixing bacteria.

Slash-and-Burn Agriculture in the Tropics

We discussed the basic concepts of shifting agriculture (slash-and-burn) in Chapter 13. Because this method of production is most commonly practiced in tropical regions around the world, it is important to look at how the climate of the tropics influences the sustainability of these systems. Recall that the basic steps to shifting agriculture were the clearing of a forested region, followed by the burning of the felled vegetation to shift nutrients from plants to soil, followed by cultivation until nutrient levels become too low to maintain good productivity levels.

In the humid tropics, the nutrients that enter the soil following burning are fairly loosely held by soil particles, leaving them quite susceptible to leaching. Initially after burning, both calcium and potassium levels in the soil increase as they are incorporated into the soil from the vegetation. As the base saturation increases as a result of this incorporation, Al^{3+} saturation decreases as Al^{3+} occupies fewer exchange sites in the soil. After burning, potassium declines more rapidly than does calcium, since calcium is held more tightly by the soil. However, over time, base saturation will continue to decrease as plants take up nutrients and as rainfall leaches them from the exchange sites, and the bases are often replaced by aluminum or hydrogen ions. As this occurs, crop yields begin to decrease dramatically until the land can no longer be worked productively. In general, the more rainfall an area receives, the faster nutrient status will decline.

In addition to changes in nutrient status, the water infiltration rate of soils that have been cleared of natural vegetation often declines markedly. Infiltration declines due to compaction because rain will often break up aggregates and cause changes in soil structure. Changes in structure will, in turn, affect the number and size of pore spaces in the soil, causing water to be held more tightly and infiltration to be much slower. If this occurs, water cannot penetrate the soil and will instead run off the surface, taking soil particles with it. Much shifting cultivation in the tropics is done on slopes and mountainsides, so this erosion can be a very serious factor, carrying away both soil particles and precious nutrients that are already quite scarce.

Effects of Temperature

As temperature increases, the rates of many biological, physical, and chemical processes in the soil increase as well. Within a limited range of temperatures, every 10°C increase in temperature will cause the rates of many processes to double. For this reason, rates of processes such as decomposition and mineralization are very high in most tropical regions. However, because of the direct solar radiation received in these areas, and the seasonality of rainfall, temperatures of soils may get warm enough to cause problems, particularly at the soil surface. When soils become too warm, beneficial microorganisms can be killed, and plants themselves may be seriously damaged by the heat. This can be prevented by mulching soils well, as discussed in Chapter 14, which will not only decrease soil temperature, but will also help to maintain adequate soil moisture.

Because temperature tends to be rather constant year-round in tropical regions, there is no severe winter freeze to break pest cycles, as occurs in temperature regions. Weed control is also more difficult in the tropics because the combination of more light, more diversity, more heat, and more animals to move seeds around favors weed establishment and growth. The same is true for many plant diseases. Conditions in the tropics are perfect for dissemination and establishment of plant diseases. For all of these reasons, integrated pest management is a particularly important strategy in tropical agroecosystems (see Chapter 11 for further discussion of the benefits of IPM).

Choosing Plants for Tropical Systems

Choices of plants for cropping cycles in temperate systems are based largely on fluctuations in temperature throughout the year. In most temperate regions, farmers add a soil-regenerating crop to their rotation, simply because cold weather forces them away from the cash crops that typically dominate their systems. In the tropics, where monthly fluctuations in temperature are minimal, farmers can grow a variety of cash crops year-round. The rotation cropping choices they make are often determined by fluctuations in rainfall, rather than by changes in temperature.

Of course, different regions in the tropics experience different rainfall patterns, and farmers must make cropping choices based on the specific pattern in their region. Many of the typical crops that are grown in the tropics can tolerate drought for one or two months of the year, while others require a good amount of moisture spread evenly throughout the growing season. With knowledge of the moisture needs of plants, growers can select a rotation pattern to match expected rains throughout the year. With knowledge of the light needs of plants, growers can choose which crops will work well together in an intercrop.

An incredible number of crop plants can be grown successfully in the tropics. This success rate is related directly to the wide variety of climatic conditions in the region and to the common cultivation practices used. Climatic conditions vary so widely in the tropics (arid, humid, high altitude, lowlands) that almost any crop from anywhere in the world can be grown there. For instance, cool-weather crops like potatoes and alfalfa can be grown in the high-altitude regions of the tropics. There is no danger of frost, so many perennial crops that are limited in distribution by cold weather can be grown in lowland tropical areas (e.g., bananas, cocoa, mango, papaya, etc.).

The history of plant cultivation in tropical regions of the world goes back 10,000–12,000 years, and all centers for early agricultural production lie there. For this reason, plants grown in this zone have faced centuries of selection and adaptation, making them very well suited for the region. Cultivation practices for many rural farmers have changed only gradually in generations, so the plants themselves are well-adapted to the methods used and thrive under typical management strategies.

Tropical Crops

There are a variety of familiar temperate crops that are also very important in the tropics and subtropics. These include corn, peanut (groundnut), soybean (soya bean), and many fruits (e.g., watermelon, strawberries, citrus, cantaloupe) and vegetables (e.g., tomato, onion, asparagus).

Peanut (*Arachis hypogeae*) and soybean (*Glycine max*) are particularly important in tropical rotations because both are leguminous plants, and can therefore increase the nitrogen content of the soil if the residues are worked into the soil following harvest. In addition, both can tolerate some periods of dryness, and both are high in protein, which makes them excellent foods for both humans and animals (see Application 18-1). Peanut is an annual plant that is produced for oil, nut, cake, and meal. While it needs warm temperatures for germination and growth, it does not need much sun, which makes it an ideal legume to include in mixed cropping systems. Peanut does best in well-drained, light soils with a pH between 6.2 and 7.5. Since the entire plant is taken from the field at harvest, the quantity of nutrients removed from the field with each peanut crop is high and, for this reason, peanut should always be grown in rotation to lower fertilizer requirements. Soybean is an annual plant that is very important in many areas of South America, Africa, and Asia. It is most productive in warm (optimum = 24–25°C), humid areas, but is very sensitive to photoperiod, although some soybean varieties that bloom and produce well under tropical daylength conditions have been developed. Some soybean cultivars do quite well in acidic or alkaline soils, but most prefer a pH of 6–6.5. Soybean is used in the production of many oils and margarines, as well as for the production of protein-rich soya milk and tofu.

While many temperate crops are equally important in the tropics, some crops that are grown in temperate zones are much more important in this region. Sorghum, sweetpotato, and cowpea are all produced in the United States, but each of them plays a much more important role in tropical systems.

Sorghum (*Sorghum bicolor*), which is used as a fodder in the United States, is an important food for both humans and animals in tropical Africa, India, and China. It is usually used for flat breads and mash, and also plays a role in brewing beer. Sorghum (Figure 18-1) is a good source of starch, oil, fodder protein, wax, and dyes. Dried stems and roots of the plants can be used as fuel, and the stems are sometimes used for construction of houses. Sorghum is often grown for a single harvest, but some cultivars may produce a second, or *ratoon,* crop. It prefers high temperatures (optimum = 27–28°C), and has good heat resistance, making it an excellent crop for very hot areas. Water requirements vary with cultivar; some are drought tolerant and others handle wet soils well. On average, plants need about 500–600 mm of rain for highest yields, and prefer a soil pH between 5–8.5. Sorghum is quite good at extracting nutrients from the soil, and fertilizer may not be necessary. However, it can remove nutrients from the soil that would be required by subsequent crops.

Sweetpotato (*Ipomoea batatas*) is one of the most important crops in China, and is also produced throughout Asia, Africa, and the Americas. It is an important source of vitamin A in many developing nations, and the overall nutritive value is quite high. Sweetpotato grows best at moderately warm temperatures (optimum = greater than 18°C), so it has a fairly wide range of distribution, all the way from lowland areas to 2500 meters above sea level. Most cultivars can survive extended periods of drought, but maximum productivity occurs when rainfall is spread evenly throughout the growing period. Sweetpotato prefers full sunlight for highest yields, and does best in light, well-drained soils. The crop cannot tolerate acidic soils (pH<5) or salinity. All parts of the sweetpotato plant are used in most regions where they are produced. In addition to the edible root, the leaves are often eaten by humans as a high-protein vegetable, and all parts of the plant can be used for animal food.

Cowpea (*Vigna unguiculata*) is one of the most important crop plants in West Africa. It is a good source of protein and starch, and is used not only for feeding humans, but also as a fodder plant and a ground cover to protect the soil. As a legume, cowpea adds nitrogen to the soil through its association with nitrogen-fixing bacteria. The plant makes very few demands in terms of soil or climate, so it is ideal for an area with poor soil and unstable

Application 18-1 Meeting Protein Needs With Plant Sources

Globally, we produce plenty of protein from plant sources alone that could be used to meet the protein needs of the world's population. As with all other foods, the problem is distribution and use rather than production. For this reason, diets are protein deficient in many countries of the world. This is especially common in the humid tropics, because the staple diets in this region consist primarily of starches (cassava, plantain, corn), and contain very little protein. Animal foods are not very common, particularly to the rural poor. Consequently, there is increasing interest in growing food legumes in tropical regions. Not only do these legumes have the advantage of being high in protein (soybean is rich in all essential amino acids), but they can also be grown in many nutrient-poor soils because they establish symbiotic associations with nitrogen-fixing bacteria. They also provide a good rotation crop to be grown with cereals, and their residues make a valuable, high-quality animal feed. Unfortunately, there are a few constraints to legume production in the tropics. While they do very well in poor soils once they are established, germination rates are often quite low, which can lead to poor stands. Legumes are also less efficient at intercepting light than are cereals, and are often more susceptible to pests and diseases. Nevertheless, if a good stand of legumes can be established, they can go a long way toward alleviating protein deficiencies in many areas of the tropical world (Table 18-2).

Animal sources of protein are superior to plant sources (with the exception of soybean) because they contain sufficient amounts of all essential amino acids. Plants, on the other hand, tend to be low in one or more essential amino acids (cereals are deficient in lysine; roots, tubers, and starchy fruits are deficient in methionine; legumes are deficient in methionine and cysteine), but these deficiencies can be balanced by combining two or more different plant foods in a diet. Combining foods like rice and beans, or corn and beans, can be a delicious and inexpensive way of obtaining all essential amino acids needed in a diet, without ever consuming animal products directly. Many "typical" diets in the humid tropics consist of these kinds of food combinations, indicating that people have been eating complementary proteins for centuries, before we ever knew what essential amino acids were!

Table 18-2 Average nutritional composition of legumes and animal products in tropical nations.

Crop	Water (%)	Energy (kcal/100 g)	Protein (%)	Carbohydrate (%)	Fat (%)
Legumes					
Cowpeas	12	340	23	61	1.6
Beans	12	340	20	63	2.2
Pigeon peas	12	340	20	63	1.2
Peanuts	12	520	21	22	43.0
Soybeans	12	390	35	31	18.0
Animal products					
Fish		73	17	0	0.5
Lean beef		202	19	0	14.0
Chicken egg		158	13	0.5	11.5

Adapted from Aykroyd, Doughty, and Walker 1982; O'Hair 1984; Platt 1976.

Figure 18-1 The seed head of sorghum is used to produce many different products, including oil, starch, and fodder.

climate conditions. Cowpea is particularly well adapted to West Africa because it is very tolerant of drought and can do quite well on acidic soils.

Tropical Fruits

There are many other crops that are produced only in tropical regions of the world, due to demands for high temperatures and intolerance of frost, or both. Perhaps the most famous and economically important exports among tropical crops of the world are the fruits.

Banana/plantain (*Musa* spp.) (Figure 18-2) is cultivated throughout the tropics and in some subtropical areas of the world. The main exporters to temperate regions are in Central and South America; production in other tropical areas is mainly for local consumption. Banana is a herbaceous perennial that cannot tolerate frost conditions. The plant requires a very warm climate, with plenty of sunshine and rainfall that is evenly distributed throughout the year. Soils used for banana cultivation should be well-drained, rich in organic matter, and should not be compacted. Banana prefers a soil pH between 5–7; fertilizer addition is almost always necessary for maximum productivity. While weeds can be a problem in banana plantations, it is not recommended that the soil be worked after planting because further cultivation can damage the feeder roots of the plant that grow close to the soil surface. Instead, mulching is often used for weed control. Propagation of banana is by suckers, which come off of underground rhizomes of existing plants. Plantain is a very important food in many tropical areas of the world, and is actually a staple in many areas of Africa and the Americas. Cultivation of plantain is identical to that of banana, but the fruit itself is usually cooked (fried, boiled, roasted), rather than eaten fresh. In some countries, the leaves and other waste of banana and plantain are used for animal fodder after the fruits have been harvested.

Citrus (*Citrus* spp.) (Figure 18-3) is produced in all parts of the world, especially subtropical regions, including the southeastern United States. The crop is grown for fresh fruit, juice, and concentrates, as well as for many oils and dried products. All citrus fruit trees require good sun and evenly distributed rainfall, but many cultivars can tolerate up to two months of drought and even very mild frost. Soil used for citrus production should be deep and well drained, with a pH between 5 and 7; citrus usually needs good soil fertility for good productivity. Propagation of citrus is usually accomplished by grafting onto rootstocks that are well adapted for the climate and soils that they are being grown in; many varieties can be grafted onto one tree. Most healthy citrus trees usually produce well for twenty to forty years before they need to be replaced.

Guava (*Psidium guajava*) (Figure 18-4) is grown primarily in the tropics of Central and South America. It can be grown anywhere from the humid tropics to subtropical areas, as long as no severe frost occurs. It is a small tree that is usually kept pruned to shrub size for ease of harvest, and is vegetatively propagated by methods such as budding, grafting, and layering. Guava prefers deep soil with plenty of water and a high level of nutrients. It produces a fruit with a thick flesh of uniform color (ranging from white to red, depending on cultivar) that is very high in vitamin C. While the fruit is seldom eaten raw, it is a popular addition to canned goods, and is commonly used for juice and nectar. The leaves of guava have been shown to have medicinal use for digestive problems.

Mango (*Mangifera* spp.) (Figure 18-5) is easily the most important tropical fruit after banana, but it is not often exported because it bruises readily. Mango is grown in all warm countries. India is the major producer, but mango can also be grown in the cooler subtropics because it can tolerate light frost and wet conditions. The trees are drought tolerant, and

Figure 18-2 Bananas are a commonly grown fruit in tropical regions. A sucker, removed from the base of a clump of bananas, can produce bananas within seven to nine months after planting under tropical lowland conditions. Production takes longer at higher elevations and in more temperate climates.

Figure 18-3 Limes in south Florida. Limes are grown in all tropical and many subtropical areas of the world, including California and Florida in the United States.

a dry period during flowering can actually encourage flowering and fruit setting. Propagation of mango is done by grafting and budding. Mango is most commonly grown for fresh or canned fruit, or for use in juices. Anthracnose is a common and important fungal disease of mango, which appears as black spots on the fruits.

Papaya (*Carica papaya*) is distributed throughout the humid tropics and subtropics, but is most commonly grown in Central and South America and in Southeast Asia. Papaya cannot tolerate frost and is very sensitive to wind. The fruit is easily bruised. It needs a great deal of water, distributed evenly throughout its growing season; if dry periods occur, mulch is recommended to retain as much moisture as possible in the soil. Papaya is a **dioecious** plant; both male and female plants are required for fertilization. Some cultivars contain hermaphrodites, with both male and female flowers on the same plant. Papaya is cultivated by seed, and plants are usually kept in production for three to five years only. Almost all papaya is consumed locally as fresh fruit, although some juice is exported. Residues can be used as animal fodder, and leaves are occasionally cooked and used as a vegetable.

Pineapple (*Ananas comosus*) is grown in almost all tropical regions, but is perhaps most common in South America. It is a very hardy plant that can handle daily variations in temperature; short, mild frost; shading; and long periods of drought. Pineapple is one of the few xerophytic crop plants in the world, using CAM metabolic processes that allow the plants to fix carbon at night and close their stomates during the day. Despite its drought tolerance, pineapple still requires approximately 1000–1500 mm of rain during its development. While

Issues in Tropical Agriculture ■ 371

Figure 18-4 Guava is a popular fruit in Central and South America, but is relatively uncommon in the United States.

Figure 18-5 Mango is an extremely popular tropical fruit that turns various shades of red as it matures. While these fruits bruise very easily, the demand in the United States is increasing, so they are becoming a more common sight in produce sections of large grocery stores.

yield in dry areas can often be improved with irrigation, pineapple is very sensitive to excessive moisture. For this reason, soil used for pineapple cultivation should be well drained. It should have a pH between 5 and 6, and may require fertilizer for good productivity. Pineapple prefers high temperatures without a great deal of fluctuation throughout the year, but cooler temperatures at night are good for flowering. It can be grown under trees, but full sun gives higher yields. If grown under full sun, the fruit must be protected from sunburn if it is not standing upright. Pineapple is vegetatively propagated by using lateral shoots and suckers from intact plants. Many of the fruits are consumed fresh locally, while others are exported for canning and juicing. Crop remnants of pineapple plants make good animal fodder.

Other common fruits grown in tropical regions include avocado (*Persea americana*), cantaloupe (*Cucumis melo*), date palm (*Phoenix dactylifera*), fig (*Ficus carica*), grapes (*Vitis* spp.), jackfruit (*Artocarpus heterophyllus*) (Figure 18-6), litchi (*Litchi chinensis*), mamey sapote (*Calocarpum sapota*), pome fruits, pomegranate (*Punica granatum*), starfruit (*Averrhoa carambola*), stone fruits, strawberries (*Fragaria* spp.), and watermelon (*Citrullus lanatus*).

Root and Tuber Crops

Edible aroids (*Colocasia* spp. and *Xanthosoma* spp.) include a variety of related plants with "elephant-ear" leaves (Figure 18-7). They are called by many different common names in different geographical areas. Species in the genus *Colocasia* are called taro, dasheen, or eddo, while those in the genus *Xanthosoma* are known as cocoyam, malanga, or tannia. All of these plants are cultivated for their large, starchy corms and related structures. Some species are adapted to dryland conditions while others thrive in extremely wet habitats.

Cassava (manioc) (*Manihot esculenta*) (Figure 18-8) is a very common crop in Central and South America and Africa. It gives the highest yields of all tropical root crops, yet requires only minimal cultivation. Cassava is a short-lived perennial shrub that prefers very warm temperatures and rainfall levels between 1000–2000 mm/year. Nevertheless, it is a drought-tolerant plant, so it can be grown in regions with seasonal rains. It can also tolerate wet conditions, provided the soil is well drained. Cassava prefers full sun, and does best under short-day light regimes. The crop is vegetatively propagated by cuttings, which are usually planted within the top 15 cm of soil. Deeper cultivation causes the roots to develop deeper in the profile, which makes harvesting more difficult. Cassava also has its own built-in storage mechanism; roots can stay in the ground for several years without damage. In addition to the starch provided for human diets, cassava plants can also serve as animal fodder, and the leaves of the plant are used in some regions as a vegetable.

Yam (*Dioscorea* spp.) is grown throughout the tropical and subtropical world, and is a staple food in many countries of tropical west Africa. Almost all yam cultivars are climbing plants. The yams are the storage organs of the plant, which are grown underground. Yams prefer the hot, wet tropics, but tolerate some drought and can be grown in areas with some months of dryness. The plant is vegetatively cultivated using small tubers or pieces of tubers, which are commonly planted on ridges. Yams have very thin stems, so stakes are usually used to support the shoots and leaves of the plant. Because of this, proper cultivation of the plant requires a great deal of manual labor.

Other important root and tuber crops of the tropics include sweetpotato and taro (as discussed previously), potato (particularly at high altitudes), and arrowroot (*Maranta arundinacea*). The nutritional compositions of these important root and tuber crops are given in Table 18-3.

374 ■ Chapter 18

Figure 18-6 Individual jackfruits can weigh between 9 and 27 kg, and the record is 50 kg (Purseglove 1974). Would you park under this tree?

Figure 18-7 Malanga or tannia (*Xanthosoma* spp.) requires substantial sunlight for good productivity, although it can tolerate light shade. The corms of these plants provide a good supply of dietary starch.

Pulses

Most species of pulses are adapted to a wide variety of ecological conditions, including the climate of the tropics. Perhaps the most important factors in regard to tropical cultivation of these plants is that they have extremely low water requirements and many are legumes, which are able to improve soil fertility.

Lentil (*Lens esculenta*) is one of the most nutritious pulses grown anywhere in the world, and is a dietary staple in many parts of the Middle East and Southeast Asia. It prefers slightly cooler temperatures than those preferred by some pulses, and is commonly grown at high altitudes in tropical regions. Lentils provide a particularly good protein source, and the crop has very high economic value as an export. The crop residue makes excellent animal fodder.

Pigeon pea (*Cajanus cajan*) is a very important pulse in India, and is grown in many other tropical and subtropical regions. It is a perennial plant that does well under short-day light regimes. It is a drought-tolerant plant with a deep root system, so it can be grown under semiarid conditions. Pigeon pea makes an excellent addition to tropical diets because of its

Figure 18-8 The leaves and branches of this cassava shrub provide a high-quality fodder for animals, while the roots can be eaten as an excellent source of dietary starch.

Table 18-3 Average nutritional composition of major root and tuber crops in tropical nations.

Crop	Water (%)	Energy (kcal/100 g)	Protein (%)	Carbohydrate (%)	Fat (%)
Cassava	64	142	1.1	34	0.3
Cocoyam	64	137	2.2	32	0.2
Potato	78	82	1.8	19	0.1
Sweetpotato	71	115	1.5	27	0.3
Taro	74	98	2.0	22	0.1
Yam	73	103	2.5	23	0.3

Adapted from O'Hair 1984; Platt 1976.

high protein content (about 20% protein), and the residues provide good fodder for animals. Pigeon pea is commonly used as a green manure to add nitrogen to soils.

Broadbean (*Vicia faba*) is a common crop in the Middle East and Southeast Asia. It can be grown in almost all tropical regions, and is a dietary staple in many countries because of its high protein content (20–35% protein). It is also an excellent animal fodder.

Chickpea (garbanzo bean) (*Cicer arietinum*) is grown primarily in subtropical and humid tropical regions for local consumption, but it also has excellent export value. Like other pulses, the residues provide good animal fodder.

Other Common Tropical Crops

Cacao (cocoa) (*Theobroma* spp.) is grown in Africa and South America, and its production is steadily increasing in Southeast Asia. This perennial tree crop grows best under humid tropical conditions with warm temperatures (25–28°C) and moderate rainfall (1500–2000 mm/yr) evenly distributed throughout the year. While cocoa is most productive under full sun conditions, it does not need full sunlight and does very well as a secondary crop under shade trees. If grown under full sun, the nutrient level of the soil needs to be carefully managed because the higher yields require more nutrients. A good way to maintain soil nutrient level without a great deal of inorganic fertilizer inputs is to return the cocoa bean shells to the soil as mulch, which will also serve to increase the organic matter content of the soil. The soil in which cocoa is grown should be well drained, but with good water-holding capacity. Since it does well in fairly acidic soils, the pH should lie between 4 and 7.5. Since many cocoa roots are found in the upper 20 cm of the soil, mechanical cultivation around the trees must be limited to prevent damage. While cocoa flowers throughout the year under warm and humid conditions, there is still a distinct seasonality to harvesting the beans because flowering is physiologically halted while fruits are developing. Pollination of the flowers is necessary for fruit development to occur, and this is often done mechanically to ensure high yields.

Coconut (*Cocos nucifera*) (Figure 18-9) is a perennial tree crop that is grown in all parts of the tropics. It prefers an average annual temperature of about 27°C without major fluctuations in day/night temperatures. For this reason, coconut is commonly grown at sea level, rather than at altitude. As long as sufficient groundwater is available, coconut can grow well even in dry regions of the tropics. Coconut palms are a common sight near the oceans of

Figure 18-9 Coconut is grown throughout the tropics, and is a common sight along beaches because it can tolerate slightly saline soils. The coconut itself can be used for a variety of purposes, and is often sold as a drink in its own natural container at local markets.

tropical countries. The plant prefers full sun, can tolerate soil salinity, and some cultivars need chloride for growth and development. As with cacao, coconut husks can be returned to the soil to maintain soil fertility, or legumes can be planted around the trees to increase soil nutrient levels and productivity of the coconuts. While most coconuts are produced locally for regional consumption, high-value export products from coconut include oil, flakes, milk, and whole coconuts. In addition, the plants can be used as animal fodder, and the leaves and trunks are often used locally as building materials.

Coffee (*Coffea* spp.) is a common perennial bush crop in Africa, Central America, South America, and Southeast Asia. It is well adapted to cooler temperatures (15–24°C) in the tropics, so it does best in the highlands and some subtropical regions. However, it cannot tolerate cold temperatures and is killed by frost. The crop is drought tolerant. Yield is actually improved with two to three dry months, so it is an excellent crop for areas with seasonal rainfall. Like cocoa, coffee tolerates shade very well, but yield is higher in full sun provided enough nutrients are available. Coffee should be grown in deep, well-drained soil with a pH between 6–6.5. The crop is usually grown from seed, but can occasionally be vegetatively propagated. Before use, the harvested berries must be processed, and the residues of the processing make a good fertilizer/mulch for the bushes. While the coffee may be consumed locally, it is in very high demand around the world, so much is exported to other countries. The crop residues of coffee plants can be used as fuel or animal fodder, and the leaves of the bushes are used to make tea in some areas of the world.

Rice (*Oryza* spp.) is grown in the warmer regions of almost all continents, but most of the world rice crop is produced in Asia. Some cultivars are grown in wet conditions (paddies), while others have been developed for upland (dry) conditions. However, the water needs of dryland rice are still quite high (more than 800 mm/season, evenly distributed), so irrigation is often necessary for high yields. Wetland rice is one of the most important agricultural crops in the world because it thrives in flooded, low-oxygen conditions. Very few plants can do this, so many of the agricultural lands on floodplains in Southeast Asia would be completely unproductive if rice was not grown there. Fields planted to wetland rice must be leveled initially and the ground must be compacted prior to cultivation to allow for even water distribution on the surface of the soil. Rice prefers high temperatures (30–32°C), and cannot survive even a very mild frost. Yield is directly related to the amount of sunlight a crop receives, so rice grown in a region that is often cloudy shows reduced yield compared to rice grown in full sun. Soil pH should fall between 4.5 and 8, but rice has few soil demands beyond that, though fertility levels should be fairly high. Rice is the only arable crop that has been grown for thousands of years as a monoculture without serious consequences for the land. It can be grown in rotation with sugarcane, pasture crops, legumes, and cotton, and many growers rotate rice production with fish culture, which adds further nutrients to the soil. Rice residues make a high-quality animal fodder, and the straw itself has many uses, including the production of rice paper.

Sesame (*Sesamum indicum*) is grown in many areas that receive seasonal summer rainfall. The largest producers of sesame are currently Southeast Asia and some countries of Africa. Most cultivars of sesame are day-neutral, but some are short-day plants. They tend to be less demanding than peanut in terms of soil demands and labor, but yields tend to be low. Sesame is usually planted in dry regions on poor soils that cannot be used for other crops. Fertilizer is seldom added to the soil. Oil from the seeds is extracted by pressing, and the resulting product is the highest-priced of all edible plant oils. The seeds themselves are used locally or exported to make tahini or to use in baking. The leaves are sometimes cooked as a vegetable.

Sugarcane (*Saccharum* spp.) (see Figure 17-1) is grown throughout the tropics and subtropics. Different cultivars are adapted to very different conditions. While some cultivars are tolerant of cool temperatures, sugarcane generally prefers a warm temperature (25–26°C) because crop growth slows when it is cooler. Sugarcane has a fairly high rainfall requirement (1500–1800 mm/yr), but a two- to three-month dry season before harvest causes growth to halt and increases the sugar content of the plant. Productivity is best on heavy, fertile soils because the nutrient requirements of cane are quite high. However, nutrients can usually be returned to the soil easily because the sugar extraction process simply removes sugar (not nutrients), leaving behind residues that can be returned to the soil as a highly nutritious organic mulch. Sugarcane is vegetatively propagated from stem cuttings and, like rice, can be grown as a monoculture for many years without harming the soil. Much cane is consumed locally (juice, alcohol production, etc.), but much of the extract is exported for further processing into sugar, molasses, and other products. As with most tropical crops we have discussed, the plant residues provide good animal fodder.

Many fiber crops are commonly produced in the tropics, including abaca (*Musa textilis*), cotton (*Gossypium* spp.), hemp (*Cannabis sativa*), jute (*Corchorus* spp.), kenaf (*Hibiscus cannabinus*), ramie (*Boehmeria nivea*), and sisal (*Agava sisalana*). There is still a good world market for plant fibers, which have a very low price compared to the price of animal and synthetic fibers. In addition, they are good at absorbing moisture and are biodegradable. While cotton provides most of the plant fibers used in the world market, other crops are of local importance.

Other cereal crops commonly grown in the tropics are corn (*Zea mays*), wheat (*Triticum* spp.), barley (*Hordeum vulgare*), and rye (*Secale cereale*). Millet (*Panicum* spp.) is an excellent cereal choice for tropical regions. It is a heat-loving plant that can tolerate low moisture conditions and poor soil. Oats and rye are occasionally grown for animal fodder in subtropical regions and in the highlands of tropical regions. See Table 18-4 for a comparison of the nutritional composition of some common cereal crops.

Sugar plants such as sugar beets (*Beta vulgaris*) and some palms are also grown in some areas of the tropics, as are oil producers like castor (*Ricinus communis*), oil palm (*Elaeis guineensis*), olive (*Olea europaea*), safflower (*Carthamus tinctorius*), sunflower (*Helianthus*

Table 18-4 Average nutritional composition of cereal crops in tropical nations.

Crop	Water (%)	Energy (kcal/100 g)	Protein (%)	Carbohydrate (%)	Fat (%)
Corn meal	12	360	9.5	73	4.0
Pearl millet	10	330	13.0	65	5.0
Rice					
brown	12	360	7.5	76	2.0
white	12	360	6.7	81	0.4
Sorghum	11	340	10.0	73	3.0
Wheat					
whole grain	12	350	12.2	70	2.5
white flour	12	350	10.9	78	1.2

Adapted from O'Hair 1984; Platt 1976.

annuus), and tung (*Aleurites montana*). Numerous vegetables (e.g., tomato [*Lycopersicon esculentum*], onion [*Allium cepa*], asparagus [*Asparagus officinalis*]), amaranth (*Amaranthus* spp.), nuts, spices, fibers, waxes (jojoba, *Simmondsia chinensis*), gums (guar, *Cyamopsis tetragonoloba*), dyes, and fodders are all grown in tropical and subtropical areas, and are of great regional importance either as a food for local consumption, or as an export product. A few examples are mentioned here; for others, see texts such as Altieri (1995) and Gliessman (1998).

Tropical Cropping Systems That Work

Various cropping systems used throughout the world were discussed in Chapter 13, and most of these systems have been used for centuries in tropical agriculture. Still, as with all agroecosystems, management and establishment of the systems must be well adapted to individual regions and even to individual farms in order to be economically profitable, sustainable, and feasible.

Multicultivos in El Salvador

An example of a successful tropical cropping system can be seen in the Zapotitan Valley of El Salvador. For years, this area produced a good yield of corn and beans, but the government of El Salvador wanted to find a way to increase vegetable production in the area, while still maintaining corn and bean production. In order to accomplish this, a system known as *multicultivos* was developed, which is an intercrop/relay crop of a number of vegetables with beans and corn. The field is initially planted to beans, corn, and radishes. The radishes are harvested first, while the beans and corn continue to grow. Beans are harvested next, followed by corn. As the corn stalks double over in the field to dry, cucumbers are planted. As they grow, they begin to twine around the dried corn stalks. After the cucumbers are harvested, the corn stalks are cut and used as a mulch on the soil. Cabbage is planted next, followed by corn where the beans had initially been grown to take advantage of the nitrogen added to the system by the legumes. Once the cabbage matures and the corn is harvested, pole beans are planted, which again use drying corn stalks as support. Once the pole beans are harvested, the cropping system can be repeated.

Double-planting the corn helps to keep yields high (about 90% of the original yields before this system was implemented), but allows for a good portion of the field to produce other crops of equal or higher value on the local market. In addition, the diversity of the system helps to reduce disease and pest problems.

This system is well adapted for tropical systems. The many crops grown in sequence rely on continuously warm weather, and the nutrients provided by one crop are used by the following crops. Because of the intensive nature of the system, a great deal of labor is involved. Most of this must be done by hand rather than by machinery, thus providing jobs in the area. The high yield and increased productivity of the land leads to increased profits as well.

Deep Ditch and High Bed System in China

Another example of a well-adapted tropical cropping system is seen along the Pearl River Delta in China (Luo and Han 1990). Because the water table in this region is so high, farmers dig ditches up to 1.5 meters deep in their fields, mounding the soil on either side of the ditches to form beds up to 7 meters wide. The water stays in the ditches, and the mounds of soil remain dry enough for vegetable or fruit tree production.

The ditches can be used for rice production, or as an aquatic system for production of snails or fish. Because the structure of the system is quite stable, the beds can be used for long-term fruit production (oranges, bananas, litchis, starfruit, etc.), or can be planted to vegetables such as potatoes, cabbage, beans, or peanuts. In some cases, farmers build structures that extend out over the ditches, and use these as supports for peas or some other vine plant. These plants can anchor in the soil of the raised bed, but then extend spatially out over the water, thereby increasing the light-use efficiency of the system.

Alley Cropping in Tropical Africa

In tropical Africa, farmers have traditionally included tree species grown together with annual or field crops (Okigbo 1990). This traditional approach to agriculture has developed into the slightly more sophisticated alley cropping systems seen today. A common tree crop grown in rows is *Leucaena*, a fast-growing leguminous tree that is often grown to improve soil nutrient levels and to provide fuelwood for local families. The tree rows are commonly planted approximately 4 meters apart, and annual crops are planted between the rows of *Leucaena*. The trees are occasionally pruned, with leaves and branches applied to the annual crop as a mulch, and larger stumps burned as fuel for cooking. The mulch is nutrient rich and is particularly high in nitrogen. When added to the annual crop, the mulch can significantly increase productivity.

Summary

Principles of tropical agriculture can be very different from those in temperate agriculture because the growing season is year-round and is affected more by changes in moisture than by changes in temperature. Tropical soils provide their own set of challenges to the producers in this region. They tend to be highly weathered and fairly low in nutrients. Choice of crops in the tropics is therefore based largely on fluctuations in rainfall and nutrient requirements of plants. A large number of plants can be grown very successfully in the tropics, including many fruit varieties, root and tuber crops, vegetables, pulses, and crops grown for sugar, oil, or fiber. Many common products used in the temperate regions of the world, including cocoa, coffee, and coconut are grown only in tropical areas. Because of the great diversity in crops that can be grown in these regions, and because of the varying degrees of expertise, input availability, and economic status of each individual farmer, cropping systems also vary greatly, and should be carefully adapted to each individual situation.

Topics for Review and Discussion

1. Compare and contrast agricultural production systems in temperate regions of the world with those in tropical regions. What do farmers in each region need to consider before choosing cropping systems and crops to include in their systems?
2. Many of the tropical regions of the world are home to some of the poorest countries in the world. From what you understand about agroecology and economics (Chap-

ter 17), why do you think that many of these regions remain underdeveloped and overcultivated?
3. Examine the scientific literature for other successful growing systems used in the tropics. What makes them so well adapted to the region in which they are used?
4. Choose one crop that is commonly grown only in the tropics. Describe the conditions necessary for high productivity of that crop. How might you use that crop in an intercropping situation? What would be the benefits of using such a crop?
5. Examine the produce department of a local grocery store in the middle of winter. What products come from tropical and subtropical areas of the world? Which products would remain available if trade with other nations was abruptly stopped?

Literature Cited

Altieri, M. A. 1995. *Agroecology: The Science of Sustainable Agriculture*. Boulder, Colo.: Westview Press.
Aykroyd, W. R., J. Doughty, and A. Walker. 1982. Legumes in human nutrition. *FAO Food and Nutrition Paper No. 20*. Rome: FAO.
Gliessman, S. R. 1998. *Agroecology: Ecological Processes in Sustainable Agriculture*. Chelsea, Mich: Ann Arbor Press.
Luo, S. M., and C. R. Han. 1990. Ecological agriculture in China. In *Sustainable Agricultural Systems,* eds. C. A. Edwards, R. Lal, P. Madden, R. H. Miller, and G. House, 299–322. Delray Beach, Fla.: St. Lucie Press
O'Hair, S. K. 1984. Farinaceous crops. In *Handbook of Tropical Food Crops,* ed. F. W. Martin, 109–137. Boca Raton, Fla.: CRC Press.
Okigbo, B. N. 1990. Sustainable agricultural system in tropical Africa. In *Sustainable Agricultural Systems*. eds. C. A. Edwards, R. Lal, P. Madden, R. H. Miller, and G. House, 323–352. Delray Beach, Fla.: St. Lucie Press.
Platt, B. S. 1976. *Tables of Representative Values of Foods Commonly Used in Tropical Countries*. London: Her Majesty's Stationery Office.
Purseglove, J. W. 1974. *Tropical Crops: Dicotyledons*. New York: Wiley.

Bibliography

Edwards, C. A., R. Lal, P. Madden, R. H. Miller, and G. House, eds. 1990. *Sustainable Agricultural Systems*. Delray Beach, Fla.: St. Lucie Press
Purseglove, J. W. 1972. *Tropical Crops: Monocotyledons*. New York: Halsted Press Division, Wiley.
Rehm, S., and G. Espig. 1991. *The Cultivated Plants of the Tropics and Subtropics: Cultivation, Economic Value, Utilization*. Weikersheim, West Germany: Margraf.
Smith, N. J. H. 1992. *Tropical Forests and Their Crops*. Ithaca, N.Y.: Comstock Pub. Associates
Weischet, W., and C. N. Caviedes. 1993. *The Persisting Ecological Constraints of Tropical Agriculture*. New York: Wiley.

Chapter 19

Human Population Growth

Key Concepts

- Exponential nature of population growth
- Pressures on cropland, environment, and resources
- Providing food to increasing populations

With few exceptions, populations continue to grow to record levels in most countries. In many locations, this growth is placing an increasing strain on natural resources, the environment, and food production systems. The challenge for agriculture remains to provide an adequate food supply for today's large population and the even larger populations anticipated in the future.

Population Growth–Past, Present, and Future

Concern over increasing world population growth is not a new issue. In 1798, Thomas Malthus worried whether a food supply that was only increasing linearly could keep up with a population that was growing geometrically (Figure 19-1). However, the fears of Malthus were not realized because, with the aid of the plow, many new sites were opened to agricultural production in the 1800s. Population size escalated more rapidly during the 1900s, especially after death rates and infant mortality declined following the discovery and use of antibiotics and other improvements in medicine and sanitation. Population increases were accommodated by rapid increases in crop yields and food production, which were brought about mainly through variety improvement and use of synthetic fertilizers and pesticides.

World population growth from 1650 to 2000 (estimated) is illustrated in Figure 19-2. This pattern of population growth resembles the exponential growth curve discussed in Chapter 6. A population of 6 billion people was reached in 1999. As of that year, approximately

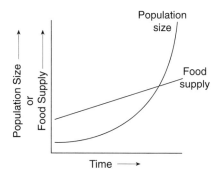

Figure 19-1 The Malthusian dilemma: An arithmetic increase in food production cannot keep pace with a geometric increase in population.

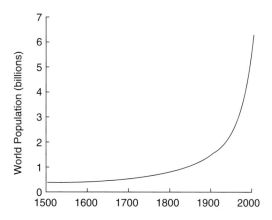

Figure 19-2 World population growth from 1500 to 2000.

80,000,000 people (about the population of Germany) were being added to the earth's population each year (Brown, Flavin, and French 1999). This is equivalent to adding a medium-sized U.S. city, such as Rochester, New York, or Akron, Ohio, to the world population *each day*. During the period from 1990 to 1995, the average annual growth rate of world population was about 1.7%. During this period, however, the annual growth rate varied widely among various countries of the world–from less than zero in Hungary, to 1.0% in the United States, to 2.1% in Mexico, to more than 3% in a number of countries in Africa and the Middle East (World Resources Institute 1994).

Future projections of world population growth vary widely, particularly those that attempt to predict growth far into the future. It is possible to make hypothetical forecasts of population growth using the exponential growth equation (see Application 19-1) or other

Application 19-1 *Projecting Future Population Growth*

We can make simple forecasts of future population growth using the exponential growth equation:

$$N_t = N_0 e^{rt}$$

Rather than the true intrinsic rate of increase of a human population, we can consider r as the effective growth rate of the population at a particular point in time. If we use a growth rate of 1.7% per year, then $r = 0.017$. Based on a 1999 population size of 6.0 billion, we can forecast a population size in 2020 (time = 21 years) of:

$$N_t = (6.0)e^{(0.017)(21)} = (6.0)e^{0.36} = 8.6 \text{ billion}$$

We can develop forecasts using other growth rates (e.g., $r = 0.015$; $r = 0.016$) as well. Growth rates have been declining very slightly during the last two decades of the 1900s. Many forecasters consider anticipated declines in growth rate in making their projections, which will therefore be somewhat less than the 8.6 billion people we predicted for 2020 in this example.

---

simple models. Most forecasts that use more detailed demographic data project a population size of 8.0–8.5 billion between the years 2020–2025. There is much speculation over what the carrying capacity of the earth may be. Some forecasters estimate a leveling off at 10–15 billion people in the second half of the twenty-first century.

Pressures From Population Growth

Several estimates suggest that cropland available in the world has increased slightly in the last decades of the twentieth century. One source (FAO 1997) estimated a 3.4% increase in world cropland between 1980 and 1995, while another (World Resources Institute 1994) estimated a 1.5% increase between 1979–1981 and 1989–1991. The changes in available cropland varied by region (Figure 19-3). Nevertheless, the overall average increase in world cropland per year is relatively small: 3.4%/15 years = 0.2% per year. Most estimates place the amount of cropland currently in use at 1.45 to 1.50 billion hectares. If this amount remains constant at 1.5 billion hectares, the amount of cropland available per person will decrease as population increases (Figure 19-4). If population grows from 6 billion to 8 billion, the amount of cropland will have to increase by 33%, to 2 billion hectares, to support a ratio of cropland per person (0.25 ha) equivalent to the ratio available to 6 billion people in 1999, assuming that productivity and consumption remain constant.

Despite any net gain in world cropland over time, extensive losses of cropland still occur each year. Some land is lost to urbanization, especially in countries such as China, which is rapidly expanding its industrial base. In the United States, an estimated 168,000 hectares of cropland were lost to urbanization each year between 1982–1992 (Gardner 1997). About

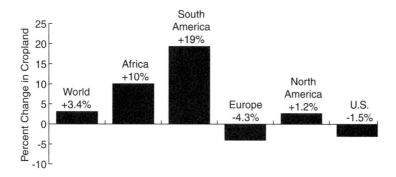

Figure 19-3 Changes in amount of cropland between 1980 and 1995.
Calculated from data presented by FAO 1997

one-third of world cropland in use has suffered some degree of degradation. Water erosion is by far the most common cause of soil degradation, but wind erosion, salinization and other forms of chemical degradation, and physical soil degradation (compaction, etc.) are also important contributors to the problem. The impact of salinization can be especially severe because of its disproportionate effect on highly productive, irrigated land. It is estimated that a few million hectares are lost to production annually to all forms of soil degradation, with 1.5–2.5 million hectares lost to salinization alone (Gardner 1997). However, in spite of losses to soil degradation and urbanization, the supply of cropland has been relatively stable, and has even increased slightly in recent years at the expense of natural ecosystems (see next section).

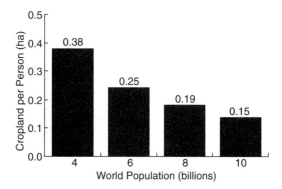

Figure 19-4 World cropland available per person, if cropland is constant at 1.5 billion hectares and population increases from 4 to 10 billion people.

Pressures on the Environment

As increased population pressure results in demand for increased agricultural production and increased land area cultivated, adverse effects of agriculture such as erosion or pollution from fertilizers and pesticides are expected to increase proportionately unless appropriate measures are taken to limit these problems. Methods for minimizing adverse environmental impacts of agricultural practices have been discussed throughout this book. However, in some situations, agriculture may directly displace natural ecosystems to meet increased demand for cropland.

The losses to cropland mentioned in the previous section are often balanced by gains at the expense of natural ecosystems. We can imagine a kind of succession as land is converted into agroecosystems and then into urban ecosystems (Figure 19-5). If the amount of agricultural land available must remain constant or increase to feed an expanding population, then agricultural land lost to urbanization must be replaced by land from another source. Natural areas provide this source of new agricultural land. Of course, natural ecosystems can be converted directly to urban systems, bypassing the intermediate agricultural stage (Figure 19-6). In any case, the movement of land through the "succession" shown in Figure 19-5 is mostly in one direction (although some abandoned agricultural areas do succeed to natural ecosystems), leaving a net loss in land available for natural ecosystems.

The natural ecosystems most capable of supplying cropland are permanent pastures (grasslands) and forests. However, much of the land that can support grasses cannot support a wide variety of other crops, and so much new farmland comes from conversion of forests. On a world basis, the land area in forests declined by 1.9% from 1981 to 1990, with a change of –3.6% in tropical countries (World Resources Institute 1996). The destruction of tropical forests continues to proceed at a rapid pace. Estimates of the rate of loss of tropical forests vary for different countries, but in general the highest loss rates today are in Latin America and Africa. Between 1981–1983 and 1991–1993, Nigeria, Ivory Coast, Mauritius, Nicaragua,

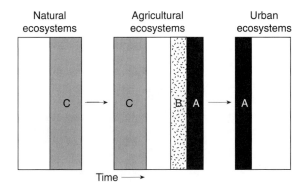

Figure 19-5 Land succession through a transition from natural to agricultural to urban ecosystems. If the amount of agricultural land remains constant, then losses to urban ecosystems (A) or degradation (B) must be balanced by gains (C) at the expense of natural ecosystems.

Figure 19-6 Forest land being converted directly to urban housing in central Florida, one of the most rapidly growing regions in the United States.

and Paraguay each lost more than 20% of their forested area (World Resources Institute 1996), an average of over 2% per year during that ten-year period. Greatest losses in terms of land area occurred in Brazil and Indonesia.

Destruction of habitat is an important cause of species extinction. The numbers of species facing extinction is expected to increase as forest habitat is destroyed, particularly tropical rain forest habitat that harbors large numbers of species. Estimates of projected extinction rates vary widely and are the subject of much controversy. Potentially useful products, sources of genetic diversity for crops, and other benefits may be lost along with these habitats. In addition, the full impact of forests in world hydrology and climate regulation is not well understood. Any disruption of climate and weather patterns will have serious consequences for agriculture.

Pressures on Supplies of Natural Resources

The proportion of cropland under irrigation is increasing, making it possible to supply fresh fruits and vegetables even from arid landscapes. The competition for water between agriculture and growing urban populations is escalating, particularly in arid and semiarid regions. In California, for example, large urban populations will demand an increasing proportion of the water needed to irrigate an agricultural production region that supplies vegetables and fruit to much of the United States. The challenges of water conservation in agriculture have already been examined in Chapter 15. Irrigated agriculture in the future will require efficient delivery systems (e.g., drip irrigation instead of overhead or furrow) and vigilance against salinization. Frequency of drought and limitations to water availability could increase in more humid climates too, if world climate patterns and hydrology are disrupted by increased CO_2 production or removal of forest vegetation.

Many conventional agricultural practices in industrialized countries are highly dependent on energy use, particularly energy for fuel and for manufacture of fertilizers and pesticides. Despite recent and ongoing discoveries of new reserves, petroleum remains a nonrenewable resource. Assuming that consumption remains at a constant level, known petroleum reserves in 1993 were projected to be adequate to supply world needs for forty years (World Resources Institute 1996). However, as world population increases from 6 billion to 8 billion, consumption will not remain constant because demand is expected to increase by a minimum of 33%. However, an increase of only 33% in energy consumption to supply 8 billion people is an extremely unlikely scenario. In 1991, the United States, with slightly less than 5% of the world's population, was responsible for 25% of the world's energy consumption (World Resources Institute 1994). If the rest of the world were to consume energy at the same rate as the United States, world energy production will need to be multiplied *five times* to meet the demand. Of course, such a level of consumption will take some time to develop and may never be enjoyed by many of the world's people. However, energy consumption rates are escalating throughout the world as the global economy expands. Urbanization, industrialization, and automobile production and use are increasing in many countries. Projected energy consumption will put an increasing strain on finite reserves of petroleum and natural gas. No one knows how long the discovery of new reserves of these fossil fuels can keep ahead of increased consumption and demand. At some point, energy conservation and increased reliance on alternative sources of energy will be critical.

Resources other than water and energy could be affected by increasing population. Loss of soil and soil fertility through erosion and other forms of degradation have already been mentioned. Phosphate reserves are unevenly distributed throughout the world. They are being depleted at different rates in different countries, sometimes through internal consumption and sometimes through export. The same is true of any materials that are used in agriculture, such as micronutrients, and for metals and minerals used in industrial processes. Reserves of some of these materials are vast, but reserves of others may become increasingly limited as world population rises.

Feeding a Growing Population

Different strategies can be used alone or in combination to provide future populations with an adequate food supply. Two of these—improved food distribution and limited population growth—involve political and sociological issues that are far beyond the scope of this text. The others—increased crop yields, reduced crop losses, and increased cropland supply—depend on our knowledge and progress in the agricultural sciences, although their implementation also depends on political and social factors.

Limited Population Growth

Limiting population growth is an obvious approach to limiting demand on the world food supply and resources. Birthrates vary throughout the world, and are still very high in some

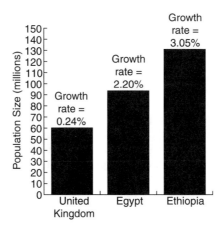

Figure 19-7 Projected population sizes in 2025 for three countries with 1995 populations estimated at about 58 million. Annual growth rate for 1995 is shown above each bar.
Based on data presented in tables by World Resources Institute 1994

countries, while other countries have achieved growth rates near zero. Figure 19-7 gives an impression of the impact that different growth rates can have on the future population sizes projected for three countries. Ethiopia receives food aid today, yet its projected population growth is by far the greatest of the three countries. The impact of increasing population growth will be felt very unevenly in different countries of the world due to the differential rates of population growth.

Improved Food Distribution

Many people believe that providing an adequate food supply to the world population is more a political problem than a production problem. World food may be adequate today, but distribution of that food to people in various parts of the world remains extremely uneven. Part of this problem is due to infrastructure and varying abilities to move perishable food to markets within a country.

Much of the food distribution problem has its roots in complex political and economic issues. International trade practices, along with land ownership patterns within countries, may determine the distribution of food to local populations. If the most fertile land in a country is used for production of crops for export, local farmers may have to subsist on marginal lands, such as highly erodible hillsides or plots cleared from forest. Impoverished rural farmers may receive such a minimal share of any export dollars flowing into their country that they cannot afford prime agricultural land or the food produced on it. Even in the United States and Europe, small farmers have been losing control over the food supply, as the scale of food production becomes larger and more industrial.

Increased Crop Yields

Yields of many (but not all) crops are increasing slightly each year; increases of about 1–2% per year are typical for many crops. However, these rates of increase are much less than those enjoyed during the period from 1950–1980, when rapid gains in yield resulted from increased energy input and variety improvement. Growth in production and yields per hectare have slowed greatly for most crops, particularly since the late 1980s and early 1990s. For example, Brown (1997) observed that grain production slowed considerably and nearly leveled off in the 1990s, with grain production per capita actually declining slightly as population increased. With yields per hectare nearly leveling off or showing only slight gains, there is little reason to expect or rely on the type of gains observed in the middle to third quarter of the twentieth century. Nevertheless, slight improvements in yield might be anticipated through variety improvement, precision farming, and reduction of losses to pests and diseases. Some locations with below-average yields still have considerable growth potential in yield. Opportunities exist to increase yields per hectare in time and space through multiple cropping systems. If these factors could combine to produce an overall yield increase of 1.5% per year, in twenty years, the 30% gain in yield should be nearly sufficient to offset a 33% increase in population from 6 billion to 8 billion.

Reduced Crop Losses

Significant proportions of many crop yields are lost to pests, diseases, or weather damage in the field. Postharvest losses are also high in many areas. Every loss that can be prevented means additional food available from the same land area. Knowledge and application of appropriate pest management practices (see Chapters 10 and 11) are essential in limiting crop losses.

Increased Cropland Supply

Although cropland can be increased by converting forests and other natural areas, supplies of these resources are limited and the adverse impacts of their destruction are numerous. The need for additional cropland can be reduced through conservation of existing cropland, particularly through practices that limit erosion and other forms of soil degradation. In some countries, zoning laws protect agricultural land from losses to urban ecosystems.

There is one vast reserve of cropland that is available for feeding expanding populations, however. In some countries, such as the United States, large amounts of fertile farmland are used for the production of crops to feed animals for meat production. As discussed in Chapter 15, this is an inefficient method of providing food to a higher-level omnivore. If the food chain was shortened, these lands could be used to produce plant food that a human population could eat directly. Thus, each producing hectare could feed more people. This approach is not possible on those lands that are so marginal that they can only support grasses or pasture. Some countries have adopted this measure as a way of significantly increasing the number of people supported per hectare, when land resources become severely limited.

Sustainable Agriculture

It is possible that an adequate food supply for 8 billion people can be achieved through a combination of the approaches outlined in the last section. In the next few decades, however, rapid population growth will place especially severe pressure on the environment and natural resources. These pressures come from both agricultural and urban systems. Yet agriculture, as currently practiced in many parts of the world, is dependent on natural resources, both renewable and nonrenewable. The depletion of nonrenewable resources poses the greatest potential threat to agricultural production as we know it today. Conservation of our existing resources is essential if agricultural production is to be sustained far into the future.

Concept of Sustainable Agriculture

The terms "sustainable agriculture" and "sustainability" have been variously defined and widely used in the 1990s. Most definitions of *sustainable agriculture* include the following elements:

- Maintenance of an adequate food supply for all people;
- Conservation of nutrients and resources;
- Minimal impact on the environment and natural ecosystems;
- Social and economic acceptance; and
- Intergenerational or even indefinite stability.

This approach is particularly attractive because of its emphasis on conservation of resources and the environment—two areas that face particular pressure from population growth. The various chapters in this book have examined methods of improving crop yields and production, alternative agricultural practices that conserve nutrients and resources while lessening environmental impact, and the integration of agroecosystems and agricultural production systems into a socioeconomic and political framework. The most difficult part of the sustainable agriculture concept is probably the extremely long-term view associated with it. Most economic decisions in agriculture and other businesses are made over a short term, such as on an annual or quarterly basis. Economic and conservation plans based on longer intervals such as five years are much less frequent, even in research programs. It is extremely difficult to project economic and resource trends 20, 50, 100, or more years in the future. Yet this must be done to some extent if natural resources and adequate agricultural production are to be available for future generations. Sustainable agriculture programs and practices attempt to provide this guarantee.

Limitations to Sustainable Agriculture

Most people probably support and aspire to the goals of sustainable agriculture, as outlined in the last section. The challenge is realizing these goals in practice. A closer examination reveals some areas of potential conflict.

Like the ecological niche, sustainable agriculture has many dimensions. A long list of physical and chemical resources must be maintained, and the maintenance of biological diversity includes a vast number of individual species. Agricultural production must be sustained too. Will the sustainability of production be in conflict with the conservation of any species or natural resources on the long list that must also be sustained? For an agricultural practice to be used and accepted, it must be economically viable, so profitability of agricultural enterprises must also be sustained. We have already discussed examples of resources that are undervalued economically, and so we cannot be sure that sustainability of profits and all resources will be parallel and not in conflict. Economics and costs will change over time as well. As natural resources become more limited and costly, and as economic costs are attached to environmental degradation, the use of resource-conservative agricultural practices should become more widespread and profitable. Another area of conflict is whether natural ecosystems are better conserved by an intensive agricultural practice (with risk of pollution) that produces high yield per unit of land or by a less intensive practice (with minimal pollution) that requires more land to produce an equivalent yield.

Precise goals will need to be established in designing sustainable agricultural systems. Exactly what level of production and profits must be sustained? What levels of each of the various resources should be maintained? What level of human population can be sustained, and for how long? The last question is particularly difficult if the goal is a moving target. A world agricultural system designed to support 6 billion people needs to be altered to sustain 8 billion or 10 billion. Obviously, these questions require long-term anticipation and planning to successfully continue sustainability over the long term.

Sustainable Agriculture in the Future

Although we may be able to make rough predictions about what may happen in twenty years, the concept of sustainable agriculture demands that we look far beyond that point, and forecasts then become much more difficult. Particularly serious issues may arise toward the middle of the twenty-first century as we approach the carrying capacity of the earth, which cannot be easily determined. For example, is a population size of 10, 12, or 15 billion beyond the intersection point on the Malthusian graph shown in Figure 19-1? Can we even determine what the carrying capacity of the earth is? Recall from Chapter 6 that a population may experience some fluctuations before stabilizing at a carrying capacity. In terms of the human population, such fluctuations can arise from war, disease, natural disaster, or some other catastrophe. Food scarcity issues (Brown 1997) could increase in importance as we approach the carrying capacity, particularly if the approach is by trial and error or left to circumstance. Population growth can be stabilized either through a decrease in the birthrate or an increase in the death rate. A gradual decline in birthrate or fertility over time would provide the smoother transition toward the carrying capacity.

At present or increased consumption rates, supplies of various resources could be depleted or severely threatened by 2050. Pressures on water have already been discussed, and reserves of phosphates, petroleum, and tropical forest may be depleted or near depletion by that point. In 1996, the projected life expectancies of world reserves of natural gas, copper, zinc, and tin were all less than sixty-five years (World Resources Institute 1996). In contrast, reserves of coal, bauxite (principal sources of aluminum), and iron ore were projected to last for 200–250 years (World Resources Institute 1996). Yet even 200–250 years is a relatively short period of time if our goal is to sustain agriculture indefinitely. Clearly, efficient recycling and development of effective alternatives are critical for long-term sustainability.

Looking back 1000 or more years, we can find examples of agricultural systems that were sustained for long periods of time. It is therefore possible to think that similar levels of stability could be achieved by sustainable systems in the future. An important difference between ancient civilizations and future ones is the per capita availability of land suitable for agriculture. As land becomes more limited, the need to stabilize population size becomes more important.

Of course, it is impossible to predict technological developments in the future. New energy sources may be discovered and come into widespread use. It is possible that the nature of the human food supply may change, perhaps with populations becoming more dependent on things like the intensive aquaculture of algae than on traditional agricultural products. Technological changes may indeed change the face of agriculture because they are usually readily adapted into agricultural systems. However, we need to be prepared for unanticipated consequences of technological advances; for instance, some types of pollution remained unrecognized for many years.

Outlook for Sustainable Agriculture

Many people are discouraged by the problems anticipated from future population growth and its ramifications. If a food crisis or environmental catastrophe is avoided today, would it not just occur anyway in twenty years or fifty years, as population grows even larger? Despite a long history of concern over global starvation, this catastrophe has not been reached. Forecasts from the 1970s that the world could not support a population of 6 billion have proved false. Today, agricultural systems have met that level of demand and should keep pace at least for several years into the near future.

Sustainability into the future will depend on the resolve of the world's people and governments to anticipate and solve these difficult problems. World population is already large enough to be capable of consuming many resources to the point of depletion for immediate benefit. However, to do so would ignore our responsibility to future generations and possibly even to our survival as a species. There are various strategies for coping with depletion of resources, increase of pollutants, or other potentially catastrophic changes. One approach is to accept the change as inevitable and adapt to it, whether it is a result of climate change or loss of some natural resource. Another strategy is to proactively anticipate the change and take measures to avoid or reduce its effects by addressing its cause.

In building sustainable systems, it is important that agriculturists and other scientists recognize and anticipate trends in agricultural production, resource usage, accumulation of pollutants, and biodiversity issues. Anticipation of such trends is critical—since environmental and food catastrophes may not occur suddenly, but may instead build up gradually over time. The limitations and conflicts that can occur in the development of sustainable agricultural systems must also be anticipated and resolved in a manner that will result in minimal long-term impact. As agriculture evolves and becomes more sustainable, the emphasis in management of agricultural systems should shift from reacting to problems and crises toward planning and developing systems in which problems and crises are less likely to occur. Improved knowledge and understanding of the structure and function of agroecosystems will be essential for designing sustainable systems. The successful development of sustainable agricultural systems is perhaps the most critical challenge facing agriculture in the early part of the twenty-first century. This is a critical step in helping world population to make a smooth transition to the carrying capacity likely to be reached in this century.

Summary

World population is growing at a nearly exponential pace. As world population increases, more pressure is placed on available cropland, the environment, and supplies of natural resources, particularly nonrenewable resources. Strategies for providing a food supply to expanded future populations include limiting population growth, improving food distribution, increasing crop yields, reducing crop losses, and increasing cropland for human food production by shortening food chains. Several of these strategies are largely political and sociological in nature. The agricultural sciences have a particularly important mission in improving or maintaining yields and limiting losses. However, the challenge for agriculture is to accomplish this through agroecosystems that are increasingly sustainable and conservative of limited resources that are being consumed at an accelerating pace.

Topics for Review and Discussion

1. What are some methods that could be used to maintain an adequate food supply for future generations?
2. Using the data and methods provided in Application 19-1, compare projections of future population growth if the growth rate is reduced to 1.5% per year ($r = 0.015$) or 1.0% ($r = 0.010$).
3. Using the data and methods provided in Application 19-1, calculate a projection of population growth for the year 2050. What percent increase does this represent over the present-day population? Over the population size projected for 2020?
4. Of the pressures on land, environment, and resources outlined in this chapter, which do you think will have the most immediate impact on world agriculture, and why?
5. Much cropland is lost each year to erosion and salinization. Discuss some ways in which this "lost" land could be rehabilitated and gradually brought back into production.

Literature Cited

Brown, L. R. 1997. Facing the prospect of food scarcity. In *State of the World 1997*, eds. L. R. Brown, C. Flavin, and H. F. French, 23–41. New York: W. W. Norton and Company.

Brown, L. R., C. Flavin, and H. F. French, eds. 1999. *State of the World 1999*. New York: W. W. Norton and Company.

FAO. 1997. *FAO Yearbook Production*. Vol. 50. Rome: Food and Agriculture Organization of the United Nations.

Gardner, G. 1997. Preserving global cropland. In *State of the World 1997*, eds. L. R. Brown, C. Flavin, and H. F. French, 42–59. New York: W. W. Norton and Company.

World Resources Institute. 1994. *World Resources 1994–95*. New York: Oxford University Press.

———. 1996. *World Resources 1996–97*. New York: Oxford University Press.

Bibliography

Brown, L. R., C. Flavin, and H. French, eds. 1997. *State of the World 1997*. New York: W. W. Norton and Company.

Buringh, P. 1989. Availability of agricultural land for crop and livestock production. In *Food and Natural Resources,* eds. D. Pimentel and C. W. Hall, 69–83. San Diego, Calif.: Academic Press.

Cleaver, K. M., and G. A. Schreiber. 1994. *Reversing the Spiral: The Population, Agriculture, and Environment Nexus in Sub-Saharan Africa*. Washington, D.C.: The World Bank.

Crews, T. E., C. L. Mohler, and A. G. Power. 1991. Energetics and ecosystem integrity: The defining principles of sustainable agriculture. *American Journal of Alternative Agriculture* 6: 146–149.

Edwards, C. A., R. Lal, P. Madden, R. H. Miller, and G. House, eds. 1990. *Sustainable Agricultural Systems*. Delray Beach, Fla.: St. Lucie Press.

FAO. 1996. *The Sixth World Food Survey.* Rome: Food and Agriculture Organization of the United Nations.

Gliessman, S. R. 1998. *Agroecology: Ecological Processes in Sustainable Agriculture.* Chelsea, Mich.: Ann Arbor Press.

Goodland, R., and H. Daly. 1996. Environmental sustainability: Universal and non-negotiable. *Ecological Applications* 6: 1002–1017.

Greenland, D. J., and I. Szabolcs. 1994. *Soil Resilience and Sustainable Land Use.* Wallingford, U.K.: CAB International.

Murdoch, W. 1990. World hunger and population. In *Agroecology,* eds. C. R. Carroll, J. H. Vandermeer, and P. M. Rosset, 3–20. New York: McGraw-Hill.

Myers, M. 1989. Loss of biological diversity and its potential impact on agriculture and food production. In *Food and Natural Resources,* eds. D. Pimentel and C. W. Hall, 49–68. San Diego, Calif.: Academic Press.

Paddock, W. C. 1992. Our last chance to win the war on hunger. In *Advances in Plant Pathology,* Vol. 8, eds. J. H. Andrews and I. Tommerup, 197–222. London: Academic Press.

Pimentel, D. 1989. Ecological systems, natural resources, and food supplies. In *Food and Natural Resources,* eds. D. Pimentel and C. W. Hall, 1–29. San Diego, Calif.: Academic Press.

Reid, T. R. 1998. Feeding the planet. *National Geographic,* October 1988: 56–75.

Spedding, C. R. W. 1996. *Agriculture and the Citizen.* London: Chapman and Hall.

Appendix I

Organizations Related to Agricultural Ecology

Following is a selection of some pertinent organizations, along with their Internet addresses and mailing addresses, which were current at our date of publication. These organizations and websites can provide additional information on many of the topics presented in this book.

The Agricultural Institute of Canada
141 Laurier Avenue West, Suite 1112
Ottawa, Ontario K1P 5J3 Canada
www.aic.ca

American Agricultural Economics Association
415 South Duff, Suite C
Ames, IA 50010–6600
www.aaea.org

American Association for the Advancement of Science
1200 New York Ave, NW
Washington D.C. 20005
www.aaas.org

American Crop Protection Association
1156 Fifteenth St, N.W., Suite 400
Washington D.C. 20005–1704
www.acpa.org

American Dairy Science Association
1111 N. Dunlap Ave.
Savoy, IL 61874
www.adsa.uiuc.edu

American Farm Bureau
225 Touhy Ave.
Park Ridge, IL 60068
www.fb.com

American Institute of Biological Sciences
1313 Dolley Madison Blvd., Suite 402
McLean, VA 22101
www.aibs.org

American Phytopathological Society
3340 Pilot Knot Road
St. Paul, MN 55121–2097
www.scisoc.org

American Seed Trade Association
601 13th Street, #570 South
Washington, D.C. 20005–3807
www.amseed.com

American Society of Agricultural Engineers
2950 Niles Road
St. Joseph, MI 49085–9659
www.asae.org

American Society of Agronomy
677 South Segoe Road
Madison, WI 53711–1086
www.agronomy.org

American Society of Animal Science
1111 North Dunlap Avenue
Savoy, IL 61874
www.asas.org

American Society for Horticultural Science
600 Cameron Street
Alexandria, VA 22314–2562
www.ashs.org

Australian Society of Horticultural Science
AuSHS, Private Bag 15
South Eastern Mail Centre, Victoria 3176
Australia
www.aushs.org.au

Council for Agricultural Science and Technology
4420 West Lincoln Way
Ames, IA 50014–3447
www.cast-science.org

Crop Science Society of America
677 South Segoe Road
Madison, WI 53711–1086
www.crops.org

Ecological Society of America
2010 Massachusetts Avenue NW, Suite 400
Washington, D.C. 20036
www.esa.sdsc.edu

Entomological Society of America
9301 Annapolis Road
Lanham, MD 20706–3115
www.entsoc.org

Greenpeace International
Keizersgracht 176
1016 DW Amsterdam, The Netherlands
www.greenpeace.org

Henry A. Wallace Institute for Alternative Agriculture
9200 Edmonston Road, Suite 117
Greenbelt, MD 20770–1551
www.hawiaa.org

National Agricultural Library
10301 Baltimore Avenue
Beltsville, MD 20705–2351
www.nalusda.gov

National Audubon Society
700 Broadway
New York, NY 10003
www.audubon.org

National Soil Survey Center
MS 32 Room 152
100 Centennial Mall North
Lincoln, NE 68508–3866
www.statlab.iastate.edu:80/soils/nssc

National Wildlife Federation
8925 Leesburg Pike
Vienna, VA 22184
www.nwf.org

The Nature Conservancy
1815 North Lynn Street
Arlington, VA 22209
www.tnc.org

Poultry Science Association
1111 Dunlap Avenue
Savoy, IL 61874
www.psa.uiuc.edu

Sierra Club
85 Second Street, Second Floor
San Francisco, CA 94105–3441
www.sierraclub.org

Society of Nematologists
3012 Skyview Drive
Lakeland, FL 33801–7072
www.ianr.unl.edu/son

Soil Science Society of America
677 South Segoe Road
Madison, WI 53711–1086
www.soils.org

Soil and Water Conservation Society
7515 NE Ankeny Road
Ankeny, IA 50021
www.swcs.org

Sustainable Agriculture Research and Education Program
University of California
One Shields Avenue
Davis, CA 95616–8716
www.sarep.ucdavis.edu

United States Geological Survey
807 National Center
Reston, VA 20192
www.h2o.er.usgs.gov

Weed Science Society of America
P.O. Box 1879, 810 E. 10th Street
Lawrence, KS 66044–8897
ext.agn.uiuc.edu/wssa

The Wildlife Society
5410 Grosvenor Lane, Suite 200
Bethesda, MD 20814–2197
www.wildlife.org

Following is a listing of Internet addresses for some other relevant organizations, including several important government sites. Internet addresses were current at our date of publication.

Organization	Internet Address
Agricultural and Agri-Food Canada	aceis.agr.ca
American Society of Parasitologists	www.museum.unl.edu/asp
British Society for Plant Pathology	www.bspp.org.uk
Canadian Phytopathological Society	res.agr.ca/lond/pmrc/cps
Consultant Group on International Agricultural Research (homepage leads to directory of international research centers)	www.cgiar.org
Cooperative State Research Education and Extension Service	www.reeusda.gov/new/csrees
International Society for Environmental Ethics	www.cep.unt.edu/ISEE.html
International Society for Horticultural Science	www.ishs.org
International Union of Forestry Research Organizations	iufro.boku.ac.at
National IPM Network	www.reeusda.gov/ipm
United States Agency for International Development	www.usaid.gov
United States Department of Agriculture	www.usda.gov
United States Environmental Protection Agency	www.epa.gov
United States Fish and Wildlife Service	www.fws.gov
USDA Forest Service	www.fs.fed.us
USDA Natural Resources Conservation Service	www.nrcs.usda.gov
USDA-ARS National Soil Tilth Laboratory	www.nstl.gov
USDA-ARS Tree Fruit Research Lab	www.tfrl.ars.usda.gov
World Wildlife Fund	www.worldwildlife.org

Appendix II

Scientific Names of Some Common Crop Plants*

Common Name	Scientific Name
Abaca	*Musa textilis*
Alfalfa	*Medicago sativa*
Almond	*Prunus amygdalus*
Amaranth	*Amaranthus* spp.
Apple	*Malus domestica*
Apricot	*Prunus armeniaca*
Arrowroot	*Maranta arundinacea*
Artichoke	*Cynaria scolymus*
Artichoke, Jerusalem	*Helianthus tuberosus*
Asparagus	*Asparagus officinalis*
Avocado	*Persea americana*
Banana	*Musa* spp.
Barley	*Hordeum vulgare*
Bean, Adzuki	*Phaseolus angularis*
Bean, broad (Faba)	*Vicia faba*
Bean, common (French, green)	*Phaseolus vulgaris*
Bean, lima	*Phaseolus lunatus*
Bean, mung	*Phaseolus aureus*
Beet	*Beta vulgaris*
Beet, sugar	*Beta vulgaris*
Blackberry	*Rubus* spp.
Blackeye pea	*Vigna unguiculata*
Blueberry	*Vaccinium corymbosum*
Brinjal	*Solanum melongena*
Broadbean	*Vicia faba*
Broccoli	*Brassica oleracea* var. *botrytis*
Brussels sprouts	*Brassica oleracea* var. *gemmifera*
Buckwheat	*Fagopyrum esculentum*
Cabbage	*Brassica oleracea* var. *capitata*

Cacao (cocoa)	*Theobroma* spp.
Canola	*Brassica napus*
Carrot	*Daucus carota*
Cashew	*Anacardium occidentale*
Cassava	*Manihot esculenta*
Castor	*Ricinus communis*
Cantaloupe	*Cucumis melo*
Cauliflower	*Brassica oleracea* var. *botrytis*
Celery	*Apium graveolens*
Cherry	*Prunus cerasus* (also *Prunus* spp.)
Chickpea (garbanzo bean)	*Cicer arietinum*
Chinese cabbage	*Brassica chinensis* var. *pekinensis*
Cinnamon	*Cinnamomum zeylanicum*
Citrus	*Citrus* spp.
Clover	*Trifolium* spp.
Coconut	*Cocos nucifera*
Coffee	*Coffea arabica* (also *Coffea* spp.)
Collard	*Brassica oleracea* var. *acephala*
Corn	*Zea mays*
Cotton	*Gossypium hirsutum* (also *Gossypium* spp.)
Cowpea	*Vigna unguiculata*
Cranberry	*Vaccinium macrocarpon*
Cucumber	*Cucumis sativus*
Dasheen	*Colocasia esculenta*
Date palm	*Phoenix dactylifera*
Eggplant	*Solanum melongena*
Endive	*Chicorium endiva*
Faba bean	*Vicia faba*
Fig	*Ficus carica*
Flax	*Linum usitatissimum*
Garlic	*Allium sativum*
Ginger	*Zingiber officinale*
Gram, black	*Phaseolus mungo*
Gram, green	*Phaseolus aureus*
Grape	*Vitis vinifera*
Grapefruit	*Citrus paradisi*
Groundnut	*Arachis hypogaea*
Gourd	*Cucurbita* spp. (also *Luffa* spp.)
Guar	*Cyamopsis tetragonoloba*
Guava	*Psidium guajava*
Hemp	*Cannabis sativa*
Hops	*Humulus lupulus*
Jackfruit	*Artocarpus heterophyllus*
Jojoba	*Simmondsia chinensis*
Jute	*Corchorus* spp.
Kale	*Brassica oleracea* var. *acephala*
Kenaf	*Hibiscus cannabinus*
Kholrabi	*Brassica oleracea* var. *gongylodes*
Leek	*Allium porrum*
Lemon	*Citrus limon*
Lentil	*Lens esculenta*
Lettuce	*Lactuca sativa*
Lima bean	*Phaseolus lunatus*

Scientific Names of Some Common Crop Plants

Lime	*Citrus aurantifolia*
Litchi	*Litchi chinensis*
Lucerne	*Medicago sativa*
Lupine	*Lupinus* spp.
Maize	*Zea mays*
Mamey sapote	*Calocarpum sapota*
Mango	*Mangifera indica*
Millet, common	*Panicum miliaceum*
Millet, pearl	*Pennisetum americanum*
Mint	*Mentha* spp.
Mung bean	*Phaseolus aureus*
Mustards	*Brassica* spp.
Oat	*Avena sativa*
Oil palm	*Elaeis guineensis*
Okra	*Hibiscus esculentus*
Olive	*Olea europaea*
Onion	*Allium cepa*
Orange, sweet	*Citrus sinensis*
Papaya	*Carica papaya*
Parsley	*Petroselinum crispum*
Parsnip	*Pastinaca sativa*
Passionfruit	*Passiflora edulis*
Pea	*Pisum sativum*
Peach	*Prunus persica*
Peanut	*Arachis hypogaea*
Pear	*Pyrus communis*
Pecan	*Carya illinoensis*
Pepper, black	*Piper nigrum*
Pepper, most types	*Capsicum annuum* (also *Capsicum* spp.)
Pigeonpea	*Cajanus cajan*
Pineapple	*Ananas comosus*
Pistachio	*Pistachio vera*
Plantain	*Musa* spp.
Plum	*Prunus domestica* (also *Prunus* spp.)
Pomegranate	*Punica granatum*
Potato	*Solanum tuberosum*
Pumpkin	*Cucurbita pepo* (also *Cucurbita* spp.)
Radish	*Raphanus sativus*
Ramie	*Boehmeria nivea*
Rapeseed	*Brassica napus*
Raspberry	*Rubus* spp.
Rhubarb	*Rheum rhaponticum*
Rice	*Oryza sativa*
Rutabaga	*Brassica napobrassica*
Rye	*Secae cereale*
Safflower	*Carthamus tinctorius*
Sage	*Salvia officinalis*
Sesame	*Sesamum indicum*
Sisal	*Agave sisalana*
Sorghum	*Sorghum bicolor*
Soybean (soya)	*Glycine max*
Spinach	*Spinacia oleracea*
Squash, summer	*Cucurbita pepo*

Squash, winter	*Cucurbita* spp.
Starfruit	*Averrhoa carambola*
Strawberry	*Fragaria* spp.
Sugarbeet	*Beta vulgaris*
Sugarcane	*Saccharum* spp.
Sunflower	*Helianthus annuus*
Sweetpotato	*Ipomoea batatas*
Tannia	*Xanthosoma* spp.
Taro	*Colocasia esculenta*
Tea	*Camellia sinensis*
Tobacco	*Nicotiana tabacum*
Tomato	*Lycopersicon esculentum*
Tung	*Aleurites montana*
Turnip	*Brassica rapa*
Vetch	*Vicia* spp.
Watermelon	*Citrullus lanatus*
Wheat	*Triticum aestivum* (some *T. durum*)
Yam	*Dioscorea* spp.
Zucchini	*Cucurbita pepo*

*No list of food plants is ever complete; these are some of the more common cultivated crops. Some crops include multiple species and hybrids; these are shown as "spp."; for example, banana = *Musa* spp.

Appendix III

Conversions Between English and Metric Systems of Measurement

English Unit	Metric Equivalent
Length	
1 mile (mi)	= 1.61 kilometers (km)
1 yard (yd)	= 0.914 meters (m)
1 foot (ft)	= 0.305 meter (m)
1 inch (in)	= 2.54 centimeters (cm) = 25.4 millimeters (mm)
Area	
1 acre (A)	= 0.405 hectare (ha)
1 square mile (mi^2)	= 2.59 square kilometers (km^2)
1 square yard (yd^2)	= 0.836 square meter (m^2)
1 square inch (in^2)	= 6.45 square centimeters (cm^2)
Volume	
1 bushel (bu)	= 35.2 liters (L)*
1 gallon (gal)	= 3.78 liters (L)
1 quart (qt)	= 0.946 liter (L)
1 pint (pt)	= 473 milliliters (ml) = 0.473 L
	= 473 cubic centimeters (cm^3 or cc)
Mass	
1 ton (t)	= 0.907 metric tonnes (mt) = 0.907 megagrams (Mg)
1 pound (lb)	= 0.454 kilograms (kg)
1 ounce (oz)	= 28.4 grams (g)

Yield

1 pound per acre (lb/A) = 1.12 kilogram per hectare (kg/ha)*
1 ton per acre (t/A) = 2.24 metric tons per hectare (mt/ha) = 2.24 Mg/ha

Pressure

1 pound per square inch (psi) = 6.90 kilopascals (kPa) = 70.3 g/cm^2

Temperature

Farenheit (F) to Celsius (C): $C = 5/9 (F - 32)$
Celsius (C) = Farenheit (F): $F = 9/5 \, C + 32$

See Application 2–4 for units of energy measurement.

*Conversion of bushels (a measure of volume) to kilograms (a measure of mass) varies with crop. A specific conversion factor must be known for the crop. For example, 1 bu of corn containing 15.5% moisture weighs 56.0 lb = 25.4 kg.

Appendix IV

Glossary of Common Terms

abiotic decomposition a decomposition process that does not require a living organism

abiotic factors nonliving or physical components of a system

accuracy how close a measurement is to reality

actinomycetes a group of bacteria with a characteristic filamentous or branched appearance

action threshold pest density at which treatment is necessary to prevent economic injury

active dispersal organisms choose where they will move

active ingredient the actual toxic compound in a pesticide formulation

adjuvant a material mixed with a spray to improve its performance

adsorption attachment of chemical molecules to the surface of soil particles

aerobic respiration respiration that uses oxygen gas

aggregated distribution individual is located near other individuals in a field or site

aggregative response predators congregate in areas of high prey density

agroecosystem an agricultural ecosystem

agroforestry an intercropping system that combines cultivation of trees with other trees, livestock, or annual and perennial crops

algae simple plants and plant-like organisms common in aquatic and soil habitats (some classified as Plantae, others as Protista)

alleles genes that occupy the same locus on different chromosomes

allelopathy indirect effects of one plant on another through modifications in the environment

alley cropping an intercropping system in which annual crops are grown in between more permanent crops

amendment a material mixed into soil, often to improve soil fertility or structure

amoebae common predatory protozoa

amphimixis reproduction originating from cross fertilization

anaerobic respiration respiration that occurs when oxygen is absent or limited

annual plant that grows and reproduces in a single season

antibiosis production of compounds that inhibit growth of microorganisms

antihelminthic any drug used to treat parasitic worm infections

aquifer layers of porous underground rock containing groundwater

assimilated energy energy equivalent of any consumed material that is actually digested

augmentation the increase in population densities of natural enemies in a site

autocidal control sterile male technique for pest management

autotroph (see **producer**)

bacteriology the study of bacteria

bacterivore an organism that feeds on bacteria

biennial plant that grows vegetatively in the first year, then reproduces during the second

bioaccumulation cumulative buildup of toxic elements or compounds in the body of an organism

bioassay culture of an organism in living plants or animals

biodegradation breakdown of organic molecules (including pesticides) by microorganisms

biodiversity number of species present

biogeochemistry the study of the movement of elements between organisms and different parts of the physical environment

biological control the management of a pest by a living organism

biological degradation (see **biodegradation**)

biomagnification increase in concentration of toxic materials in the higher consumers within a food chain

biomass mass or weight of living tissue that remains in an organism after water is removed

biotic decomposition a decomposition process that requires a living organism

biotic factors biological components of a system

block (see **replication**)

blue-green algae close relatives of bacteria included in the kingdom Monera

C:N ratio the ratio of carbon to nitrogen in a substance

carrying capacity the upper limit to population size that a given environment will support

carryover the persistence of a herbicide residue in soil where it may affect a later crop

casting an earthworm deposit

cation exchange capacity the total number of exchangeable cations that a soil can absorb, expressed in centimoles of charge per kilogram of soil

centipede a myriapod with one pair of legs per body segment

chromosomes structures in cells that contain the genes

class a taxonomic group of related orders

classical biological control introduction of a natural enemy to manage a non-native pest

clumped distribution (see **aggregated distribution**)

coefficient of variation the ratio of the standard deviation to the mean, expressed as a percent

cohort a population of individuals of similar age

Collembola an abundant group of soil insects

community all populations in a given area

compost a material resulting from decomposition

composting the natural breakdown of organic materials to produce a humus-like substance

conservation tillage general term for practices in which tillage operations are reduced or eliminated

consumed energy energy equivalent of any material that is eaten by a consumer

consumer in food webs, an organism that obtains its carbon and energy from another organism

consumer subsidy equivalent difference between international and domestic price

conventional tillage general term for practices in which crop residues are buried and the soil surface is cleaned and smoothed prior to planting

cosmopolitan worldwide in distribution

cover crop a lower-value crop grown in a season that is less favorable for cash crop production

critical period time during crop growth when weeds will affect yield

crop damage threshold pest population level below which no measurable yield loss occurs

cross protection inoculation with a mild strain of virus to protect against a more pathogenic strain

cultural eutrophication acceleration of eutrophication by human activities

CV (see **coefficient of variation**)

dauer a resistant stage of a nematode

day neutral plant flowering not affected by day length

degree days (see **heat units**)

denitrification loss of nitrogen from soil as N_2 or other forms, usually under conditions of poor aeration

desertification conversion of arid or semiarid lands to deserts

desiccant a chemical applied to a crop plant before harvest to kill crop foliage and weeds

determinate plant in which vegetative growth stops at flowering

deterministic model that does not include statistical variation

detritus small fragments of decomposed litter that are no longer recognizable

digestible energy energy content of the material that is actually digested when used as a feed by an animal

dioecious both male and female are required for fertilization

disease progress curve a graph of disease incidence or severity over time

diversity richness and evenness of species in a community

diversity index a numerical measure of diversity

division a taxonomic group of related classes of plants

dolomitic limestone limestone that contains Mg^{2+} as well as Ca^{2+}

dormancy an interruption in the development of a plant or other organism

drainage basin (see **watershed**)

earthworms segmented annelid worms common in soil

ecological niche combination of physical and biological factors that define the role of an organism in an ecosystem

economic injury level pest density at which the value of the loss prevented equals the cost of treatment

ecosystem all biotic and abiotic factors in a defined area

ecotone a transition zone between two different kinds of ecosystems

ectomycorrhizae mycorrhizal fungi that grow around and between plant root cells but do not enter cells

ectoparasite a parasite that does not enter the host

eluviation leaching within a soil horizon

enchytraeids small, white or gray-colored relatives of earthworms

endomycorrhizae mycorrhizal fungi that grow directly into plant root cells

endoparasite a parasite that enters the host

endophyte a fungus that lives inside of a plant

epidemic an outbreak of a disease or pest population in an area

erosion movement of soil, including its removal from one location and deposition at another

eutrophication aging process that results from nutrient enrichment of lakes and ponds

evapotranspiration collective term including both evaporation and transpiration

evenness relative abundance of individuals among the various species present in a community

exponential growth population growth in an unlimited, favorable environment

facilitation modification of the environment by an intercrop

facultative parasite an organism that feeds on dead tissue or organic matter, but may also feed on living plant or animal tissue under certain conditions

family a taxonomic group of related genera

fecundity number of offspring produced per female

field capacity soil moisture level at which soil macropores are drained but micropores are filled with water

flatworms unsegmented worms belonging to the phylum Platyhelminthes

focus initial point of infection or colonization

foliar referring to aboveground plant parts

food chain the hierarchy of producers, primary consumers, secondary consumers, and higher-level consumers

food web a diagram that shows the organisms present in an ecosystem, with arrows indicating direction of food or energy transfer

fragmentation breakup of habitat into small, isolated patches

functional response the change in the rate of prey consumed per predator in relation to the prey density

fundamental niche niche occupied in the absence of competition or restraints on resources

fungivore an organism that feeds on fungi

gene the basic unit of heredity

generalist an organism that feeds on a variety of food sources

genotype genetic composition of an organism

genus a taxonomic group of related species

germplasm general term for genes from crop cultivars, varieties, lines, wild species, and other sources that could be used in crop improvement

global extinction all members of a species disappear from the earth

greenhouse effect a gradual warming of the earth due to increased absorption of heat by CO_2 and other gases

gross primary production the total amount of energy captured by plants during photosynthesis

guild a group of species that use the same food resource

hairworms specialized worm parasites of some insects

hard pan a compacted layer of subsurface soil

harvest index proportion of the crop yield that is actually used

heat units number of days or hours of accumulated temperature above some threshold

herbivorous plant-feeding

heterotroph (see **consumer**)

horizontal resistance general resistance to a range of pathogens or isolates

host an organism infected by a parasite

humus a dark-brown mixture of various substances and compounds resulting from decomposition

hydrological cycle the movement of water through and between the ecosystems and environments of the earth

hydrolysis reaction with water

hyperparasitism parasitism of a parasite

hyphae filamentous strands that make up the body, or mycelium, of a fungus

hypothesis a guess about the outcome of an experiment under a particular set of conditions

illuviation accumulation of materials within a soil horizon

immobilization the tie-up of nitrogen in an organic form, such as in the tissues of a microorganism or plant

incidence number of plants or organisms infected by a pathogen

indeterminate plant in which some vegetative growth continues after the first flowering

individual a single organism

induced resistance resistance in response to a pathogen

inorganic nitrogen (N) nitrogen forms that do not contain carbon

inorganic nitrogen fertilizer a nitrogen fertilizer that does not contain carbon

input anything that enters an agroecosystem

instar a stage between molts

integrated pest management an approach to managing pests only when needed, through a variety of methods

integration combined use of multiple tactics in a pest management strategy

intensity quantity of light received at a particular location

intercropping the planting of two or more crops simultaneously on the same unit of land

intermediate host an additional host required by some parasites to complete the life cycle

interspecific competition between members of difference species

intraspecific competition between members of the same species

intrinsic rate of increase maximum rate of population growth that a species can biologically support

invertebrate general term for an animal without a backbone

IPM (see **integrated pest management**)

isomorphous substitution replacement of a charged atom within a mineral soil with an atom of similar size, but often varying charge

isopod a common terrestrial crustacean

k (see **carrying capacity**)

k strategist term given to a persister that devotes more energy to maintenance of a stable population size than to reproduction

key pest one that is responsible for the majority of crop damage

kingdom a taxonomic group of related phyla or divisions

landscape all of the heterogeneous ecosystems and components within a geographic region

larva an immature stage in insects with complete metamorphosis

leaching loss of soluble materials due to water movement through the soil profile

limiting factor a resource essential to the growth of an organism that is in short supply

litter a layer of debris on the soil surface formed from fragments of plant and animal material

living mulch a cover crop grown in association with a cash crop, usually for weed control

local extinction all members of a species are eliminated from a region, but can still be found elsewhere

locus the position of a gene on a chromosome

logistic growth population growth in a limited environment

long-day plant plant that flowers only when night length is shorter than some critical value

macrofauna animals large enough to see with the unaided eye

macronutrient an element or nutrient required in relatively large quantities by an organism

macroparasite parasites characterized by long generation time; often produce infective stages that are released from the host to infect new hosts

macropores spaces between soil aggregates

marginal value theorem governs the length of time that a predator remains in a patch of prey

master horizon one of the common soil horizons

mean the numerical average of a set of measurements

mesofauna animals that are intermediate in size between microfauna and macrofauna

metamorphosis change in form during development

microarthropods collective term for mites and Collembola

microbivore general term to describe feeding on small organisms such as bacteria or fungi

microfauna animals too small to observe without a microscope

micronutrient an element or nutrient required in only small or trace amounts by an organism

microparasite organism that multiplies within the host or host cells; characterized by small size, short generation time, and high reproductive rates

micropores spaces within soil aggregates

migration mass movement of members of one species from one location to another

millipede a myriapod with two pairs of legs on most body segments

mineralization the release of inorganic nitrogen from organic nitrogen compounds

mites microscopic or near-microscopic arachnids common in agricultural habitats

molting shedding of an old exoskeleton during growth

monoculture repeated production of one crop on the same piece of land over time

mulch a material applied to the soil surface

mycelium the mass of hyphae that collectively make up the body of a fungus

mycoherbicide a herbicide developed from fungi that are pathogenic to weeds

mycology the study of fungi

mycoparasites fungi that are parasitic on other fungi

mycorrhizae specialized fungi that live in close association with plant roots and aid in nutrient uptake

myriapod a general term for a segmented arthropod with many pairs of legs

N rob the immobilization of soil nitrogen by bacteria or other microorganisms

natural product a compound produced by or isolated from living organisms

net primary production (NPP) the portion of the gross primary production that is stored in plant tissue over time

net reproductive rate total number of female offspring produced per female during one generation

nitrification two-step process in which ammonium is converted to nitrate

nitrogen fixation the conversion of N_2 gas from the atmosphere into NH_3

numerical response increased reproduction leading to a larger population of predators

nutrient cycling movement and recycling of elements essential for life

nymph an immature insect stage that resembles the adult

omnivore an organism that feeds on plant or animal matter

order a taxonomic group of related families

organic matter living organisms and any substances and organic compounds that originated from living organisms

organic nitrogen nitrogen that is included in organic (carbon-containing) compounds

organic nitrogen fertilizer a carbon compound that contains organic nitrogen as the nitrogen source

output anything removed from or leaving an agroecosystem

parasite an organism that feeds on and lives in or on the prey

parasitoid an insect that lays its eggs in the body of the host, killing the host during the course of its development

parthenogenesis reproduction originating from cross fertilization

partitioning allocation of new dry matter to various plant parts

passive dispersal organisms have no choice in where they move (e.g., moved by wind or water)

pathogen an infectious agent that can reproduce or replicate

perennial plant that lives for three years or more

period of stay amount of time that a group of animals remains in a paddock in a rotational grazing system

pesticide degradation chemical or biological breakdown of pesticide into other compounds

phenology the study of the sequence in which certain biological and developmental events occur

phenotype observable trait of an organism

pheromone a chemical compound given off by one individual that produces a behavioral response in another member of the same species

photoperiodism flowering of plants in response to length of night

photosynthetic pathways series of biochemical steps in photosynthesis; differs with C3, C4, and CAM plants

phylogenetic tree a diagram used to show relationships among different phyla

phylum a taxonomic group of related classes of animals

phytophagous herbivorous

phytotoxicity injury to crop plants

planarian a free-living flatworm

plant disease a disturbance from a pathogen or environmental factor that interferes with plant physiology

plant pathology the study of causal agents of plant diseases

plinthite a hard soil type formed from iron and aluminum oxides

plot the smallest unit of a field experiment

plow layer top layer of a soil profile that has been mixed by plowing

polyculture production of several different crops on the same piece of land

polygenic resistance resistance involving many genes

polyphagous feeding on a variety of food sources

population a group of individuals of the same species

population dynamics the study of the changes in population size over time

pore space space within or between soil aggregates

precision how close a series of measurements are to each other

predator an organism that feeds on other living organisms but does not take up residence in them

primary consumer a herbivore that obtains its energy from plant producers

primary tillage initial tillage operation in preparing a site

producer a green plant that produces an organic carbon food source for an ecosystem

production energy potential energy that has been stored in biomass over time

pupa a resting stage during insect development

r (see **intrinsic rate of increase**)

r **strategist** term given to a colonizer that devotes more energy to reproduction than to maintenance

random distribution an individual has an equal probability of being located at any point within a field or site

realized niche niche occupied when modifying factors are present

recharge area land that water filters through to reach an aquifer

regular distribution locations of individuals are equally spaced throughout a field or site

relay cropping an intercropping system in which a second crop is planted during the life cycle of the first crop

replication an experimental unit containing several plots; each plot has received a different treatment

resilience rate at which a community recovers following disturbance

resistance ability of a community to avoid disturbance or ability of an organism to withstand a pest or disease (definition depends on context)

respiration energy the portion of assimilated energy that is used to maintain the life functions and activities of the consumer

rest regrowth of pasture during rotational grazing

resurgence increase in pest populations to levels greater than those present before treatment

rhizobacteria general term for bacteria that colonize the region around plant roots

rhizosphere the region of the soil that is closely associated with plant roots

richness number of species in a community

rotifers microscopic aquatic invertebrates often found in soil

salinization accumulation of salts in the soil

sample unit size of the area or item from which an individual sample is collected

scientific method a formal sequence of stems in obtaining information: observation, hypothesis, experiment, results, conclusion

scientific name the unique two-part Latin name for a species

secondary consumer a consumer that feeds on primary consumers

secondary parasite (see **hyperparasitism**)

secondary tillage tillage for seedbed preparation and routine cultivation

seed bank the reservoir of weed seeds present in the soil

septicemia a general and widespread infection

severity proportion of plant tissue with disease symptoms

shifting cultivation clearing of a natural ecosystem for temporary production of agricultural crops

short-day plant plant that flowers only when night length is greater than some critical value

simulation model predictor of population estimates over small time intervals, using current weather and environmental information

sink place where output materials leaving an agroecosystem accumulate

slash-burn agriculture (see **shifting cultivation**)

slime mold a fungus-like organism related to protozoa

slugs gastropod molluscs without shells

snails gastropod molluscs with shells

sodicity accumulation of Na^+ ions in the soil

soil horizon a distinct layer within the soil profile

soil profile a vertical section extending from the soil surface down to the parent material

soil solarization heating soil under clear plastic to kill soilborne pests

soil texture proportions of particles of different size (sand, silt, clay) within a soil

soilborne referring to disease agents present in soil

source place where an input material is obtained

specialist an organism that feeds only on a few kinds of food items

species group of closely related organisms that are reproductively isolated from other such groups

spiders predatory arachnids common in agricultural habitats
springtails (see **Collembola**)
stable age distribution proportions of all age groups in the population are the same from generation to generation
standard deviation a measure of precision among a set of measurements
standing crop energy equivalent of biomass measured at one point in time
statistical tests methods for evaluating whether differences observed in experiments likely occurred as a result of treatment or as a result of chance
stochastic model that includes sources of error and statistical variation
stocking rate number of animals supported per unit of land
strata subdivisions within a field or any other area to be sampled
strategy overall plan used to manage a pest; can include one or more tactics
subsidy an agricultural price support
subtropical a climate with seasonal differences in both temperature and rainfall, located between tropical and midlatitude zones
succession natural and directional changes in community structure over time
survivorship curve graph showing the survival of members of a cohort at various ages
susceptible organism that has no defense against a particular disease or pest organism
symbiosis a relationship that is beneficial to both organisms involved
symphylid a small, white myriapod with up to twelve pairs of legs
synthetic fertilizer a fertilizer that is produced by an industrial process
synthetic product an artificial compound produced in a laboratory
systemic a pesticide that is taken up by a plant (or other organism) and translocated to various plant parts

tactic method for managing a particular pest
tardigrades microscopic aquatic invertebrates that may also occur in soil
taxonomy the branch of biology dealing with the classification of organisms
transect a series of samples collected along a straight path
transpiration the release of water from plants to the atmosphere
treatment a difference imposed on plots in an experiment
trophic level any one layer of producers or consumers within a trophic pyramid
trophic pyramid diagram of energy transfer through producers and several levels of consumers

underpopulation slower growth rate of population when potential mates are rare

vector an organism that acquires a pathogen from an infected host and transmits it to another host
vertebrate general term for an animal with a bony or cartilaginous backbone
vertical resistance resistance to a specific pathogen or isolate
vesicular-arbuscular mycorrhizae (see **endomycorrhizae**)
volatilization loss of gases, such as NH_3, to the atmosphere

waterlogging excessive moisture in a soil, often resulting from overirrigation or poor drainage
watershed area of land that collects surface water and diverts it into bodies of water
water table upper layer of water in the aquifer
wilting point low soil moisture level at which remaining water is held within soil micropores but is unavailable to plants

Index

Abiotic decomposition, 68
Abiotic factors, 2
Acaricide, 216
Accelerated biodegradation (*see* Biodegradation)
Accuracy, 10, 222–223
Acid soil, 98–99, 292–293
Actinomycetes, 36–37
Action threshold, 228, 231–232
Active ingredient, 217
Adsorption, 258
Adjuvant, 217
Aeration, 89
Aerobic respiration, 17
  decomposition, 68–69
Africa, cropping systems, 382
Aggregated distribution (*see* Distribution, aggregated)
Agriculture
  experimentation in, 6–11
  history of, 4–6
  sustainable, 392–395
Agroecology, 1
  information sources, 12–13
Agroecosystem
  animals in, 330–336
  carbon cycle, 17–19
  crop diversity, 148–149
  definition of, 2–3
  earthworms in, 42
  energy, 20–30
  environmental interactions, 250–252
  food webs in, 34–36, 46–51
  full-industrial, 28
  inputs and outputs to, 15–16
  intensive, 28, 266–268

Agroecosystem, *continued*
  landscape, 248–252
  metals, 262–263
  nitrogen in, 57–64
  pesticides, 257–261
  phosphorus in, 64–66
  potassium in, 67
  predators and parasites, 166–178
  pre-industrial, 28
  semi-industrial, 28
  stability of, 247–248
  subsistence, 266–268
  sustainable, 392–395
  tropical, 360–364
  water cycle in, 83–85
  weeds, 139–143
Agroforestry, 273–275, 382
Alcohol from crops, 30
Aldicarb, 217
Alfisol, 97
Algae, 36, 38
Alkaline soil, 98–99, 321–322
Allele, 150–151
  frequency, 151–155
Allelopathy, 127, 131
  weeds, 142
Alley cropping, 274, 382
Aluminum, 362–363
Amaranth, 143–144
Amaranthus spp., 143–144
Ambrosia artemisiifolia, 183
Amendments (*see* Organic amendments)
Amino acids, 366
Ammonia, 57–58, 63
Ammonium, 57–59

417

Amoebae, 39
Amphimixis, 120
Anaerobic respiration, 17
    decomposition, 69
Ananas comosus, 370
Animalia, 53
Animals, domestic (*see* Livestock)
Annelids, 39, 41 (*see also* Earthworms)
Annual, 139–140
Antibiosis, 194
Antihelminthic, 210
Ants, 44
Aquaculture, 381–382
Aquatic ecosystem, 255–256
Aquifer, 309–310
Arachis hypogaea, 364
Aridisol, 96
Aroid, 373, 375
Arthropods
    classification of, 43
    soil, 41–44
Artocarpus heterophyllus, 373–374
Aschelminthes, 45
Atrazine, 261
Augmentation, 221
Autocidal control, 220
Autotroph, 17
Azolla, 59
Azotobacter, 59

Bacillus thuringiensis, 220
Bacteria
    bactericide, 193
    biodegradation, 260
    biological control, 194
    cyanobacteria, 38
    denitrifying, 58
    management of, 192–196
    nitrifying, 58
    nitrogen-fixing, 59–61
    parasitic, 175–176
    population growth, 111
    plant pathogenic, 187, 189, 191–196
    rhizobacteria, 37
    soil, 36–37
Bactericide, 193
Bacteriology, 188
Bacterivore, 39–40
Banana, 368–369
    nematode management, 199
    wind damage, 82

Barnyard grass, 145–146
Bean
    broad, 377
    garbanzo, 377
    lima, 190
    mosaic virus 190
    yield and spacing, 31
Bees, 203
Beetles
    ladybird, 168
    soil, 44
Beneficial insects (*see* Biological control)
Bermudagrass, 145
Bicarbonate, 19, 57
Biennial, 139–140
Bindweed, field, 147
Bioaccumulation, 261
Bioassay, 223
Biodegradation, 259–261
Biodiversity, 36, 148–149, 243
Biogas, 30, 305
Biogeochemistry, 16
Biological control
    advantages and disadvantages, 177–178
    classical, 221
    definitions, 220
    disease management, 193–194
    green lacewing, 168
    insect management, 174, 207
    ladybird beetles, 168
    natural, 220–221
    nematode management, 200
    predators and parasites in, 166–167, 174,
        176–178, 220–221
    strategies, 220–221
    weed management, 182, 184
Biomagnification, 261–262
Biomass, 22
    accumulation in plants, 102–103
    crops as fuel sources, 30
    partitioning in plants, 102–103
Biotic decomposition, 68–69
Biotic factors, 2
Birds as pests, 208
Block, 7
Blue-green algae, 38
Boll weevil, 234–235
Braconid wasp, 174
Breeding (*see* Plant breeding)
Broadbean, 377
Bt (*see Bacillus thuringiensis*)

Buffer strip, 249–250, 254
Bugs, 205
Burning (*see also* Shifting cultivation)
    CO_2 release, 18–19
    decomposition, 68
Burrowing nematode, 199

C3 and C4 plants, 22, 79, 182, 269
Cacao, 377
Cajanus cajan, 375
Calcium, 67
California, water conservation, 317
Calorie, 25
Carbofuran, 260–261
Carbohydrate content, 366, 377, 380
Carbon, 16–20, 299
    cycle, 17–19
    essential element, 56–57
    photosynthesis, 16
    respiration, 17
Carbon dioxide
    carbonic acid, 19
    fossil fuel, 18–20
    global warming, 19–21
    photosynthesis, 16
    production by agriculture, 19–20
    respiration, 17
Carbon monoxide, 17
Carbon-nitrogen ratio, 72
    decomposition rate, 71–73
    immobilization, 73
    organic amendments, 298–300
    organic fertilizers, 72–73
    values for various materials, 72
Carbonate, 19
Carbonic acid, 19
Carica papaya, 370
Carrion, 44
Carrying capacity, 117
    changed by agriculture, 119
    pest management, 230–232
    plant population, 128–130
Carryover, 185
Cassava, 373, 376
    ethanol source, 30
Casting, earthworm, 42
Caterpillar, 202
Cation exchange capacity, 98
Cattle, 209–210 (*see also* Livestock, pests of)
Centipede, 44
Cereal crops, 380

Chemical control (*see* Pesticides)
Chemical processes
    photosynthesis, 16
    respiration, 17
Chenopodium album, 146
Chickpea, 377
Chilling days, 104–105
Chilling hours, 104–105
China
    cropping systems, 381–382
    water issues, 311–313
Chromosome, 149–151
Cicer arietinum, 377
Citrus, 368, 370
Class, 53
Classification
    organisms, 53
    pesticides, 216–218
    soils, 96–97
Clay, 90–92
Climate
    change and global warming, 20, 81–82
    effect on decomposition, 70
    factors determining, 77–81
    tropics, 361–362
    types, 76–77
Clumped distribution (*see* Distribution, aggregated)
C:N ratio (*see* Carbon-nitrogen ratio)
Cocoa, 377
Coconut, 377–379
Cocos nucifera, 377–379
Coefficient of variation, 10, 225–226
Coffee, 379
Cogongrass, 146
Cohort, 110, 115
Collembola, 41, 43
Colocasia, 373
Colonization, 132–134
    distribution patterns, 132–134
    establishment, 132–134
Community, 2
    climax, 135
Competition, 127–131
    allelopathy, 131
    ecological niche, 126–127
    intercropping, 268–273
    interspecific, 127, 129–130
    intraspecific, 127–129
    weed management, 140–142, 181–182

Compost, 71
Composting, 71
Conservation
    energy in food chains, 328–330
    genetic resources, 156, 159
    human population growth, 388–390
    natural enemies, 221
    natural resources, 355–356
    nutrients, 287–305
    phosphorus, 356
    soil, 88, 290
    sustainability, 392–395
    tillage (see Tillage, conservation)
    water, 291, 312–320
Consumer
    carbon, 17
    demand, 351
    economic choices, 351, 354–355
    feeding terminology, 40
    primary, 24
    secondary, 24
Consumer subsidy equivalent, 346
Convolvulus arvensis, 147
Corn
    ethanol source, 30
    pesticides, 215
    plant density, 108
    U.S. yields, 28
Cosmopolitan, 41
Cotton
    Bt cultivar, 220
    pest management, 234–235
    pesticides, 215
Cover crops, 278–280
    examples, 280
    green manures, 302–305
    living mulch, 280
    management, 279
Cowpea, 279, 365–366, 368
    determinate characteristics, 108
    powdery mildew, 189
Crabgrass, large, 147
Critical period, 182
Crop (see also Plant, individual crop names)
    genetic diversity, 162
    losses, 214, 226–227, 392
    management, 230, 234
    scientific name, 403–406
    transgenic, 220
    tropical, 364–381
Crop damage threshold, 226–227
Crop rotation, 276–280

Crop rotation, *continued*
    cover crops, 278–280
    crop sequence, 276–277
    green manures, 303–304
Cropland, 386–387, 392
Cropping systems, 266–284
    agroforestry, 273–275
    cover crops, 278–280
    crop rotation, 276–280
    integration of, 283
    intercropping, 268–276
    living mulches, 280
    multiple cropping, 268–280
    relay cropping, 274, 276
    shifting cultivation, 280–283
    tropical, 381–382
Cross protection, 194
Cultivar, 148–162
    development, 155–156
    genetic variation, 149–151, 162
    hybrids, 156
    pest resistance, 159–161
Cultural practices, 221–222
    disease management, 193, 195–196
    insect management, 204–208
    nematode management, 199–200
    weed management, 184, 187
CV (see Coefficient of variation)
Cyanobacteria, 38
Cynodon dactylon, 145
Cyperus esculentus, 143
Cyperus rotundus, 143
Cyst, 198
Cyst nematode (see Nematode, cyst)

Dairies
    nitrates, 253
Data
    analysis, 8–11
    obtaining, 7–8
Dauer, 174
Day length, 77–79
DDT, 260–261
Decomposition, 68–73
    abiotic, 68
    biotic, 68–69
    composting, 71
    factors affecting, 70–73
    organisms involved in, 45–46, 69–70
    recycling of nitrogen, 58, 62–64
    recycling of phosphorus, 65–66
    sequence of events, 69

Deforestation, 388–389
Degradation
    biological, 259–260
    pesticide, 259–261
Degree days, 103–105
Demand, 351
Denitrification, 58
Desertification, 333–334
Desiccant, 186
Determinate plant, 105, 108
Deterministic model, 121
Detritus, 69
Digitaria sanguinalis, 147
Dioecious, 370
Dioscorea, 373
Disease (*see* Plant disease)
Disease progress curve, 191–192
Disease triangle, 191
Dispersal, 131–132, 224
    active, 131
    passive, 131
    pathogens, 131–132
    weed seeds, 140
Dispersion (*see* Distribution)
Distribution
    aggregated, 132–133, 223–224
    clumped, 132, 223–224
    food, 391
    patterns, 132
    random, 132, 223–224
    regular, 132–133, 223–224
    sampling, 223–226
Diversity
    benefits to agroecosystems, 148–149
    crop breeding, 162
    genetic, 149, 162
    Green Revolution, 157–158
    indices, 243–247
    insect species, 249
    levels of, 248–250
    limitations of, 245, 247
    maintenance of, 248–250, 271
    measurement of, 244–247
    regional, 162–163
    Shannon, 245–247
    Simpson, 245–247
    species, 148–149, 243–247
    stability, 247–248
Division, 53
Dolomitic limestone, 67
Dominance, 244
Dormancy, 109

Drainage
    basin, 309
    systems, 323–324
Drought, 310
Dry matter (*see* Biomass)
Dynamics (*see* Population dynamics)

Earth
    precipitation patterns, 80–81
    seasons, 77–78
    temperature patterns, 79–80
Earthworms
    benefits to agroecosystems, 42
    soil, 36, 39, 41–42
Echinochloa crusgalli, 145
Echinochloa colonum, 145
Ecological efficiency, 26
Ecological niche (*see* Niche)
Ecology, 1
Economics
    consumer choice, 354–355
    environment, 354
    injury level, 228, 231–232
    pesticide use, 214–215
    pest management, 226–229
    resource conservation, 354–355
    small farmers, 353–354
    subsidies, 346–353
    supply and demand, 351
    threshold, 227–229
    tropical agriculture, 360–361
    water use, 317
Ecosystems, 2–3
    health, 148–149
    interconnection of, 50–51, 252, 388
Ecotone, 249–250
Ectomycorrhizae, 66–67
Ectoparasite, 196
Egypt
    population growth, 391
    water issues, 325
Elements, essential, 55–57
Eleusine indica, 146
Eluviation, 95
Enchytraeids, 36, 41
Endomycorrhizae, 66–67
Endoparasite, 196
Endophyte, 38
Energy, 16–17, 20–30
    agricultural inputs, 26–27
    agroecosystem, 20–30

Energy, *continued*
 assimilated, 23–24
 conservation tillage, 294
 consumed, 23–24
 consumption rates, 344–346
 content of materials, 22–23, 366, 377, 380
 digestible, 22–23
 efficiency of transfer, 26
 fertilizer manufacture, 62–63
 food chain efficiency, 328–330
 input and yield, 28–29
 photosynthesis, 16, 20–21
 production, 23–24
 pyramid, 24–25
 respiration, 17, 23–24
 solar radiation, 16, 20–21, 361
 thermodynamics, 21–22
 units, 25
Entisol, 96
Environment (*see also* Light, pH, Physical factors, Soil, Temperature, Water)
 agrochemicals, 218–219, 252–263
 economics, 354–355
 fertilizers, 253–256
 interactions with agroecosystems, 250–252
 metals, 262–263
 pesticides, 256–263
 plant disease, 191–192
 population growth, 388–389
 weeds, 182
Epidemic, 160
Eradication, 215–216, 235
Erosion, 86–88
 conservation tillage, 290
 water, 87
 wind, 87–88
Error, 225–226
Establishment, 132–134
Ethanol from crops, 30
Ethiopia
 population growth, 391
Eutrophication, 255–256
Evaporation, 85, 291
Evapotranspiration, 85
 factors influencing, 85
Evenness, 244–247
Exclusion (*see* Prevention methods)
Experimentation, 6–11
 analysis of data, 8–11
 conclusions, 11
 obtaining data, 6–8

Exponential growth, 111–117, 386
Export subsidy, 347
Extinction, 134
 global, 134
 human population affecting, 389
 local, 134

Facilitation, 270
Facultative parasite, 48
Family, 53
Fat content, 366, 377, 380
Fecundity, 115–116
Feeding habits, 34–36, 40
 generalist, 36
 nematodes, 39–40
 soil organisms, 36–48
 specialist, 36
Fertilizer
 conservation tillage, 291–293
 environment, 253–256
 green manure, 303–304
 inorganic, 62–63
 leaching, 253–256
 nitrogen, 59–60, 62–63
 NPK analysis, 68
 organic, 59–60, 62–63
 salts, 63, 99
 soil pH, 99
 synthetic, 62–63
 use in U.S., 29
Fiber crops, 380
Field capacity, 93
Field experiment (*see* Experimentation)
Filtration, vegetative, 254
Fire (*see* Burning, Shifting cultivation)
Fixation (*see* Nitrogen fixation)
Flatworms, 45, 209
Flies, 44
Flooding, 310
Flowering and photoperiod, 78–79, 105–106
Focus, 132
Foliar, 192
Food chain, 24, 34
 biomagnification, 261–262
 shortening, 328–330
Food supply, human, 330, 390–392
Food web
 above ground, 34–36, 50–51
 below ground, 46–51
 cyclical nature of, 49
 interconnection, 50–51

Forecasting, 121, 251, 386
Formulation of pesticide, 217
Fossil fuels
    carbon dioxide, 19
    population growth, 389–390
Fragmentation, 251
Fruit crops
    pest management, 236
    tropical, 368–373
Full-industrial agroecosystem, 28
Fumigant, 186, 237–238
Functional response (*see* Predators, responses to prey)
Fungi, 38, 53, 188
    biological control of weeds, 184
    management, 192–196
    parasites, 176
    plant disease, 187–196
    soil, 36, 38, 48
Fungicide, 193–194, 215
Fungivore, 39–40

Gene, 150–151
    resistance, 159–161
    gene bank, 156
Generalist food habits, 36, 163, 172–173
Genetic drift, 151–155
Genetic engineering, 150
    crop cultivars, 220
Genetic variation
    conservation, 156, 159
    crop breeding, 162
    crop cultivars, 149–151
Genotype, 151
Genus, 53
Germplasm, 150
Global warming, 20, 81–82
Glycine max, 364
Goosegrass, 146
Gramineae as C4 plants, 22
Grazing
    carbon cycle, 17–19
    continuous, 337
    creep, 342
    management, 336–342
    period of stay, 341
    rest periods, 338–341
    rotational, 337–342
    set stocking, 337
    Voisin, 338–342
Green manure, 302–305

Green manure, *continued*
    examples, 304
    fertilizer, 303–304
    future outlook, 304–305
    legumes, 303–305
    nitrogen fixation, 303
Green revolution, 157–158
Greenhouse effect, 20
Groundnut (*see* Peanut)
Guava, 368, 371
Guild, 36
Gypsum, 322

Hairworms, 45
Hard pan, 95
Hardy-Weinberg Law, 151–155
Harvest index, 30
Heat units, 103–105
Herbicide, 184–186, 214–215
    desiccant, 186
    nonselective, 185
    selective, 185
    tolerance, 220
Herbivore, 40, 162–164
Herbivorous, 34, 40
Herbivory, 162–164
Heredity, 149–151
Heterodera (*see* Nematode, cyst)
Heterotroph, 17
Heterozygous, 150–154
History
    agriculture, 4–6
    desertification, 333
Histosol, 96
Homozygous, 150–154
Horizon (*see* Soil horizon)
Horsehair worms, 45
Host, 166, 191, 209
Host plant resistance (*see* Resistance)
Human population (*see* Population growth, human)
Humus, 69
Hybrids, 156
Hydrogen
    essential element, 56–57
    photosynthesis, 16
Hydrogen sulfide, 17
Hydrological cycle, 84–85
Hydrolysis, 259
Hyperparasite, 175, 194

Hyphae, 38
Hypothesis, 6

Illuviation, 95
Immobilization, 57, 292
    C:N ratio, 72–73
Imperata cylindrica, 146
Import quota, 347
Inceptisol, 96
Incidence of disease, 191
Indeterminate plant, 105, 108
Individual, 2
Industrialized agriculture (*see* Agroecosystem, intensive)
Inoculation with *Rhizobium,* 60
Insecticides, 205, 207, 214–215
Insects, 201–208
    agricultural importance, 204
    beneficial, 202–204
    conservation tillage, 294
    crop diversity, 249
    intercropping, 270–271
    life cycles, 201–202
    management, 204–208
    orders, 204
    parasites of, 39, 173–175
    pests, 202–208
    predators, 167–173
    soil, 41, 43–44
    virus vectors, 190
Instar, 201
Integrated pest management (*see* Pest management, integrated)
Intensive agriculture (*see* Agroecosystem, intensive)
Interactions
    agriculture-environment, 250–252
    multiple pests, 232–233
Intercropping, 268–276
    advantages, 269–273
    agroforestry, 273–275
    alley cropping, 274, 382
    competition, 269
    disadvantages, 283–284
    examples, 381–382
    insect management, 270–271
    land equivalent ratio, 271–273
    legumes, 270–271
    mechanisms, 269–271
    relay cropping, 274, 276
Intermediate host, 209
International justice, 355–357

International trade, 346–352
Internet, 12
Interspecific competition (*see* Competition, interspecific)
Intraspecific competition (*see* Competition, intraspecific)
Intrinsic rate of increase (*see* r)
Invertebrate, 39, 54
IPM (*see* Pest management, integrated)
Ipomoea batatas, 365
Irrigation (*see also* Water)
    conservation, 313–317
    drip, 317–318
    efficiency of, 314–317
    micro, 317
    overhead, 316
    surface, 315–316
Isolate, 191
Isomorphous substitution, 98
Isopods, 44

Jackfruit, 373–374
Johnsongrass, 146
Joule, 25
Journals, scientific, 12
Jungle rice, 145–146
Justice, 355–357

K (*see* Carrying capacity)
K strategist, 120–121
    succession, 135–136, 181
Key pest, 233
Kingdom, 53

Lacewings, 168
Ladybird beetles, 168
Lambsquarter, 146
Land (*see* Cropland)
Land equivalent ratio, 271–273
Landscape, 243, 248–252
    agriculture-environment interactions, 250–252
    diversity in, 248–250
Large crabgrass, 147
Larva, 201–202
Late blight, 155, 236–237
Laterite, 362
Leaching, 58
    nitrates, 253–254
    pesticides, 258
Leafminer, 206, 221

Legumes
    cover crops, 278–280
    green manure, 303–305
    intercrop, 270–271
    nitrogen fixation, 60
    protein content, 366
    pulses, 375, 377
    rotation, 276–277
Lens esculenta, 375
Lentil, 375
Leucaena, 274–275, 382
Life table, 115–117
Light (*see also* Solar radiation)
    energy, 16, 20–21
    intensity, 21, 79, 361
    visible spectrum, 21
Lime (fruit), 370
Limestone, 67
Liming of soil, 67, 99
Limiting factors, 83
Litter, 41, 69
Livestock, 328–342
    agroecosystems with, 330–336
    desertification, 333–334
    food chain, 328–330
    grazing management, 336–342
    manure, 332
    parasites, 176
    pests, 209–210
    production systems, 330–333, 335–336
Locus, 150–151
Logistic growth, 117–120
Lotka-Volterra model, 167, 169–170, 177

Macrofauna, 39
Macronutrient, 55–56, 64–68
Macroparasites, 175
Macropores, 90–92
Maize (*see* Corn)
Malanga, 373, 375
Malthus, 384–385
Mammals as pests, 208
Management (*see also* Pest management)
Management, crop, 230, 234
    grazing, 336–342
    tropical agroecosystems, 362–364
Mangifera, 368
Mango, 368, 370, 372
Manihot esculenta, 373
Manioc, 373
Manure, 332

Marginal value theorem, 173
Master horizon, 94
Maturity groups, 106–107
Mean, 8–9, 223–226
Mechanical management of weeds, 184, 186–187
Meloidogyne (*see* Nematode, root-knot)
Mesofauna, 39
Metals addition to agroecosystems, 262–263, 302
    toxicity, 56
Metamorphosis, 201
Methane, 17, 30, 69
Methyl bromide, 237–238
Microarthropods, 39
Microbivore, 40
Microfauna, 39
Microhabitat, 224
Micro-irrigation (*see* Irrigation, micro)
Micronutrient, 55–56, 67–68
Microparasites, 175
Micropores, 90–92
Migration, 131
Mildew, 189
Millipedes, 44–45
Mineral soil (*see* Soil, mineral)
Mineralization, 57–58
    nitrogen cycling, 58
Minimum tillage (*see* Tillage)
Mites
    agricultural importance, 204
    oribatid, 41
    soil, 36, 41, 46
Models
    deterministic, 121
    examples for various crops, 122
    exponential growth, 111–115
    interspecific competition, 129–130
    intraspecific competition, 127–129
    logistic growth, 117–120
    Lotka-Volterra, 167, 169–170
    predator-prey systems, 167, 169–171
    simulation, 121
    stochastic, 121
    succession, 135–136
    types, 121
    validation, 121
Mollisol, 97
Molting, 201
Monera, 53
Monoculture, 266–268
Muck, 91

Mulch, 295–298
    living, 280, 296
    plastic, 237–238, 296, 298, 318
    synthetic, 296
Multicultivos, 381
Multiple cropping (*see* Cropping systems, multiple cropping)
Musa, 368
Mycelium, 38
Mycoherbicide, 184
Mycology, 188
Mycoparasites, 194
Mycoplasma, 188–189
Mycorrhizae, 38, 66–67
    types of, 66–67
Myriapod, 44

Natural enemies (*see* Biological control)
Natural product, 216
Natural resources (*see also* Conservation, Energy, Soil, Water)
    conflict over, 324–326, 355–357
    human population pressures, 389–390
    subsidized, 347
Natural selection, 149
    plant breeding, 155–156
Nematodes
    burrowing, 199
    cyst, 198, 236
    damage and symptoms, 196–198
    distribution, 133
    ectoparasites, 196
    endoparasites, 196
    feeding habits, 39–40
    foliar, 197
    insect parasites, 39, 174
    management, 198–201
    parasites, 176
    plant parasites, 188, 196–201
    root-knot, 165, 196–198
    soil, 39–40
    soybean cyst, 236
Nematomorpha, 45
Net reproductive rate, 115–116
Niche
    competition, 127–130
    ecological, 126–127
    factors determining, 126–127
    fundamental, 126–127
    intercropping, 268–270
    realized, 127
Nigeria, 274
Nile River, 325

Nitrate, 57
    dairies, 253
    environment, 253–255
    fertilizer sources, 62
    leaching, 253–254
    nitrification, 58
    organic amendments, 299–301
    water, 253–254
Nitrification, 58
Nitrifying bacteria, 58
Nitrite, 57
    nitrification, 58
Nitrogen
    agroecosystems, 57–64
    conservation tillage, 291–293
    cycle, 62–64
    fertilizer, 59–60, 62–63, 68
    fixation, 59–61, 303
    gas, 57–60, 62–64
    green manures, 298–302
    inorganic, 57, 62–63
    leaching, 253–254
    mineralization, 57–58
    organic sources, 57, 62–63, 298–301
    rob, 73
    water, 253–254
Nitrous oxide, 57
No tillage (*see* Tillage)
Nutrient
    conservation of, 287–305
    conservation tillage, 291–293
    deficiencies, 188
    essential, 55–57
    organic amendments, 298–302
    shifting cultivation, 363
    tropical soils, 362–363
Nutrient cycling, 55–68
    conservation tillage, 291–293
    decomposition, 68–73
    food webs, 49
    nitrogen cycle, 62–64
    phosphorus cycle, 65–66
    soil, 89
Nutsedge, 143, 145
    purple, 143, 145
    yellow, 143, 145
Nymph, 201

Oil crops, 380–381
Omnivore, 35, 40
Onion, 129
Optimal foraging theory, 171–173
Order, 53

Organic amendments, 295–305
　　examples, 298
　　green manures, 302–305
　　metals in, 262–263
　　nematode management, 199–200
　　nitrogen source, 298–305
　　pest management, 302
Organic matter, 68
　　O horizon, 94
Organic soil (*see* Soil, organic)
Organisms
　　classification of, 53–54
　　soil, 36–46
Organochlorines, 261–262
Oryza, 379
Oxisol, 97
Oxygen
　　essential element, 56–57
　　photosynthesis, 16
　　respiration, 17
　　soil, 89

Paddock, 339–342
Papaya, 370
Parasites, 166–167, 173–176
　　animal, 176, 209–210
　　biological control agents, 174, 176–178
　　crop plants, 175–176
　　facultative, 48
　　hyperparasites, 175
　　insect, 39, 174
　　primary, 175
　　secondary, 175
Parasitoid, 166
　　Braconid wasp, 174
Parthenogenesis, 120
Partitioning of biomass, 102–103
Pasture management (*see* Grazing)
Pathogens
　　agents of plant disease, 187–191
　　dispersal, 131–132
　　establishment of, 132–133
　　isolates, 191
Peanut, 61, 364–365
Perennial, 139–140
Persistence of pesticides, 259–261
Pest (*see* Bacteria, Birds, Fungi, Insects, Mites, Nematodes, Pest management, Rodents, Viruses, Weeds)
Pest management, 214–239
　　bacteria, 192–196
　　biological control, 166–168, 174, 176–178, 220–221

Pest management, *continued*
　　birds, 208
　　cost/benefit, 229
　　cultural practices (*see* Cultural practices)
　　decision making, 222–230
　　dynamic, 228–229
　　ecological basis, 230–232
　　economics, 226–229
　　examples, 234–239
　　fungi, 192–196
　　insects, 204–208
　　integrated, 222, 233–239
　　integration, 215, 222, 233–239
　　mammals, 208–210
　　multiple pests, 232–233
　　nematodes, 198–201
　　organic amendments, 302
　　plant pathogens, 192–196
　　preemptive, 216
　　resistance, 159–161, 219–220
　　responsive, 216
　　rodents, 208
　　sampling, 222–226
　　scouting, 228, 230
　　solarization, 195
　　static, 228–229
　　strategies, 215–216
　　tactics, 215–222
　　thresholds, 226–229
　　viruses, 192–196
　　weeds, 139–143, 180–187
Pesticides, 216–219 (*see also* Bactericide, Fumigant, Fungicide, Herbicide, Insecticides)
　　active ingredient, 217
　　advantages, 218
　　bioaccumulation, 261–262
　　classification, 216–218
　　degradation, 259–261
　　disadvantages, 218–219, 256–257
　　disease management, 193–194
　　environment, 252–263
　　expectations, 219
　　formulation, 217
　　inorganic, 260–261
　　insect management, 207
　　leaching, 258
　　mode of action, 217–218
　　names, 217
　　natural, 216
　　nematode management, 200
　　non-target effects, 218–219, 256–257
　　persistence of, 259–261

Pesticides, *continued*
   resistance to, 219
   runoff, 258
   sales, 214–215
   synthetic, 216
   systemic, 218
   weed management, 184
pH, 98–99, 321–322
   liming, 67
   phosphorus availability, 65
   tropical soils, 362
Phenology, 105–106
Phenotype, 151
Pheromone, 205
Phosphorus
   agroecosystems, 64–66
   aquatic ecosystems, 255–256
   cycle, 65–66
   fertilizer, 68
   mycorrhizae, 66–67
   natural resource, 356
   soil chemistry, 65–66
   soil pH, 65
Photoperiod, 78
   flowering, 78–79, 105–107
   soybean, 106–107
Photosynthesis, 16–26
Photosynthetic pathway, 22
Phylogenetic tree, 54
Phylum, 53
Physical factors
   agroecosystem, 76–99
   ecological niche, 126–127
Physical methods
   disease management, 193–195
   insect management, 207
   nematode management, 200
   weed management, 184
Phytophagous, 40
Phytophthora infestans, 155, 236–237
Phytotoxicity, 185
Pigeonpea, 292, 375, 377
Pigweed, 144
Pineapple, 370
   CAM plant, 22, 370
Planarians, 45
Plant
   breeding, 155–156, 162
   C3 and C4, 22, 79, 182
   CAM, 22, 79
   competition, 127–131
   cultivars, 148–165
   day-neutral, 79

Plant, *continued*
   defense to herbivory, 162–164
   determinate, 105, 108
   grazing, 337–338
   growth, 102–106
   indeterminate, 105, 108
   long-day, 78–79
   parasitic, 176, 181, 188
   population, 30, 107–108, 128–130
   short-day, 78
   spacing, 30–31, 129
   succession, 134–137
Plant disease, 187–196
   causative agents, 187–191
   conservation tillage, 293–294
   environment, 191–192
   incidence, 191
   isolates, 191
   management, 192–196
   noninfectious agents, 188
   resistance, 159–161
   severity, 191
   triangle, 191
Plant pathogen (*see* Plant disease)
Plant pathology, 188
Plantae, 53
Plantain, 368
Plinthite, 95
Plot, experimental, 7
Plow layer, 95
Plow pan, 95
Politics
   animal production, 330
   conflict, 357
   economics, 346–357
   environment, 354–355
   international, 355–357
   population growth, 384–396
   resource conservation, 355
   small farmers, 353–357
   subsidies, 346–353
   water issues, 311–313, 324–326
Pollinator, 203–204
Polyculture, 266–268 (*see also* Cropping systems)
   problems with, 283–284
Polyphagous, 36
Population, 2
   genetics, 151–155
   human, 384–396
   plant, 30
   world, 384–386
Population dynamics, 101–122
   pest management, 230–232

Population dynamics, *continued*
    plant disease, 191–192
    plants, 107–111
    predator-prey systems, 167, 169–170
    weeds, 109–111
Population growth
    density-dependent, 117–120
    density-independent, 111–115, 386
    examples, 122
    exponential, 111–115, 386
    human, 384–396
    logistic, 117–120
    models, 121–122
Pore space, 90–93
Portulaca oleracea, 146
Postemergence, 185
Potassium
    agroecosystems, 67
    fertilizer, 68
Potato
    ethanol source, 30
    famine, 155, 236
    late blight, 155, 236–237
    rotation, 277
Powdery mildew, 189
Precipitation
    patterns and climate, 76–77, 80–81
    tropics, 361
Precision, 10, 223–224
    agriculture, 224
    sampling, 225–226
Predators, 39–40, 166–173
    aggregative response, 171
    biological control, 176–178
    diet width, 171–173
    foraging efficiency, 173
    functional response, 170–171
    models, 167, 169–171
    numerical response, 171
    optimal foraging theory, 171–173
    responses to prey, 170–171
    soil ecosystems, 46
Preemergence, 185
Pre-industrial agroecosystem, 28
Preplant, 185
Preventive methods, 215
    disease management, 193, 195
    insect management, 207
    nematode management, 200
    weed management, 184, 187
Price (*see* Economics)
Price supports (*see* Subsidies)
Producer, 17, 25

Production, 23–24
    gross primary, 24
    net primary, 24
Protein content, 366, 377, 380
Protista, 53
Protozoa, 36, 38–39, 188
Psidium guajava, 368
Pulse crops, 375, 377
Pupa, 202
Purslane, common, 146

Quarantine (*see* Preventive methods)

r, 112–114
    estimation of, 112, 114, 117
    life tables, 117
r strategist, 120–121
    succession, 135–136, 180–181
Radopholus similis, 199
Ragweed, 183
Rainfall (*see* Precipitation)
Random distribution (*see* Distribution, random)
Random sample, 224–225
Recharge area, 310
Reclamation
    desertification, 334
    saline and sodic soils, 322
Recycling
    nitrogen, 58–64
    nutrients, 49, 89
    pesticide residues, 259–260
    phosphorus, 65
    wastewater, 319
Regular distribution (*see* Distribution, regular)
Relay cropping, 274, 276
Replication, 7–8
Resilience, 247
Resistance, 159–161, 219–220
    breaking of, 161
    carrying capacity, 119
    defense to herbivory, 162–164
    disease, 159–161
    disturbance, 247
    genetics of, 161
    horizontal, 161
    induced, 194
    nematode, 199, 236
    pest, 159–161
    pesticide, 219
    polygenic, 161
    postinfectional, 160
    preinfectional, 160

Resistance, *continued*
   tolerance, 160
   types, 160–161
   vertical, 161
Respiration, 17–20, 23–26
   aerobic, 17
   anaerobic, 17
Resurgence, 219
Rhizobacteria, 37
Rhizobium, 59–61
Rhizosphere, 37
Rice, 379
Richness, 244
Rodenticide, 216
Rodents, 208
Root crops, 373, 375–377
   nutritional composition, 377
Root-knot nematode, 165, 196–198
Rotation (*see* Crop rotation)
Rotifers, 45
Runoff, 309
   fertilizer elements, 253–256
   pesticides, 258
Ryegrass toxicity, 232

Saccharum, 380
Salinity, 320–322
Salinization, 314, 320–322
Salts, 56, 63, 99, 321–322
Salvador, cropping systems, 381
Sample, 222
   number of, 225–226
   random, 224–225
   stratified, 225
   systematic, 224–225
Sample unit, 223
Sampling, 222–226
   error, 225–226
   methods, 222–223
   number of samples, 225–226
   pattern, 224–225
   precision, 223–226
   sequential, 226
Sand, 90–92
Scientific method, 6–7
Scientific names, 53, 401–404
Scientific organizations, 398–400
Screwworm, 204–205
Scouting, 228, 230
Season
   determination, 77–78
   effect on decomposition, 71
Secondary parasite, 175

Secondary pest, 219, 221
Seed
   dispersal, 109–111, 131–132, 140
   survival, 109–111
Seed bank, 109, 140
Semi-industrial agroecosystem, 28
Septicemia, 174
Sesame, 379–380
Sesamum indicum, 379
Severity of disease, 191
Shannon index, 245–247
Shifting cultivation, 280–283, 363
   problems, 282
   soil nutrients, 281, 363
Silt, 90–92
Simpson index, 245–247
Simulation model, 121
Sink, 15–16
Sinkhole, 315
Slash and burn (*see* Shifting cultivation)
Slime molds, 39
Slugs, 44
Snails, 44
Social issues, 355–357
   conflict over resources, 357
Sodicity, 320–322
Sodium, 321–322
Solar energy (*see* Solar radiation)
Solar radiation (*see also* Light)
   photosynthesis, 16, 20–21
   seasonal trends, 77–79
   tropics, 361
Solarization, 195
Soil, 85–99
   aeration, 89
   classification, 96–97
   conservation, 88, 290
   erosion (*see* Erosion)
   horizons, 94–95
   laterite, 362
   medium for plant growth, 88–89
   mineral, 90–92
   moisture (*see* Water)
   nutrient cycling in, 55–68, 89
   organic, 90, 94, 96
   organic matter, 68–69, 90
   particle size, 90–91
   pH (*see* pH)
   pore space, 90–93
   profile, 93–95
   reaction (*see* pH)
   solarization (*see* Solarization)
   solution, 91

Soil, *continued*
    texture, 90–91
    tropical, 362–363
Soil fertility
    cation exchange capacity, 98
    conservation tillage, 291–293
    earthworms, 42
    tropical, 362–363
Soil organisms, 36–49
    classification, 36–45, 53–54
    decomposition, 45–46, 69–70
    food webs, 46–50
    predators and prey, 46
Soil structure
    conservation tillage, 291
    earthworms, 42
    porosity, 90–93
Soil water, 89, 291 (*see also* Leaching)
    holding capacity, 90–93
    movement, 90–93
Soilborne, 192
Sorghum, 365, 367
    ethanol source, 30
    rotation, 277
Sorghum bicolor, 365
Sorghum halepense, 146
Source, 15–16
Soybean, 364–365
    cyst nematode, 236
    maturity groups, 106–107
    pesticides, 215
    photoperiod, 106–107
Spacing (*see* Plant spacing)
Spatial distribution (*see* Distribution)
Specialist food habits, 36, 163, 172–173
Species, 53–54
    unidentified, 54
Species abundance curve, 244
Species diversity (*see* Diversity)
Spiders
    agricultural importance, 204
    soil, 43–44
Spodosol, 97
Springtails (*see* Collembola)
Stability, 247–248
Stable age distribution, 114
Standard deviation, 9–10, 223–226
Standing crop, 24
Statistical tests, 11
Statistics, 6–11
Sterile male release, 204–205, 207, 220
Stochastic model, 121
Stocking (*see* Grazing management)

Strata, 225
Stratified sample, 225
Stocking rate, 336
Subsidence, 315
Subsidies, 346–353
    consumer, 346
    export, 347
    natural resources, 347
    sugar, 348–350
Subsistence agriculture (*see* Agroecosystem, subsistence)
Subtropical climate, 77
Succession, 134–137
    models, 135–136
    r and k strategies, 135–137
    weeds, 137
Sugar
    crops, 380
    energy source, 16–17
    subsidies, 348–350
Sugarbeet
    ethanol source, 30
    subsidies, 348–350
Sugarcane, 348, 380
    ethanol source, 30
    subsidies, 348–350
Sulfur, 67–68
Sunlight (*see* Solar radiation)
Supply, 351
Support price, 347
Survivorship, 116
Survivorship curve, 115
Susceptible, 160
Sustainability
    agriculture, 392–395
    economics, 354
    grazing, 337–342
    human population growth, 390–393
    limitations, 393–394
    natural resources, 355
    outlook, 394–395
    phosphorus, 356
Sweetpotato, 365
Symbiosis, 59, 66
Symbiotic nitrogen fixation, 59–61
Symphylids, 44
Synthetic product, 216
Systematic sample, 224–225
Systemic pesticide, 218

Tannia, 373, 375
Tardigrades, 45
Taro, 373

Taxonomy, 54
Temperature
  conservation tillage, 291
  effect on decomposition, 70
  global warming, 20
  heat units, 103–105
  patterns and climate, 76–77, 79–80
  tropics, 361, 363–364
Termites, 41, 44
Texture (see Soil texture)
Theobroma, 377
Thermodynamics, 21–22
Threshold, 226–229
  action, 228, 231–232
  crop damage, 226–227
  economic, 227–229
  injury level, 228, 231–232
Ticks, 209
Tillage
  benefits and risks, 290–295
  conservation, 287–295
  conventional, 287
  diseases, 293–294
  energy, 295
  insects, 295
  minimum, 288
  mulch, 288
  primary, 186
  ridge, 288
  secondary, 186
  strip, 288–289
  types, 287–289
  water conservation, 320
  weeds, 142, 186–187, 293
  yield, 294–295
Tolerance, 160
Tomato, 318
Toxicity of pesticides, 217–218, 256–262
Trade (see International trade)
Transect, 224
Transgenic crops, 220
Transpiration, 84
Treatment, 7–8
Trophic levels, 24
  food webs, 35
Trophic pyramid, 24–25
Tropical agriculture (see Agroecosystem, tropical)

Ultisol, 97
Underpopulation, 120
United States
  corn yields, 28

United States, *continued*
  fertilizer use, 29
  irrigated farmland, 313
  price supports, 346
  sugar subsidies, 391
Units, conversion of, 25, 405–406
Urea, 62, 300

Validation, 121
Varieties (see Cultivars)
Vector
  animal diseases, 209
  virus, 190
Vegetable leafminer (see Leafminer)
Vertebrates, 45
  pests, 208–210
Vertisol, 96
Vesicular-arbuscular mycorrhizae, 66
Vicia faba, 377
Vigna unguiculata, 365
Viruses, 53, 190
  bean mosaic, 190
  cross protection, 194
  management, 192–196
  parasites, 176
  plant disease, 188, 190
  vector, 190
Voisin grazing (see Grazing, Voisin)
Volatilization
  nitrogen, 58
  pesticides, 258

Wasp, braconid, 174
Water, 83–85, 308–326
  agrochemicals in, 252–262, 310, 313
  conflict over, 324–326, 357
  conservation of, 312–320
  conservation tillage, 291
  cycle in agroecosystems, 84–85
  drainage, 90–93, 323–324
  erosion, 86–87
  nitrates in, 254
  phosphorus in, 255–256
  photosynthesis, 16
  politics, 311–313, 324–326
  population growth, 389–390
  respiration, 17
  soil, 84, 89–93
  sources, 309–310
  wastewater recycling, 317, 319
Water bears, 45
Water table, 310

Waterlogging, 314, 323–324
Watershed, 309
Weeds
   allelopathy, 142
   annual, 139–140
   benefits in agroecosystems, 142–143
   biennial, 139–140
   competition with crops, 130, 140–142, 180–182
   conservation tillage, 142, 293
   dispersal, 140
   ecology, 139–143
   establishment of, 133–134
   management, 139–143, 182–187
   parasitic plants, 176, 181
   perennial, 139–140
   population density, 181–182
   population dynamics, 109–111
   succession, 137
   tillage, 186
   worst weeds, 143–147
Wilting point, 93

Wind, 81–82
   erosion, 86–88
World
   cropland, 387
   population growth, 385

Xanthosoma, 373, 375

Yam, 373
Yield
   compensation, 270
   conservation tillage, 294–295
   crop damage, 226–229
   energy input, 28–29
   Green Revolution, 157–158
   intercrops, 270–273
   maximization of, 29–30
   plant spacing, 31, 129
   population growth, 391–392
   thresholds, 226–229
   weeds, 181